C0-CDM-373

HANDBOOK OF PRODUCTION SCHEDULING

***Recent titles in the* INTERNATIONAL SERIES IN OPERATIONS RESEARCH & MANAGEMENT SCIENCE**

Frederick S. Hillier, Series Editor, *Stanford University*

Maros/ *COMPUTATIONAL TECHNIQUES OF THE SIMPLEX METHOD*

Harrison, Lee & Neale/ *THE PRACTICE OF SUPPLY CHAIN MANAGEMENT: Where Theory and Application Converge*

Shanthikumar, Yao & Zijm/ *STOCHASTIC MODELING AND OPTIMIZATION OF MANUFACTURING SYSTEMS AND SUPPLY CHAINS*

Nabrzyski, Schopf & Węglarz/ *GRID RESOURCE MANAGEMENT: State of the Art and Future Trends*

Thissen & Herder/ *CRITICAL INFRASTRUCTURES: State of the Art in Research and Application*

Carlsson, Fedrizzi, & Fullér/ *FUZZY LOGIC IN MANAGEMENT*

Soyer, Mazzuchi & Singpurwalla/ *MATHEMATICAL RELIABILITY: An Expository Perspective*

Chakravarty & Eliashberg/ *MANAGING BUSINESS INTERFACES: Marketing, Engineering, and Manufacturing Perspectives*

Talluri & van Ryzin/ *THE THEORY AND PRACTICE OF REVENUE MANAGEMENT*

Kavadias & Loch/*PROJECT SELECTION UNDER UNCERTAINTY: Dynamically Allocating Resources to Maximize Value*

Brandeau, Sainfort & Pierskalla/ *OPERATIONS RESEARCH AND HEALTH CARE: A Handbook of Methods and Applications*

Cooper, Seiford & Zhu/ *HANDBOOK OF DATA ENVELOPMENT ANALYSIS: Models and Methods*

Luenberger/ *LINEAR AND NONLINEAR PROGRAMMING, 2nd Ed.*

Sherbrooke/ *OPTIMAL INVENTORY MODELING OF SYSTEMS: Multi-Echelon Techniques, Second Edition*

Chu, Leung, Hui & Cheung/ *4th PARTY CYBER LOGISTICS FOR AIR CARGO*

Simchi-Levi, Wu & Shen/ *HANDBOOK OF QUANTITATIVE SUPPLY CHAIN ANALYSIS: Modeling in the E-Business Era*

Gass & Assad/ *AN ANNOTATED TIMELINE OF OPERATIONS RESEARCH: An Informal History*

Greenberg/ *TUTORIALS ON EMERGING METHODOLOGIES AND APPLICATIONS IN OPERATIONS RESEARCH*

Weber/ *UNCERTAINTY IN THE ELECTRIC POWER INDUSTRY: Methods and Models for Decision Support*

Figueira, Greco & Ehrgott/ *MULTIPLE CRITERIA DECISION ANALYSIS: State of the Art Surveys*

Reveliotis/ *REAL-TIME MANAGEMENT OF RESOURCE ALLOCATIONS SYSTEMS: A Discrete Event Systems Approach*

Kall & Mayer/ *STOCHASTIC LINEAR PROGRAMMING: Models, Theory, and Computation*

Sethi, Yan & Zhang/ *INVENTORY AND SUPPLY CHAIN MANAGEMENT WITH FORECAST UPDATES*

Cox/ *QUANTITATIVE HEALTH RISK ANALYSIS METHODS: Modeling the Human Health Impacts of Antibiotics Used in Food Animals*

Ching & Ng/ *MARKOV CHAINS: Models, Algorithms and Applications*

Li & Sun/ *NONLINEAR INTEGER PROGRAMMING*

Kaliszewski/ *SOFT COMPUTING FOR COMPLEX MULTIPLE CRITERIA DECISION MAKING*

Bouyssou et al/ *EVALUATION AND DECISION MODELS WITH MULTIPLE CRITERIA: Stepping stones for the analyst*

Blecker & Friedrich/ *MASS CUSTOMIZATION: Challenges and Solutions*

Appa, Pitsoulis & Williams/ *HANDBOOK ON MODELLING FOR DISCRETE OPTIMIZATION*

***** A list of the early publications in the series is at the end of the book ****

HANDBOOK OF PRODUCTION SCHEDULING

Edited by

JEFFREY W. HERRMANN
University of Maryland, College Park

Jeffrey W. Herrmann (Editor)
University of Maryland
USA

Library of Congress Control Number: 2006922184

ISBN-10: 0-387-33115-8 (HB) ISBN-10: 0-387-33117-4 (e-book)
ISBN-13: 978-0387-33115-7 (HB) ISBN-13: 978-0387-33117-1 (e-book)

Printed on acid-free paper.

Printed in the United States of America.

9 8 7 6 5 4 3 2 1

springer.com

Dedication

This book is dedicated
to my family, especially
my wife Laury and
my parents Joseph and Cecelia.
Their love and encouragement
are the most wonderful
of the many blessings
that God has given to me.

Contents

Contributing Authors

Milind Dawande
University of Texas at Dallas

Jörg Thomas Dickersbach
SAP AG

John W. Fowler
Arizona State University

Jeffrey W. Herrmann
University of Maryland, College Park

Jayant Kalagnanam
IBM T. J. Watson Research Center

Stephan Kreipl
SAP AG

Mark Kuchta
Colorado School of Mines

Ho Soo Lee
IBM T. J. Watson Research Center

Emmett J. Lodree, Jr.
Auburn University

Bart MacCarthy
University of Nottingham

Michael Martinez
Colorado School of Mines

Scott J. Mason
University of Arkansas

Kenneth N. McKay
University of Waterloo

Lars Mönch
Technical University of Ilmenau

Alexandra Newman
Colorado School of Mines

Bryan A. Norman
University of Pittsburgh

Michele E. Pfund
Arizona State University

Michael Pinedo
New York University

Chandra Reddy
IBM T. J. Watson Research Center

Oliver Rose
Technical University of Dresden

Stuart Siegel
IBM T. J. Watson Research Center

Mark Trumbo
IBM T. J. Watson Research Center

Guilherme E. Vieira
Pontifical Catholic University of Paraná

Vincent C.S. Wiers
Eindhoven University of Technology

Preface

The purpose of the book is to present scheduling principles, advanced tools, and examples of innovative scheduling systems to persons who could use this information to improve production scheduling in their own organization.

The intended audience includes the following persons:

- Production managers, plant managers, industrial engineers, operations research practitioners;
- Students (advanced undergraduates and graduate students) studying operations research and industrial engineering;
- Faculty teaching and conducting research in operations research and industrial engineering.

The book concentrates on real-world production scheduling in factories and industrial settings, not airlines, hospitals, classrooms, project scheduling, or other domains. It includes industry case studies that use innovative techniques as well as academic research results that can be used to improve real-world production scheduling.

The sequence of the chapters begins with fundamental concepts of production scheduling, moves to specific techniques, and concludes with examples of advanced scheduling systems.

Chapter 1, "A History of Production Scheduling," covers the tools used to support decision-making in real-world production scheduling and the changes in the production scheduling systems. This story covers the charts developed by Henry Gantt and advanced scheduling systems that rely on sophisticated software. The goal of the chapter is to help production schedulers, engineers, and researchers understand the true nature of production scheduling in dynamic manufacturing systems and to encourage them to consider how production scheduling systems can be improved even more.

Chapter 2, "The Human Factor in Planning and Scheduling," focuses on the persons who do production scheduling and reviews some important results about the role of these persons. The chapter presents guidelines for designing decision support mechanisms that incorporate the individual and organizational aspects of planning and scheduling.

Chapter 3, "Organizational, Systems and Human Issues in Production Planning, Scheduling and Control," discusses system-level issues that are relevant to production scheduling and highlights their importance in modern manufacturing organizations.

Chapter 4, "Decision-making Systems in Production Scheduling," looks specifically at the interactions between decision-makers in production scheduling systems. The chapter presents a technique for representing production scheduling processes as complex decision-making systems. The chapter describes a methodology for improving production scheduling systems using this approach.

Chapter 5, "Scheduling and Simulation," discusses four important roles for simulation when improving production scheduling: generating schedules, evaluating parameter settings, emulating a scheduling system, and evaluating deterministic scheduling approaches. The chapter includes a case study in which simulation was used to improve production scheduling in a semiconductor wafer fab.

Chapter 6, "Rescheduling Strategies, Policies, and Methods" reviews basic concepts about rescheduling and briefly reviews a rescheduling framework. Then the chapter discusses considerations involved in choosing between different rescheduling strategies, policies, and methods.

Chapter 7, "Understanding Master Production Scheduling from a Practical Perspective: Fundamentals, Heuristics, and Implementations," is a helpful discussion of the key concepts in master production scheduling and the techniques that are useful for finding better solutions.

Chapter 8, "Coordination Issues in Supply Chain Planning and Scheduling," discusses the scheduling decisions that are relevant to supply chains. The chapter presents practical approaches to important supply chain scheduling problems and describes an application of the techniques.

Chapter 9, "Semiconductor Manufacturing Scheduling and Dispatching," reviews the state-of-the-art in production scheduling of semiconductor wafer fabrication facilities. Scheduling these facilities, which has always been difficult due to the complex process flow, is becoming more critical as they move to automated material handling.

Chapter 10, "The Slab-design Problem in the Steel Industry," discusses an interesting production scheduling problem that adds many unique constraints to the traditional problem statement. This chapter presents a heuristic solution based on matching and bin packing that a large steel plant uses daily in mill operations.

Chapter 11, "A Review of Long- and Short-Term Production Scheduling at LKAB's Kiruna Mine," discusses the use of mathematical programming to solve large production scheduling problems at one of the world's largest mines. The chapter discusses innovative techniques that successfully reduce solution time with no significant decrease in solution quality.

Chapter 12, "Scheduling Models for Optimizing Human Performance and Well-being," covers how scheduling affects the persons who have to perform the tasks to be done. The chapter includes guidelines on work-rest scheduling, personnel scheduling, job rotation scheduling, cross-training, and group and team work. It also presents a framework for research on sequence-dependent processing times, learning, and rate-modifying activities.

The range of the concepts, techniques, and applications discussed in these chapters should provide practitioners with useful tools to improve production scheduling in their own facilities.

The motivation for this book is the desire to bridge the gap between scheduling theory and practice. I first faced this gap, which is discussed in some of the chapters of this book, when investigating production scheduling problems motivated by semiconductor test operations and developing a job shop scheduling tool for this setting.

It has become clear that solving combinatorial optimization problems is a very small part of improving production scheduling. Dudek, Panwalkar, and Smith (*Operations Research*, 1992), who concluded that the extensive body of research on flowshop sequencing problems has had "limited real significance," suggest that researchers have to step back frequently from the research and ask: "Will this work have value? Are there applications? Does this help anyone solve a problem?"

More generally, Meredith (*Operations Research*, 2001) describes a "realist" research philosophy that yields a body of knowledge that is not connected to reality. Unfortunately, this describes scheduling research too well. To avoid this problem, Meredith instructs us to validate models against the real world and with the managers who have the problems.

Therefore, in addition to the practical goal stated above, it is my hope that this book, by highlighting scheduling research that is closely tied to a variety of practical issues, will inspire researchers to focus less on the mathematical theory of sequencing problems and more on the real-world production scheduling systems that still need improvement.

Jeffrey W. Herrmann

Acknowledgments

A book such as this is the result of a team effort. The first to be thanked, of course, must be authors who contributed their valuable time to produce the interesting chapters that comprise this volume. I appreciate their effort to develop chapters, write and revise the text, and format the manuscripts appropriately. Thanks also to Charles Carr, Ken Fordyce, Michael Fu, Chung-Yee Lee, and Vincent Wiers for their useful comments about draft versions of the chapters.

I would like to thank Fred Hillier for inviting me to edit this handbook. It is an honor to be a part of this distinguished series. My thanks also go to Gary Folven and Carolyn Ford, who taught me a great deal about editing and producing a handbook. It has been a pleasure working with them.

In addition to everything else she does to care for our family, my wife Laury provided useful and timely editorial assistance. I am grateful for her excellent help.

Chung-Yee Lee, my advisor, has been and continues to be an exceptional influence on my development as a researcher, teacher, and scholar. I appreciate his indispensable support and guidance over the last 15 years.

Finally, I am indebted to my family and to the wonderful friends, colleagues, teachers, and students whom I have known. All have added something valuable to my life.

Chapter 1

A HISTORY OF PRODUCTION SCHEDULING

Jeffrey W. Herrmann
University of Maryland, College Park

Abstract: This chapter describes the history of production scheduling in manufacturing facilities over the last 100 years. Understanding the ways that production scheduling has been done is critical to analyzing existing production scheduling systems and finding ways to improve them. The chapter covers not only the tools used to support decision-making in real-world production scheduling but also the changes in the production scheduling systems. This story goes from the first charts developed by Henry Gantt to advanced scheduling systems that rely on sophisticated algorithms. The goal of the chapter is to help production schedulers, engineers, and researchers understand the true nature of production scheduling in dynamic manufacturing systems and to encourage them to consider how production scheduling systems can be improved even more. This chapter not only reviews the range of concepts and approaches used to improve production scheduling but also demonstrates their timeless importance.

Key words: Production scheduling, history, Gantt charts, computer-based scheduling

1. INTRODUCTION

This chapter describes the history of production scheduling in manufacturing facilities over the last 100 years. Understanding the ways that production scheduling has been done is critical to analyzing existing production scheduling systems and finding ways to improve them.

The two key problems in production scheduling are, according to Wight (1984), "priorities" and "capacity." In other words, "What should be done first?" and "Who should do it?" Wight defines *scheduling* as "establishing the timing for performing a task" and observes that, in a manufacturing firms, there are multiple types of scheduling, including the detailed scheduling of a shop order that shows when each operation must start and complete. Cox *et al.* (1992) define *detailed scheduling* as "the actual assignment of starting and/or completion dates to operations or groups of operations to show when these must be done if the manufacturing order is to be completed on time." They note that this is also known as *operations scheduling*, *order scheduling*, and *shop scheduling*. This chapter is concerned with this type of scheduling.

One type of dynamic scheduling strategy is to use dispatching rules to determine, when a resource becomes available, which task that resource should do next. Such rules are common in facilities where many scheduling decisions must be made in a short period of time, as in semiconductor wafer fabrication facilities (which are discussed in another chapter of this book).

This chapter discusses the history of production scheduling. It covers not only the tools used to support decision-making in real-world production scheduling but also the changes in the production scheduling systems. This story goes from the first charts developed by Henry Gantt to advanced scheduling systems that rely on sophisticated algorithms. The goal of the chapter is to help production schedulers, engineers, and researchers understand the true nature of production scheduling in dynamic manufacturing systems and to encourage them to consider how production scheduling systems can be improved even more. This review demonstrates the timeless importance of production scheduling and the range of approaches taken to improve it.

This chapter does not address the sequencing of parts processed in high-volume, repetitive manufacturing systems. In such settings, one can look to JIT and lean manufacturing principles for how to control production. These approaches generally do not need the same type of production schedules discussed here.

Although project scheduling will be discussed, the chapter is primarily concerned with the scheduling of manufacturing operations, not general project management. Note finally that this chapter is not a review of the production scheduling literature, which would take an entire volume.

For a more general discussion of the history of manufacturing in the United States of America, see Hopp and Spearman (1996), who describe the changes since the First Industrial Revolution. Hounshell (1984) provides a detailed look at the development of manufacturing technology between 1800 and 1932. McKay (2003) provides a historical overview of the key concepts behind the practices that manufacturing firms have adopted in modern times, highlighting, for instance, how the ideas of just-in-time (though not the term) were well-known in the early twentieth century.

The remainder of this chapter is organized as follows: Section 2 discusses production scheduling prior to the advent of scientific management. Section 3 describes the first formal methods for production scheduling, many of which are still used today. Section 4 describes the rise of computer-based scheduling systems. Section 5 discusses the algorithms developed to solve scheduling problems. Section 6 describes some advanced real-world production scheduling systems. Section 7 concludes the chapter and includes a discussion of production scheduling research.

2. FOREMEN RULE THE SHOP

Although humans have been creating items for countless years, manufacturing facilities first appeared during the middle of the eighteenth century, when the First Industrial Revolution created centralized power sources that made new organizational structures viable. The mills and workshops and projects of the past were the precursors of modern manufacturing organizations and the management practices that they employed (Wilson, 2000a). In time, manufacturing managers changed over the years from capitalists who developed innovative technologies to custodians who struggle to control a complex system to achieve multiple and conflicting objectives (Skinner, 1985).

The first factories were quite simple and relatively small. They produced a small number of products in large batches. Productivity gains came from using interchangeable parts to eliminate time-consuming fitting operations. Through the late 1800s, manufacturing firms were concerned with maximizing the productivity of the expensive equipment in the factory. Keeping utilization high was an important objective. Foremen ruled their shops, coordinating all of the activities needed for the limited number of products for which they were responsible. They hired operators, purchased materials, managed production, and delivered the product. They were experts with superior technical skills, and they (not a separate staff of clerks) planned production. Even as factories grew, they were just bigger, not more complex.

Production scheduling started simply also. Schedules, when used at all, listed only when work on an order should begin or when the order is due. They didn't provide any information about how long the total order should take or about the time required for individual operations (Roscoe and Freark, 1971). This type of schedule was widely used before useful formal methods became available (and can still be found in some small or poorly run shops). Limited cost accounting methods existed. For example, Binsse (1887) described a method for keeping track of time using a form almost like a Gantt chart.

Informal methods, especially expediting, have not disappeared. Wight (1984) stated that "production and inventory management in many companies today is really just order launching and expediting." This author's observation is that the situation has not changed much in the last 20 years. In some cases, it has become worse as manufacturing organizations have created bureaucracies that collect and process information to create formal schedules that are not used.

3. THE RISE OF FORMAL SYSTEMS

Then, beginning around 1890, everything changed. Manufacturing firms started to make a wider range of products, and this variety led to complexity that was more than the foremen could, by themselves, handle. Factories became even larger as electric motors eliminated the need to locate equipment near a central power source. Cost, not time, was the primary objective. Economies of scale could be achieved by routing parts from one functional department to another, reducing the total number of machines that had to purchased. Large move batches reduced material handling effort. Scientific management was the rational response to gain control of this complexity. As the next section explains, planners took over scheduling and coordination from the foremen, whose empire had fallen.

3.1 The production control office

Frederick Taylor's separation of planning from execution justified the use of formal scheduling methods, which became critical as manufacturing organizations grew in complexity. Taylor proposed the production planning office around the time of World War I. Many individuals were required to create plans, manage inventory, and monitor operations. (Computers would take over many of these functions decades later.) The "production clerk" created a master production schedule based on firm orders and capacity. The "order of work clerk" issued shop orders and released material to the shop (Wilson, 2000b).

Gantt (1916) explicitly discusses scheduling, especially in the job shop environment. He proposes giving to the foreman each day an "order of work" that is an ordered list of jobs to be done that day. Moreover, he discusses the need to coordinate activities to avoid "interferences." However, he also warns that the most elegant schedules created by planning offices are useless if they are ignored, a situation that he observed.

Many firms implemented Taylor's suggestion to create a production planning office, and the production planners adapted and modified Gantt's charts. Mitchell (1939) discusses the role of the production planning department, including routing, dispatching (issuing shop orders) and scheduling. Scheduling is defined as "the timing of all operations with a view to insuring their completion when required." The scheduling personnel determined which specific worker and machine does which task. However, foremen remained on the scene. Mitchell emphasizes that, in some shops, the shop foremen, who should have more insight into the qualitative factors that affect production, were responsible for the detailed assignments. Muther (1944) concurs, saying that, in many job shops, foremen both decided which work to do and assigned it to operators.

3.2 Henry Gantt and his charts

The man uniquely identified with production scheduling is, of course, Henry L. Gantt, who created innovative charts for production control. According to Cox *et al.* (1992), a *Gantt chart* is "the earliest and best known type of control chart especially designed to show graphically the relationship between planned performance and actual performance." However, it is important to note that Gantt created many different types of charts that represented different views of a manufacturing system and measured different quantities (see Table 1-1 for a summary).

Gantt designed his charts so that foremen or other supervisors could quickly know whether production was on schedule, ahead of schedule, or behind schedule. Modern project management software includes this critical function even now. Gantt (1919) gives two principles for his charts:

1. Measure activities by the amount of time needed to complete them;
2. The space on the chart can be used to represent the amount of the activity that should have been done in that time.

Gantt (1903) describes two types of "balances": the *man's record*, which shows what each worker should do and did do, and the *daily balance of work*, which shows the amount of work to be done and the amount that is done. Gantt's examples of these balances apply to orders that will require many days to complete.

The daily balance is "a method of scheduling and recording work," according to Gantt. It has rows for each day and columns for each part or each operation. At the top of each column is the amount needed. The amount entered in the appropriate cell is the number of parts done each day and the cumulative total for that part. Heavy horizontal lines indicate the starting date and the date that the order should be done.

The man's record chart uses the horizontal dimension for time. Each row corresponds to a worker in the shop. Each weekday spans five columns, and these columns have a horizontal line indicating the actual working time for each worker. There is also a thicker line showing the the cumulative working time for a week. On days when the worker did not work, a one-letter code indicates the reason (e.g., absence, defective work, tooling problem, or holiday).

Table 1-1. Selected Gantt charts used for production scheduling.

Chart Type	Unit	Quantity being measured	Representation of time	Sources
Daily balance of work	Part or operation	Number produced	Rows for each day; bars showing start date and end date	Gantt, 1903; Rathe, 1961
Man's Record	Worker	Amount of work done each day and week, measured as time	3 or 5 columns for each day in two weeks	Gantt, 1981; Rathe, 1961
Machine Record	Machine	Amount of work done each day and week, measured as time	3 or 5 columns for each day in two weeks	Gantt, 1919, 1981; Rathe, 1961
Layout chart	Machine	Progress on assigned tasks, measured as time	3 or 5 columns for each day in two weeks	Clark, 1942
Gantt load chart	Machine type	Scheduled tasks and total load to date	One column for each day for two months	Mitchell, 1939
Gantt progress chart	Order	Work completed to date, measured as time	One column for each day for two months	Mitchell, 1939
Schedule Chart	Tasks in a job	Start and end of each task	Horizontal axis marked with 45 days	Muther, 1944
Progress chart	Product	Number produced each month	5 columns for each month for one year	Gantt, 1919, 1981; Rathe, 1961
Order chart	Order	Number produced each month	5 columns for each month for one year	Gantt, 1919, 1981; Rathe, 1961

Gantt's *machine record* is quite similar. Of course, machines are never absent, but they may suffer from a lack of power, a lack of work, or a failure.

McKay and Wiers (2004) point out that Gantt's man record and machine record charts are important because they not only record past performance but also track the reasons for inefficiency and thus hold foremen and managers responsible. They wonder why these types of charts are not more widely used, a fact that Gantt himself lamented (in Gantt, 1916).

David Porter worked with Henry Gantt at Frankford Arsenal in 1917 and created the first progress chart for the artillery ammunition shops there. Porter (1968) describes this chart and a number of similar charts, which were primarily progress charts for end items and their components. The unit of time was one day, and the charts track actual production completed to date and clearly show which items are behind schedule. Highlighting this type of exception in order to get management's attention is one of the key features of Gantt's innovative charts.

Clark (1942) provides an excellent overview of the different types of Gantt charts, including the machine record chart and the man record chart, both of which record past performance. Of most interest to those studying production scheduling is the *layout chart*, which specifies "when jobs are to be begun, by whom, and how long they will take." Thus, the layout chart is also used for scheduling (or planning). The key features of a layout chart are the set of horizontal lines, one line for each unique resource (e.g., a stenographer or a machine tool), and, going across the chart, vertical lines marking the beginning of each time period. A large "V" at the appropriate point above the chart marks the time when the chart was made. Along each resource's horizontal line are thin lines that show the tasks that the resource is supposed to do, along with each task's scheduled start time and end time. For each task, a thick line shows the amount of work done to date. A box with crossing diagonal lines shows work done on tasks past their scheduled end time. Clark claims that a paper chart, drawn by hand, is better than a board, as the paper chart "does not require any wall space, but can be used on a desk or table, kept in a drawer, and carried around easily." However, this author observes that a chart carried and viewed by only one person is not a useful tool for communication.

As mentioned before, Gantt's charts were adapted in many ways. Mitchell (1939) describes two types of Gantt charts as typical of the graphical devices used to help those involved in scheduling. The Gantt load chart shows (as horizontal lines) the schedule of each machine and the total load on the machine to date. Mitchell's illustration of this doesn't indicate which shop orders are to be produced. The Gantt progress chart shows (as horizontal lines) the progress of different shop orders and their due dates.

For a specific job, a *schedule chart* was used to plan and track the tasks needed for that job (Muther, 1944). Various horizontal bars show the start and end of subassembly tasks, and vertical bars show when subassemblies should be brought together. Filling in the bars shows the progress of work completed. Different colors are used for different types of parts and subassemblies. This type of chart can be found today in the Gantt chart view used by project management software.

In their discussion of production scheduling, Roscoe and Freark (1971) give an example of a Gantt chart. Their example is a graphical schedule that lists the operations needed to complete an order. Each row corresponds to a different operation. It lists the machine that will perform the operation and the rate at which the machine can produce parts (parts per hour). From this information one can calculate the time required for that operation. Each column in the chart corresponds to a day, and each operation has a horizontal line from the day and time it should start to the day and time it should complete. The chart is used for measuring progress, so a thicker line parallel to the first line shows the progress on that operation to date. The authors state that a "Gantt chart is essentially a series of parallel horizontal graphs which show schedules (or quotas) and accomplishment plotted against time."

For production planning, Gantt used an *order chart* and a *progress chart* to keep track of the items that were ordered from contractors. The progress chart is a summary of the order charts for different products. Each chart indicates for each month of the year, using a thin horizontal line, the number of items produced during that month. In addition, a thick horizontal line indicates the number of items produced during the year. Each row in the chart corresponds to an order for parts from a specific contractor, and each row indicates the starting month and ending month of the deliveries.

In conclusion, it can be said that Gantt was a pioneer in developing graphical ways to visualize schedules and shop status. He used time (not just quantity) as a way to measure tasks. He used horizontal bars to represent the number of parts produced (in progress charts) and to record working time (in machine records). His progress (or layout) charts had a feature found in project management software today: the length of the bars (relative to the total time allocated to the task) showed the progress of tasks.

3.3 Loading, boards, and lines of balance: other tools

While Gantt charts remain one of the most common tools for planning and monitoring production, other tools have been developed over the years, including loading, planning boards, and lines of balance.

Loading is a scheduling technique that assigns an operation to a specific day or week when the machine (or machine group) will perform it

(MacNiece, 1951). Loading is finite when it takes into account the number of machines, shifts per day, working hours per shift, days per week as well as the time needed to complete the order.

MacNiece (1951) also discusses *planning boards*, which he attributes to Taylor. The board described has one row of spaces for each machine, and each row has a space for each shift. Each space contains one or more cards corresponding to the order(s) that should be produced in that shift, given capacity constraints. A large order will be placed in more than one consecutive space. MacNiece also suggests that one simplify scheduling by controlling the category that has the smallest quantity, either the machines or the products or the workers. Cox *et al.* (1992) defines a *control board* as "a visual means of showing machine loading or project planning." This is also called a *dispatching board*, a *planning board*, or a *schedule board*.

The rise of computers to solve large project scheduling problems (discussed in the next section) did not eliminate manual methods. Many manufacturing firms sought better ways to create, update, visualize, and communicate schedules but could not (until much later) afford the computers needed to run sophisticated project scheduling algorithms. Control boards of various types were the solution, and these were once used in many applications. The Planalog control board was a sophisticated version developed in the 1960s. The Planalog was a board (up to six feet wide) that hung on a wall. (See Figure 1-1.) The board had numerous rows into which one could insert gauges of different lengths (from 0.25 to 5 inches long). Each gauge represented a different task (while rows did not necessarily represent resources). The length of each gauge represented the task's expected (or actual) duration. The Planalog included innovative "fences." Each fence was a vertical barrier that spanned multiple rows to show and enforce the precedence constraints between tasks. Moving a fence due to the delay of one task required one to delay all subsequent dependent tasks as well.

Also of interest is the *line of balance*, used for determining how far ahead (or behind) a shop might be at producing a number of identical assemblies required over time. Given the demand for end items and a bill-of-materials with lead times for making components and completing subassemblies, one can calculate the cumulative number of components, subassemblies, and end items that should be complete at a point in time to meet the demand. This line of balance is used on a progress chart that compares these numbers to the number of components, subassemblies, and end items actually done by that point in time (See Figure 1-2). The underlying logic is similar to that used by MRP systems, though this author is unaware of any scheduling system that use a line of balance chart today. More examples can be found in O'Brien (1969) and *Production Scheduling* (1973).

Also of interest is the *line of balance*, used for determining how far ahead (or behind) a shop might be at producing a number of identical assemblies required over time. Given the demand for end items and a bill-of-materials with lead times for making components and completing subassemblies, one can calculate the cumulative number of components, subassemblies, and end items that should be complete at a point in time to meet the demand. This line of balance is used on a progress chart that compares these numbers to the number of components, subassemblies, and end items actually done by that point in time (See Figure 1-2). The underlying logic is similar to that used by MRP systems, though this author is unaware of any scheduling system that use a line of balance chart today. More examples can be found in O'Brien (1969) and *Production Scheduling* (1973).

Figure 1-1. Detail of a Planalog control board (photograph by Brad Brochtrup).

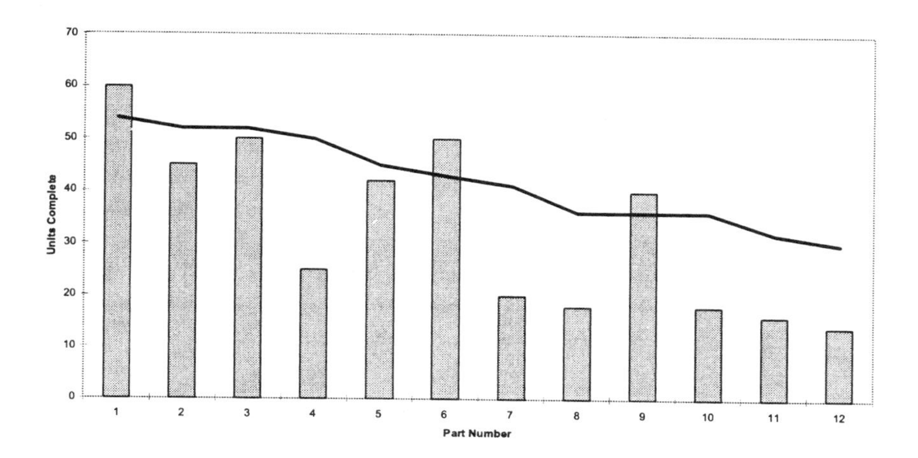

Figure 1-2. A line of balance progress chart (based on O'Brien, 1969). The vertical bars show, for each part, the number of units completed to date, and the thick line shows the number required at this date to meet planned production.

4. FROM CPM TO MRP: COMPUTERS START SCHEDULING

Unlike production scheduling in a busy factory, planning a large construction or systems development project is a problem that one can formulate and try to optimize. Thus, it is not surprising that large project scheduling was the first type of scheduling to use computer algorithms successfully.

4.1 Project scheduling

O'Brien (1969) gives a good overview of the beginnings of the critical path method (CPM) and the Performance Evaluation and Review Technique (PERT). Formal development of CPM began in 1956 at Du Pont, whose research group used a Remington Rand UNIVAC to generate a project schedule automatically from data about project activities.

In 1958, PERT started in the office managing the development of the Polaris missile (the U.S. Navy's first submarine-launched ballistic missile). The program managers wanted to use computers to plan and monitor the Polaris program. By the end of 1958, the Naval Ordnance Research Calculator, the most powerful computer in existence at the time, was programmed to implement the PERT calculations. Both CPM and PERT are now common tools for project management.

4.2 Production scheduling

Computer-based production scheduling emerged later. Wight (1984) lists three key factors that led to the successful use of computers in manufacturing:

1. IBM developed the Production Information and Control System starting in 1965.
2. The implementation of this and similar systems led to practical knowledge about using computers.
3. Researchers systematically compared these experiences and developed new ideas on production management.

Early computer-based production scheduling systems used input terminals, centralized computers (such as an IBM 1401), magnetic tape units, disk storage units, and remote printers (O'Brien, 1969). Input terminals read punch cards that provided data about the completion of operations or material movement. Based on this status information, the scheduling computer updated its information, including records for each machine and employee, shop order master lists, and workstation queues. From this data, the scheduling computer created, for each workstation, a dispatch list (or "task-to-be-assigned list") with the jobs that were awaiting processing at that workstation. To create the dispatch list, the system used a rule that considered one or more factors, including processing time, due date, slack, number of remaining operations, or dollar value. The dispatcher used these lists to determine what each workstation should do and communicate each list to the appropriate personnel. Typically, these systems created new dispatch lists each day or each shift. Essentially, these systems automated the data collection and processing functions in existence since Taylor's day.

Interactive, computer-based scheduling eventually emerged from various research projects to commercial systems. Godin (1978) describes many prototype systems. An early interactive computer-based scheduling program designed for assembly line production planning could output graphs of monthly production and inventory levels on a computer terminal to help the scheduling personnel make their decisions (Duersch and Wheeler, 1981). The software used standard strategies to generate candidate schedules that the scheduling personnel modified as needed. The software's key benefit was to reduce the time needed to develop a schedule. Adelsberger and Kanet (1991) use the term *leitstand* to describe an interactive production scheduling decision support system with a graphical display, a database, a schedule generation routine, a schedule editor, and a schedule evaluation routine. By that time, commercial leitstands were available, especially in Germany. The emphasis on both creating a schedule and monitoring its progress (planning and control) follows the principles of Henry Gantt. Similar types of systems are now part of modern manufacturing planning and control systems and ERP systems.

Computer-based systems that could make scheduling decisions also appeared. Typically, such systems were closely connected to the shop floor tracking systems (now called manufacturing execution systems) and used dispatching rules to sequence the work waiting at a workstation. Such rules are based on attributes of each job and may use simple sorting or a series of logical rules that separate jobs into different priority classes.

The Logistics Management System (LMS) was an innovative scheduling system developed by IBM for its semiconductor manufacturing facilities. LMS began around 1980 as a tool for modeling manufacturing resources. Modules that captured data from the shop floor, retrieved priorities from the daily optimized production plan (which matched work-in-process to production requirements and reassigned due dates correspondingly), and made dispatching decisions were created and implemented around 1984. When complete, the system provided both passive decision support (by giving users access to up-to-date shop floor information) and proactive dispatching, as well as issuing alerts when critical events occurred. Dispatching decisions were made by combining the scores of different "advocates" (one might call them "agents" today). Each advocate was a procedure that used a distinct set of rules to determine which action should be done next. Fordyce et al. (1992) provide an overview of the system, which was eventually used at six IBM facilities and by some customers (Fordyce, 2005).

Computer-based scheduling systems are now moving towards an approach that combines dispatching rules with finite-capacity production schedules that are created periodically and used to guide the dispatching decisions that must be made in real time.

4.3 Production planning

Meanwhile, computers were being applied to other production planning functions. Material requirements planning (MRP) translates demand for end items into a time-phased schedule to release purchase orders and shop orders for the needed components. This production planning approach perfectly suited the computers in use at the time of its development in the 1970s. MRP affected production scheduling by creating a new method that not only affected the release of orders to the shop floor but also gave schedulers the ability to see future orders, including their production quantities and release dates. Wight (1984) describes MRP in detail.

The progression of computer-based manufacturing planning and control systems went through five distinct stages each decade from the 1960s until the present time (Rondeau and Litteral, 2001). The earliest systems were reorder point systems that automated the manual systems in place at that time. MRP was next, and it, in turn, led to the rise of manufacturing resources planning (MRP II), manufacturing execution systems (MES), and now enterprise resource planning (ERP) systems. For more details about modern production planning systems, see, for instance, Vollmann, Berry, and Whybark (1997).

4.4 The implementation challenge

Modern computer-based scheduling systems offer numerous features for creating, evaluating, and manipulating production schedules. (Seyed, 1995, provides a discussion on how to choose a system.) The three primary components of a scheduling system are the database, the scheduling engine, and the user interface (Yen and Pinedo, 1994). The scheduling system may share a database with other manufacturing planning and control systems such as MRP or may have its own database, which may be automatically updated from other systems such as the manufacturing execution system. The user interface typically offers numerous ways to view schedules, including Gantt charts, dispatch lists, charts of resource utilization, and load profiles. The scheduling engine generates schedules and may use heuristics, a rule-based approach, optimization, or simulation.

Based on their survey of hundreds of manufacturing facilities, LaForge and Craighead (1998) conclude that computer-based scheduling can be successful if it uses finite scheduling techniques and if it is integrated with the other manufacturing planning systems. Computer-based scheduling can help manufacturers improve on-time delivery, respond quickly to customer orders, and create realistic schedules. Finite scheduling means using actual shop floor conditions, including capacity constraints and the requirements of orders that have already been released. However, only 25% of the firms responding to their survey used finite scheduling for part or all of their operations. Only 48% of the firms said that the computer-based scheduling system received routine automatically from other systems, while 30% said that a "good deal" of the data are entered manually, and 21% said that all data are entered manually. Interestingly, 43% of the firms said that they regenerated their schedules once each day, 14% said 2 or 3 times each week, and 34% said once each week.

More generally, the challenge of implementing effective scheduling systems remains, as it did in Gantt's day (see, for instance, Yen and Pinedo, 1994, or Ortiz, 1996). McKay and Wiers (2005) argue that implementation should be based on the amount of uncertainty and the ability of the operators in the shop to recover from (or compensate for) disturbances. These factors should be considered when deciding how the scheduling system should handle uncertainty and what types of procedures it should use.

5. BETTER SCHEDULING ALGORITHMS

Information technology has had a tremendous impact on how production scheduling is done. Among the many benefits of information technology is the ability to execute complex algorithms automatically. The development of better algorithms for creating schedules is thus an important part of the history of production scheduling. This section gives a brief overview that is follows the framework presented by Lenstra (2005). Books such as Pinedo (2005) can provide a more detailed review as well as links to surveys of specific subareas.

5.1 Types of algorithms

Linear programming was developed in the 1940s and applied to production planning problems (though not directly to production scheduling). George Dantzig invented the simplex method, an extremely powerful and general technique for solving linear programming problems, in 1947.

In the 1950s, research into sequencing problems motivated by production scheduling problems led to the creation of some important algorithms, including Johnson's rule for the two-machine flowshop, the earliest due date (EDD) rule for minimizing maximum lateness, and the shortest processing time (SPT) rule for minimizing average flow time (and the ratio variant for minimizing weighted flow time).

Solving more difficult problems required a different approach. Branch-and-bound techniques appeared around 1960. These algorithms implicitly enumerated all the possible solutions and found an optimal solution. Meanwhile, Lagrangean relaxation, column generation techniques for linear programming, and constraint programming were developed to solve integer programming problems.

The advent of complexity theory in the early 1970s showed why some scheduling problems were hard. Algorithms that can find optimal solutions to these hard problems in a reasonable amount of time are unlikely to exist.

Since decision-makers generally need solutions in a reasonable amount of time, search algorithms that could find near-optimal solutions became more important, especially in the 1980s and 1990s. These included local search algorithms such as hillclimbing, simulated annealing, and tabu search. Other innovations included genetic algorithms, ant colony optimization, and other evolutionary computation techniques. Developments in artificial intelligence led to agent-based techniques and rule-based procedures that mimicked the behavior of a human organization.

5.2 The role of representation

Solving a difficult problem is often simplified by representing it in the appropriate way. The representation may be a transformation into another problem that is easy to solve. More typically, the representation helps one to find the essential relationships that form the core of the challenge. For instance, when adding numbers, we place them in a column, and the sum is entered at the bottom. When doing division, however, we use the familiar layout of a long division problem, with the divisor next to the dividend, and the quotient appears above the bar. For more about the importance of representation in problem-solving, see Simon (1981), who discussed the role of representation in design.

Solving scheduling problems has been simplified by the use of good representations. Modern Gantt charts are a superior representation for most traditional scheduling problems. They clearly show how the sequence of jobs results in a schedule, and they simplify evaluating and modifying the schedule.

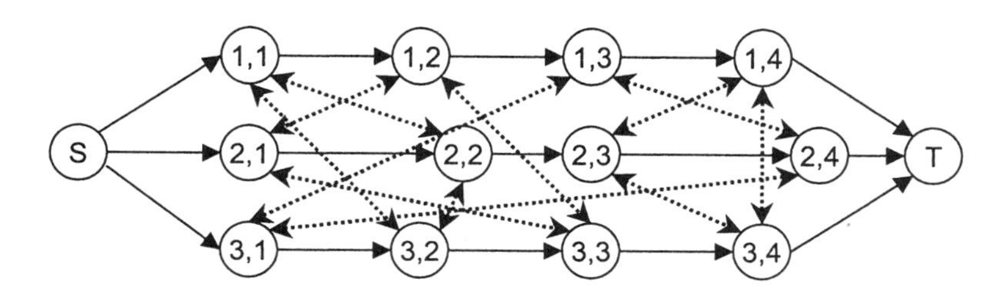

Figure 1-3. A disjunctive graph for a three-job, four-machine job shop scheduling problem.

MacNiece (1951) gives a beautiful example of using a Gantt chart to solve a scheduling problem. The problem is to determine if an order for an assembly can be completed in 20 weeks. The Gantt chart has a row for each machine group and bars representing already planned work to which he adds the operations needed to complete the order. He argues that using a Gantt chart is a much quicker way to answer the question.

Gantt charts continue to be refined in attempts to improve their usefulness. Jones (1988) created an innovative three-dimensional Gantt chart that gives each of the three key characteristics (jobs, machines, and time) its own axis.

Another important representation is the disjunctive graph, which was introduced by Roy and Sussmann (1964). The disjunctive graph is an excellent way to represent the problem of minimizing the makespan of a job shop scheduling problem (see Figure 1-3). Note that this representation represents each activity with a node. (Activity-on-arc representations have been used elsewhere.) The dashed edges in the graph represent the precedence constraints between tasks that require the same resource. Thus, these show the decisions that must be made. When the disjunctive arcs have been replaced with directed arcs, the graph provides a way to calculate the makespan. This representation also inspired many new algorithms that use this graph.

6. ADVANCED SCHEDULING SYSTEMS

Advances in information technology have made computer-based scheduling systems feasible for firms of all sizes. While many have not taken advantage of them (as discussed above), some firms have created advanced systems that use innovative algorithms. Each of these systems formulates the problem in a unique way that reflects each firm's specific scheduling objectives, and the system collects, processes, and generates information as part of a larger system of decision-making.

This section highlights the diversity of the approaches used to solve these scheduling problems. Many years of research on optimization methods have created a large set of powerful algorithms that can be applied to generate schedules, from mathematical programming to searches that use concepts from artificial intelligence.

6.1 Mathematical programming

An aluminum can manufacturing facility uses mathematical programming to create a weekly schedule (Katok and Ott, 2000). The can plant uses six production lines, each of which can make up to one million cans in an eight-hour shift. The cans are used by three filling plants. Each week the can plant must decide what to produce, where to store inventory, and how to satisfy demand (from inventory or production). A changeover is required when a production line switches from one can label to another. These changeovers are undesirable due to the scrap that is created and the downtime incurred. The problem of minimizing total production cost subject to satisfying demand and capacity constraints is a type of multi-level capacitated lot-sizing problem. It was formulated as a mixed-integer program and can be solved using GAMS in less than one minute.

One of the world's largest underground mines uses a mathematical programming approach to develop long-term production schedules (Newman et al., 2005). The mining operations, which began over 100 years ago, now yield nearly 24 million tons of iron ore each year. The production scheduling problem is to determine, for the next five years, which parts of the mine should be mined each month. Different parts of the mine contain different amounts of three ore types. The objective is to minimize the total deviation from the amount of each type of ore desired each month. The mixed-integer problem formulation includes constraints that reflect the nature of the mining operations and the resources available. Because the problem has nearly 66,000 binary variables, the scheduling system uses specialized algorithms to remove and aggregate the decision variables and add additional constraints. This resulting problem, programmed in AMPL, has 700 integer variables and can be solved using CPLEX in about five minutes.

6.2 Other solution approaches

Mathematical programming is not the only approach for solving scheduling problems. Approaches based on concepts from artificial intelligence and other areas of operations research can also be successful.

A Japanese steel plant uses a rule-based *cooperative scheduling* approach to create production schedules for three converters, nine sets of refining equipment, and five continuous casters, which together process up to 15,000 tons of steel each day (Numao, 1994). The unit of production is a 300-ton charge. Subschedules for a set of similar charges are backwards scheduled from casting, the bottleneck operation. The scheduling engine then merges the subschedules, which may be overlapping, and resolves any conflicts. The scheduling engine uses the rules in the rule base to satisfy a variety of general and domain-specific constraints. The scheduling system was designed to allow the user to modify the schedule at any point during the process, but especially after the scheduling engine merges the subschedules.

The system, implemented in a rule-based language, reduced the time needed to create a daily schedule from 3 hours to 30 minutes.

To solve the slab design problem, a different large steel plant uses a scheduling heuristic based on matching and bin packing (Dawande et al., 2005). Steel slabs are about 0.2 meters thick, 2 meters wide, and 12 meters long. They weigh between 15 and 40 tons. Steel slabs are used to create steel coils and sheets, and a single slab can be used to satisfy more than one customer order. The slab design problem is to determine the number of slabs that need to be produced, to specify each slab's size, and to assign orders to the slabs. Orders that require the same grade of steel and the same surface finish can be assigned to the same slab. The scheduling objective is to minimize the number of slabs and to minimize surplus weight. The scheduling engine (programmed using C++) can find good solutions in a few minutes.

Kumar et al. (2005) presents an innovative optimization algorithm to create cyclic schedules for robotic cells used in semiconductor manufacturing. The firm that manufactures these cells can use the algorithm to find a sequence of robot moves that maximizes that particular cell's capacity. The algorithm, which finds least common multiple cycles, uses a genetic algorithm to search the set of robot move cycles, while linear programming is used to evaluate each cycle. The algorithm requires a few minutes to find a near-optimal solution for complex robotic cells with 16 stations.

6.3 It takes a system

As mentioned earlier in this chapter, a scheduling system includes much more than the scheduling engine. Links to corporate databases are needed to extract information automatically. User interfaces are needed for the scheduling personnel to enter and update data, to view and modify schedules, and generate reports.

Sophisticated mathematical programming techniques use software that scheduling personnel do not understand. Thus, it is necessary to construct user interfaces that use terms and concepts that are familiar. These can be programmed from the ground up, or one can use common office software as the interface. For example, the can plant scheduling system mentioned above uses an Excel-based interface for entering data.

It is also important to note that developing a scheduling system requires carefully formulating a problem that includes the plant-specific constraints, validating the problem formulation, and creating specialized algorithms to find solutions using a reasonable amount of computational effort.

7. CONCLUSIONS

Since the separation that established production scheduling as a distinct production management function, the large changes in production scheduling are due to two key events. The first is Henry Gantt's creation of useful ways to understand the complex relationships between men, machines, orders, and time. The second is the overwhelming power of information technology to collect, visualize, process, and share data quickly and easily, which has enhanced all types of decision-making processes. These events have led, in most places, to the decline of shop foremen, who used to rule factories, and to software systems and optimization algorithms for production scheduling.

The bad news is that many manufacturing firms have not taken advantage of these developments. They produce goods and ship them to their customers, but the production scheduling system is a broken collection of independent plans that are frequently ignored, periodic meetings where unreliable information is shared, expediters who run from one crisis to another, and ad-hoc decisions made by persons who cannot see the entire system. Production scheduling systems rely on human decision-makers, and many of them need help.

This overview of production scheduling methods should be useful to those just beginning their study of production planning and control. In addition, practitioners and researchers should use this chapter to consider what has been truly useful to improve production scheduling practice in the real-world.

REFERENCES

Adelsberger, H., Kanet, J., 1991, "The Leitstand - A New Tool for Computer Integrated Manufacturing," *Production and Inventory Management Journal*, **32**(1):43-48.

Binsse, H.L., 1887, "A short way to keep time and cost," *Transactions of the American Society of Mechanical Engineers*, **9**:380-386.

Clark, W., 1942, *The Gantt Chart, a Working Tool of Management*, second edition, Sir Isaac Pitman & Sons, Ltd., London.

Cox, J.F., Blackstone, J.H., and Spencer, M.S., 1992, *APICS Dictionary*, American Production and Inventory Control Society, Falls Church, Virginia.

Dawande, M., Kalagnanam, J., Lee, H. S., Reddy, C., Siegel, S., and Trumbo, M., 2005, in *Handbook of Production Scheduling*, Herrmann J.W., ed., Springer, New York.

Duersch, R.R., and Wheeler, D.B., 1981, "An interactive scheduling model for assembly-line manufacturing," *International Journal of Modelling & Simulation*, **1**(3):241-245.

Fordyce, K., 2005, personal communication.

Fordyce, K., Dunki-Jacobs, R., Gerard, B., Sell, R., and Sullivan, G., 1992, "Logistics Management System: an advanced decision support system for the fourth tier dispatch or short-interval scheduling," *Production and Operations Management*, **1**(1):70-86.

Gantt, H.L., 1903, "A graphical daily balance in manufacture," *Transactions of the American Society of Mechanical Engineers*, **24**:1322-1336.

Gantt, H.L., 1916, *Work, Wages, and Profits*, second edition, Engineering Magazine Co., New York. Reprinted by Hive Publishing Company, Easton, Maryland, 1973.

Gantt, H.L., 1919, *Organizing for Work*, Harcourt, Brace, and Howe, New York. Reprinted by Hive Publishing Company, Easton, Maryland, 1973.

Godin, V.B., 1978, "Interactive scheduling: historical survey and state of the art," *AIIE Transactions*, **10**(3):331-337.

Hopp, W.J., and Spearman, M.L., 1996, *Factory Physics*, Irwin/McGraw-Hill, Boston.

Hounshell, D.A., 1984, *From the American System to Mass Production* 1800-1932, The Johns Hopkins University Press, Baltimore.

Jones, C.V., 1988, "The three-dimensional gantt chart," *Operations Research*, **36**(6)891-903.

Katok, E., and Ott, D., 2000, Using mixed-integer programming to reduce label changes in the Coors aluminum can plant, *Interfaces*, **30**(2):1-12.

Kumar, S., Ramanan, N., and Sriskandarajah, C., 2005, "Minimizing cycle time in large robotic cells," *IIE Transactions*, **37**(2):123–136.

LaForge, R.L., and Craighead, C.W., 1998, "Manufacturing scheduling and supply chain integration: a survey of current practice," American Production and Inventory Control Society, Falls Church, Virginia.

Lenstra, J.K, 2005, "Scheduling, a critical biography," presented at the Second Multidisciplinary International Conference on Scheduling: Theory and Applications, July 18-21, 2005, New York.

MacNiece, E.H., 1951, *Production Forecasting, Planning, and Control*, John Wiley & Sons, Inc., New York.

McKay, K.N., 2003, Historical survey of manufacturing control practices from a production research perspective, International Journal of Production Research, **41**(3):411-426.

McKay, K.N., and Wiers, V.C.S., 2004, *Practical Production Control: a Survival Guide for Planners and Schedulers*, J. Ross Publishing, Boca Raton, Florida. Co-published with APICS.

McKay, K.N., and Wiers, V.C.S., 2005, The human factor in planning and scheduling, in *Handbook of Production Scheduling*, Herrmann J.W., ed., Springer, New York.

Mitchell, W.N., 1939, *Organization and Management of Production*, McGraw-Hill Book Company, New York.

Muther, R., 1944, *Production-Line Technique*, McGraw-Hill Book Company, New York.

Newman, A., Martinez, M., and Kuchta, M., 2005, A review of long- and short-term production scheduling at LKAB's Kiruna mine, in *Handbook of Production Scheduling*, Herrmann J.W., ed., Springer, New York.

Numao, M., 1994, Development of a cooperative scheduling system for the steel-making process, in *Intelligent Scheduling*, Zweben, M., and Fox, M.S., eds., Morgan Kaufmann Publishers, San Francisco.

O'Brien, J.J., 1969, *Scheduling Handbook*, McGraw-Hill Book Company, New York.

Ortiz, C., 1996, "Implementation issues: a recipe for scheduling success," *IIE Solutions*, March, 1996, 29-32.

Pinedo, M., 2005, *Planning and Scheduling in Manufacturing and Services*, Springer, New York.

Porter, D.B., 1968, "The Gantt chart as applied to production scheduling and control," *Naval Research Logistics Quarterly*, **15**(2):311-318.

Production Scheduling, 1973, American Institute of Certified Public Accountants, New York.

Rondeau, P.J., and Litteral, L.A., 2001, "Evolution of manufacturing planning and control systems: from reorder point to enterprise resource planning," *Production and Inventory Management Journal*, Second Quarter, 2001:1-7.

Roscoe, E.S., and Freark, D.G., 1971, *Organization for Production*, fifth edition, Richard D. Irwin, Inc., Homewood, Illinois.

Roy, B., and Sussmann, B., 1964, "Les problems d'ordonnancement avec constraintes disjonctives." Note DS No. 9 bis, SEMA, Montrouge.

Seyed, J., 1995, "Right on schedule," *OR/MS Today*, December, 1995:42-44.

Simon, H.A., 1981, *The Sciences of the Artificial*, second edition, MIT Press, Cambridge, Massachusetts.

Skinner, W., 1985, The taming of the lions: how manufacturing leadership evolved, 1780-1984, in *The Uneasy Alliance*, Clark, K.B., Hayes, R.H., and Lorenz, C., eds., Harvard Business School Press, Boston.

Vieira, G.E., Herrmann, J.W., and Lin, E., 2003, Rescheduling manufacturing systems: a framework of strategies, policies, and methods," *Journal of Scheduling*, **6**(1):35-58.

Vollmann, T.E., Berry, W.L., and Whybark, D.C., 1997, *Manufacturing Planning and Control Systems*, fourth edition, Irwin/McGraw-Hill, New York.

Wight, O.W., 1984, *Production and Inventory Management in the Computer Age*, Van Nostrand Reinhold Company, Inc., New York.

Wilson, J.M., 2000a, "History of manufacturing management," in *Encyclopedia of Production and Manufacturing Management*, Paul M. Swamidass, ed., Kluwer Academic Publishers, Boston.

Wilson, J.M., 2000b, "Scientific management," in *Encyclopedia of Production and Manufacturing Management*, Paul M. Swamidass, ed., Kluwer Academic Publishers, Boston.

Yen, B.P.-C., and Pinedo, M., "On the design and development of scheduling systems," Proceedings of the Fourth International Conference on Computer Integrated Manufacturing and Automation Technology, October 10-12, 1994:197-204.

Chapter 2

THE HUMAN FACTOR IN PLANNING AND SCHEDULING

Kenneth N. McKay, Vincent C.S. Wiers
University of Waterloo, Eindhoven University of Technology

Abstract: In this chapter, we will review the research conducted on the human factor in planning and scheduling. Specifically, the positive and negative aspects of the human factor will be discussed. We will also discuss the consequences when these aspects are ignored or overlooked by the formal or systemized solutions.

Key words: Decision-making, scheduling decision support systems

1. INTRODUCTION

In this chapter, the term scheduling will be used for any decision task in production control that involves sequencing, allocating resources to tasks, and orchestrating the resources. Depending on the degree of granularity, scope, and authority, this view covers dispatching, scheduling, and planning. Unless noted, the basic issues and concepts discussed in this chapter apply to all three activities. This common view is also taken because in practice, it has been the authors' experience that in many cases, a single individual will be doing some planning, some scheduling, and some dispatching - all depending on the situation at hand and it is not possible to identify or specify clean boundaries between the three.

The gap between theory and practice in scheduling has been noted since the early 1960's and has been discussed by a number of researchers (Buxey, 1989; Dudek et al., 1992; Graves, 1981; MacCarthy and Liu, 1993; Pinedo, 1995; Rodammer and White, 1988; Crawford and Wiers, 2001). There are many possible reasons for the gap and each contributes to the difficulties associated with improving the effectiveness and efficiency of production. One of the reasons for the gap has been speculated to be the lack of fit between the human element and the formal or systemized element found in the scheduling methods and systems (McKay et al., 1989; McKay and Wiers, 1999). This specific lack of fit is what we will call the human factor in the success or failure of scheduling methodology.

With very few exceptions, there is some component of human judgment and decision making in the production control of real factories. The human element might be responsible for the majority of sequencing and resource allocation decisions from the initial demand requirements, or might be responsible for initial parameter setting for algorithms and software, or might be involved in interpretation and manipulation of recommended plans generated by a software tool. It is very hard to think of a situation where the mathematical algorithms and planning logic are self-installing, self-setting, self-tuning, and self-adapting to the situational context of business realities. Unfortunately, the role and contribution of the human element has been largely ignored and under-researched compared to the effort placed on the mathematical and software aspects (McKay et al., 1988).

In this chapter, we will review the research conducted on the human factor in planning and scheduling. Specifically, the positive and negative aspects of the human factor will be discussed. We will also discuss the consequences when these aspects are ignored or overlooked by the formal or systemized solutions. Section 2 presents a discussion on scheduling and sequencing from the perspective of schedule feasibility. Section 3 presents a review of the human scheduling research focusing on the scheduler as an individual. Section 4 reviews the research on the task nature of scheduling. Section 5 proposes a set of concepts for how to better design decision support mechanisms which incorporate the individual and organizational aspects of planning and scheduling. The concepts presented in Section 5 are integrated in a design model for scheduling decision support systems in Section 6. Finally, Section 7 presents our conclusions.

2. SCHEDULING AND SEQUENCING

Encapsulated in mathematical logic or software, the scheduling problem is presented as a sequencing problem. This view dates back to the early 1960's and was explicitly expanded upon in the seminal work of Conway, Maxwell, and Miller (1967). The mathematical dispatching and sequencing rules focus on what to select from the work queue at any specific resource. In academia, *Theory of Scheduling* = *Theory of Sequencing*. At a higher level, this also includes what to release into the manufacturing process. During the past four decades, various sophisticated algorithms have been developed that attempt to find the best possible sequence given a number of constraints and objectives (e.g., Pinedo, 1995). The implicit or explicit goal of the research has been to create sequences of work that can be followed, and by this following of the plan, the firm will be better off. The ultimate goal is to use information directly from a manufacturing system, run the algorithms without human intervention, and have the shop floor execute the plans as directed - without deviation. This goal has been reasonably obtained in a number of industries, such as process, single large machine equivalents, highly automated work cells, and automated (or mechanically controlled) assembly lines. For these factories, the situation is reliable and certain enough for plans to be created and followed for the required time horizon.

The acceptable time horizon is one in which a change can take place without additional costs and efforts. For example, a, b, and c is the planned schedule and instead of picking a, we decide to work on c. If this decision does not incur additional costs and efforts as the rest of the plan unfolds, then the change does not matter. It might be that changes in the plan can be made without penalty two days from now. In this case, a good schedule and scheduling situation would be one that could be created and actually followed for today and tomorrow. Here, the *Theory of Scheduling = Theory of Sequencing*. For rapidly changing job shops or industries, plans for the immediate future might be changing every half hour or so. It has also been observed in the factories which have been studied that the objective functions and constraints change almost as quickly. In unstable situations such as these, sequencing is only part of the problem and the *Theory of Scheduling ≠ Theory of Sequencing*. The ability to reschedule and perform reactive scheduling does not really solve the problem either if the changes are being made within the critical timing horizon. Reactive re-sequencing may incur additional costs and wastes within the window if the manufacturing system does not have sufficient degrees of freedom with which to deal with the situation. In this chapter, the focus is on the situations where sequencing is only part of the scheduling problem and the challenge is the short term time horizon during which any changes might be problematic.

When sequencing is only part of the problem and the human scheduler is expected to supply the remaining knowledge and skill, other issues may remain. One issue relates to the starting point provided by the human. That is, using the model, data, and algorithms in the computer system, generate a starting schedule for the human to interpret and manipulate. An extreme level is that of 100% - either accepting the proposed sequence or rejecting it. At one field site, it was reported that the schedulers started each day with deleting the software-generated plan and starting from scratch manually. This is an example of 100% rejection. An automated work cell capable of reliable and predictable operation without human intervention might be an example of 100% acceptance. The quality of a starting plan might be considered the degree of acceptance - how much of the plan is acceptable.

The human might reject a plan, or part of a plan, for two major reasons. First, the plan (or sequence) is not feasible and is not operational. These are the decisions that can be considered 0-1 decisions - e.g., "That cannot happen." Second, operationally feasible sequences may be rejected because they are considered suboptimal or not desirable. These sequences could be executed on the shop floor, but it would not be a good way to use the firm's resources. One sequence is preferred over another. The feasible and infeasible criteria, and the preferred and not preferred criteria can be examined by what is included in the traditional models and methods.

In the *Theory of Sequencing*, what is a feasible plan? First, one test of feasibility is whether or not a resource is planned to do something it cannot do. That is, if a machine can only drill one thing at a time, is only drilling assigned, and is only one item scheduled at a time? This is basic feasibility. Second, another test of feasibility is a time pattern of availability; a machine can only be scheduled work when it is possible to schedule work. Third, precedence constraints can dictate order of tasks and how the tasks relate to each other. Fourth, assuming forward loading is being performed, work cannot start before it is available to be worked on. If a sequence satisfies these four conditions, it is generally assumed to be mathematically feasible. The parameters into the mathematical structure include information such as: routings, sequencing relationships between tasks, machine capability, machine availability, set up criteria, processing times, earliest start dates, due dates, possibly penalties and possibly yield. Once it is possible to generate a feasible plan given the input parameters, the next task is to create a good sequence. In mathematical terms, a good plan is one that would attempt to maximize (or minimize) one or more quantitative metrics that can be derived from the interpretation of the plan. The mathematical approaches either use heuristics and algorithms to guide the creation of better schedules while considering feasibility (e.g., traditional OR) or use methods for generating multiple feasible sequences and then have methods for selecting the better sequences. There are reasonably clear definitions and understanding about what *feasible* and *better* means in mathematical sequencing research. Feasible and better are usually explicitly discussed in the research publications. Other, less obvious, assumptions are not. Ten key, implicit assumptions we have observed in mathematical formulations are:

- the relatively small set of facts about the scheduling problem handled by the mathematical model or algorithm are sufficient to capture the main characteristics of the problem.

- the objectives or measurement metrics do not vary over the time horizon being sequenced.
- the feasibility constraints defining the problem do not vary over the time horizon.
- time can be modeled as an abstract time series with t(i), and t(j) and that sequencing is not dependent upon a Monday effect, a shift effect, a holiday effect, or a time of year effect.
- any routing or processing requirements are also independent of state or context for the planning horizon.
- the resource capability and output is largely independent of time (after learning is achieved).
- resources are largely insensitive to the work (in terms of causing problems to the machine).
- work is largely insensitive to the state of the machine upon which the work is performed.
- work can actually be late - numbers of late jobs, degree of lateness being common objectives in the scheduling research.
- operations, routings, quantities, and such do not dynamically change based on the current state of manufacturing.

When these assumptions do not hold, the ability of the mathematical model or software system to create a feasible plan is challenged. As the number of invalid assumptions increases, the more difficult it will be to create a feasible plan. These assumptions illustrate why the *Theory of Scheduling* is not always the same as the *Theory of Sequencing* (McKay and Wiers, 1999). When an assumption fails, either additional logic or manual intervention is required to bridge the gap - creating a feasible schedule. In McKay (1992), these assumptions were investigated and the types of information needed to bridge the gap were analyzed in a case study. While a large portion of the enriched data could be conceptually encoded, programmed, and added into algorithms, approximately 30% of the data used to bridge the gap was not considered to be easily computerized (or legally computerized). For example, scheduling decisions were observed that depended upon knowledge regarding a crew's attitude during training. This particular fact was only important on one day, for one job, on a specific shift and was not relevant for any other decision on the planning horizon. However, it was important enough to force the pertinent job to a specific place on the schedule. In hindsight, an analyst can say that this could be programmed (ignoring practical aspects of the challenge). Another example decision involved a worker's alcohol problem and how the scheduler wanted to avoid this worker's effort on any critical job. Legally, this type of information would be a challenge to include. While this could be coded, it might open up human resource issues and lawsuits.

Hence, human interpretation and enhancement of the scheduling problem is required when the assumptions cannot be satisfied by mathematics or software. While not needed in every factory situation, many factories fail to satisfy, or come close to satisfying the implicit assumptions of sequencing theory. The next section discusses the research that has been performed on the human element in scheduling.

3. THE HUMAN ELEMENT

3.1 Introduction

The first question to ask is "What can humans do that computers and/or mathematics cannot?" The answer is that there is much uncertainty in the physical world. It is humans who are very well equipped to cope with many 'soft,' qualitative task elements as well as any creative problem solving that might be needed in order to create a feasible or better schedule. Empirical field studies have suggested that humans are superior to existing scheduling techniques and information systems regarding the following characteristics (McKay et al., 1989):

- *Flexibility, adaptability and learning.* Humans can cope with many stated, not–stated, incomplete, erroneous, and outdated goals and constraints. Furthermore, humans are able to deal with the fact that these goals and constraints are seldom more stable than a few hours.
- *Communication and negotiation.* Humans are able to influence the variability and the constraints of the shop floor; they can communicate with the operators on the shop floor to influence job priorities or to influence processing times. Humans are able to communicate and negotiate with (internal) customers if jobs are delayed, or communicate with suppliers if materials are not available as planned.
- *Intuition.* Humans are able to fill in the blanks of missing information required to schedule. This requires a great amount of 'tacit knowledge.' At the time of collecting this knowledge it is not always clear which goals are served by it.

While the human's ability to deal with uncertainty is important, it is also important to consider the human's ability to create the sequences of work that form a schedule. The seminal article by Sanderson (1989) summarizes and reviews 25 years of work done on the human role in scheduling. Two types of studies are discussed in the review: laboratory studies and field studies. Sanderson also discusses methodological and conceptual aspects of the literature reviewed. The laboratory studies summarized in Sanderson's review have mainly focused on three themes: comparing unaided humans with scheduling techniques, studying interactive systems of humans and techniques, and studying the effect of display types on scheduling performance. However, there have been very few studies replicated or performed in such a way for the results to be generalized. The tasks studied in the research have been quite varied, as have the study methods. Moreover, the research questions which mainly focused on comparisons of humans and techniques might not be as relevant today with the large number of scheduling software tools available (e.g., over one hundred scheduling tools are available, see LaForge and Craighead, 2000). In addition, the majority of field studies prior to 1990 focused on highly experienced schedulers with very little decision support.

Sanderson concludes with the observation that more and better coordinated research on the human factor in scheduling is required. The research reported in Sanderson's review were widely dispersed over a variety of research journals and the reported works were often carried out in isolation from each other. She also notes that a common research question that was addressed in much of the literature reviewed—i.e., which is better, humans or algorithms—was no longer relevant (even in 1989). Her conclusion was that humans and algorithms seem to have complementary strengths which could be combined. To be able to do this, a sound understanding of the human scheduler was considered necessary. In the following two subsections, recent literature on empirical research and cognitive scheduling models is discussed.

3.2 Recent empirical research on scheduling

Although Sanderson identified the need for extensive field studies on the scheduling task and scheduler, relatively few studies have been performed on the human scheduler in the time since. With the emergence of commercially available computer technology and scheduling decision support systems in the 1980s, a different stream of research evolved. In this stream, researchers were driven by the differences between Operations Research-based scheduling theory that formed the basis of most of the software, and the real-life scheduling process as decision makers conducted it. Inspired by the real-world complexities of the human scheduler (McKay et al., 1988; McKay et al., 1989), researchers spent much time in factories, using descriptive qualitative research methods. Accordingly, subsequent to Sanderson's review, most of the work studying the human factor in planning and scheduling has been of an exploratory and qualitative nature (see Crawford and Wiers, 2001, for a review) and has served very well the purpose of documenting actual scheduling behavior in production organizations.

Several interesting and novel results have been reported in the qualitative studies, such as the heuristics used by schedulers to avoid future problems, or the temporal aspect and the role of resource aging (McKay, 1992). A major contribution of these studies has been to move the field of study from the laboratory to the factory, with extensive qualitative empirical studies having been conducted, mostly based on ethnographic or case methodologies. Most studies have relied on relatively short observation or data collection periods to collect the empirical data. For example, one or two weeks might be spent with each factory doing an intense analysis and data collection. There have also been several longer term studies performed.

McKay (1992) performed two six month longitudinal field studies on the schedulers' cognitive skill and decision making processes. These two studies on the scheduling task are reported in the context of research on the effectiveness of the hierarchical production planning (HPP) paradigm in dealing with uncertainty. A task analysis at a printed circuit board (PCB) factory was used to identify the decisions made in response to uncertainties in the manufacturing system. The human scheduler turned out to be especially important in managing uncertainty (see also McKay et al., 1989). The field study in the PCB factory is also reported in McKay et al. (1995a). In this paper, the formal versus the informal scheduling practices are compared in the context of managing uncertainty. Several interesting aspects of the scheduling practices are mentioned in this study. First, it was observed that the scheduler worked with multiple schedules: a political schedule for the world to see, a realistic schedule, an idealistic schedule, and an optimistic schedule that was orally communicated to the line. This suggests that any field observations or field studies involving schedulers might be sensitive and aware of this possible multi-plan situation. This phenomenon has been observed in every factory situation the authors have been involved in. Second, it was observed that the scheduler did not accept the current situation as fate; instead, he endeavored to influence the amount and allocation of capacity, the amount of customer demand, the technical characteristics of machines (e.g., to minimize setups). The scheduler employed a large number of heuristics (more than one hundred) to anticipate possible problems and take precautionary measures.

Another exception to the short study is a nine year field study in which real world planning and scheduling has been researched (McKay and Wiers, 2003b). In this field study, flow shop and job shop systems were developed based on ethnographic methods for understanding the requirements and the systems have evolved with the requirements of the plant. In this effort, the schedulers' ontology or mental mapping of the problem was used to create a custom interface using their vocabulary and meta-functions. The focus of the research work was on task analysis to obtain deeper insights regarding the differences and similarities between scheduling, dispatching, and planning (McKay and Wiers, 2003a). The job shop part of the factory was structured in a hierarchical fashion with multiple individuals involved. The job shop was pulled just-in-time from the assembly area which was managed in an integrated way with one individual performing the three planning tasks. Weekly or bi-weekly visits to the plant have been made for the duration of the nine year study and have permitted many insights to be gained about the evolution and usage of scheduling technology as the plant itself evolved.

The third longer study is the work reported in Wiers (1996). The decision behavior of four production schedulers in a truck manufacturing company was investigated by means of a quantitative model. This model consisted of three parts: performance variables, action variables and disturbance variables. The results showed that schedulers who control equal production units show quite different decision behaviors. Also, a 'good' schedule turned out to be no guarantee for good performance. Moreover, some scheduling actions worked positively in the short term but negatively in the longer term. However, the methodological discussion of the case made clear that it is very difficult to construct a reliable quantitative model of production scheduling. Den Boer (1992, 1994) also conducted a quantitative field study on the decision behavior of material requirements planners. The model was based on the paramorphic representation of judgment (Hoffman, 1960) and contained four elements: performance, actions, disturbances and environment. Based on this study, Den Boer concluded that planners suffer from a lack of feedback in setting parameters such as safety time and safety stock.

This leaves the field now with a substantial number of insights from empirical studies, along with the results from earlier experimental studies and those from scheduling theory. Furthermore, in the field of cognitive psychology, advances have been made over the past decades regarding the more general problem of understanding human decision making in complex problem settings.

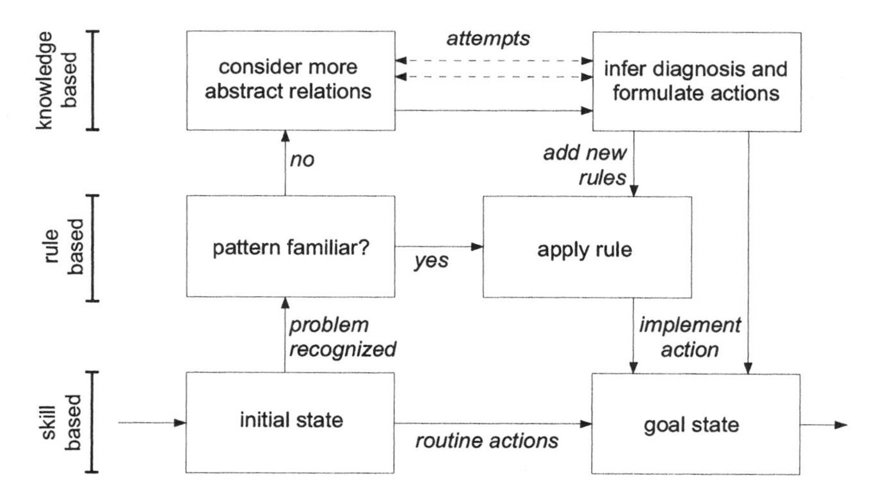

Figure 2-1. The GEMS model of human decision behavior (adapted from Reason, 1990)

3.3 The decision level - cognitive scheduling models

The area of modeling cognitive processes in complex tasks—such as the scheduling task—still appears to be in a relatively preliminary stage. In a special issue of Ergonomics about cognitive processes in complex tasks, Van der Schaaf (1993) notes that the process of developing a cognitive task model is more useful than the model itself. In an article about task allocation, Price (1985) observes that there is no universally applicable 'cookie cutter' for task allocation decisions; moreover, the ultimate configuration of tasks in a specific situation has to be determined throughout the design cycle. According to Price, covert and cognitive information processing tasks have not been adequately considered in systems design, or by human factors scientists generally. However, the decision models of Rasmussen (1986) are mentioned by Price as being helpful in this respect. The decision ladder of Rasmussen has been used by many authors to model cognitive processes in complex tasks, and is used by Sanderson (1991) and Sanderson and Moray (1990) to construct a model of human scheduling (MHS). The decision ladder model was also used by Higgins (1999) to study a scheduling situation and to create a matching user interface for scheduling technology.

The GEMS model (Generic Error Modeling System, Reason (1990)) is an adapted version of the decision ladder of Rasmussen. The GEMS model is depicted in the figure above. According to the model, humans reason with different levels of *attention* and *routine*. The more attention a task requires, the less routine the task, and vice versa. Tasks become more routine when they are repeated. The model distinguishes three levels of human information processing: skill-based, rule-based, and knowledge-based.

The GEMS model (Generic Error Modeling System, Reason (1990)) is an adapted version of the decision ladder of Rasmussen. The GEMS model is depicted in the figure above. According to the model, humans reason with different levels of *attention* and *routine*. The more attention a task requires, the less routine the task, and vice versa. Tasks become more routine when they are repeated. The model distinguishes three levels of human information processing: skill-based, rule-based, and knowledge-based.

At the *skill–based* level, the actions are carried out almost automatically, i.e., without the need for conscious reasoning. Automatic progress of the activities is checked periodically, but as long as these checks are satisfactory, control stays at the skill–based level. If a difference between the expected and real outcome is noted, control passes to the *rule–based* level. At the rule–based level there are many *if–then* rules competing to become active. The pattern of the problem is matched with the *if* part of the rules. If this succeeds, a particular (set of) rules is applied. The predominance of a certain rule depends mainly on the match between the *if* part and the environment, and the strength of the rule as a whole. If there are no rules that match the environment, reasoning passes on to the *knowledge–based* level. At the knowledge–based level, problems are identified, analyzed and solved by combining novel and existing knowledge in a new way. First, a representation of the problem and its causes is built. Second, alternative solutions for the problem are generated. Third, a solution is evaluated, selected and implemented. Knowledge about the problem solving process is stored and can be re–used if a similar problem occurs. In this way new *if–then* rules are added to the rule base.

Limitations of human information processing capabilities stem mainly from two factors: (1) bounded rationality, and (2) incomplete problem representation. Bounded rationality is caused by limited mental capacities, and therefore, large real-world problem representations do not fit into memory. Even if mental capacities were large enough to encompass the problems mentioned, then incomplete problem representation, i.e., insufficient knowledge about the problem, would still impede the full understanding of the problem and its context. The relationship between bounded rationality and limited problem representation can be compared to a beam of light that shines on a screen with information. The size of the light beam on the screen represents bounded rationality; the fact that not all information is visible within the beam of light represents incomplete problem representation (Wagenaar et al., 1990).

Using frameworks such as this might assist in understanding and designing hybrid interfaces for human schedulers. It is important for the human to be able to identify infeasible solutions or parts of a proposed schedule which are infeasible. Understanding the limits of the cognitive process should assist in designing display content and tools that will augment and strengthen areas of weakness while at the same time avoiding negative impacts.

3.4 Issues relating to the use of formal decision processes

Kleinmuntz (1990) discusses why humans still prefer to use their heads instead of decision techniques, given the fact that cognition is bounded and that techniques can help humans to increase performance. A proposed explanation is that people are unwilling to settle for techniques they know are imperfect. Possibly erroneously, people also believe that increased mental effort improves performance. According to Kleinmuntz, this is particularly true for situations where they are confident about their expertise.

The issue of trust in automation has also been studied by Muir (1994) and Muir and Moray (1996). The former paper presents a theoretical model of human trust in machines. In the latter paper, two experiments are reported that examine operators' trust in the use of automation in a simulated supervisory process control task. Results showed that operators' ratings of trust were mainly determined by their perception of its competence. Trust was reduced following any sign of incompetence in the automation, even one which had no effect on overall system performance. Another finding of Muir and Moray's experiments is that operators' trust changes very little with experience; whereas Kleinmuntz concludes that the use of decision aids decreases with the subject's *belief* in his experience. Relating this back to schedule feasibility, if schedulers do not trust the feasibility of a proposed schedule, or if the schedulers observe repeated infeasibility, the decision support aids for scheduling might not be used or used as intended.

The question of how to improve the use of decision rules is studied by Davis and Kotteman (1995). They investigated the determinants of decision rule use in a production planning task. Decision rule use can be improved by offering feedback in which actual performance is compared to performance that would have been realized if the rule had been used. However, measuring the performance of production scheduling has recently been highlighted as a very complex problem (Gary et al., 1995; Stoop, 1996). Apart from basic criteria such as the absence of possibilities for minor improvements and feasibility, it is not clear that any objective criteria can be set. While performance feedback can be given by monitoring performance over time, this is likely to be of limited value when the manufacturing environment is unstable. Davis and Kotteman (1995) indicate that a somewhat less effective measure to improve decision rule use is to explicitly describe the performance characteristics (i.e., the way a certain rule affects a certain performance) to humans, in this way making the rule more transparent. According to Norman (1988), the transparency of a decision rule is especially important in situations where critical, novel or ill–specified problems have to be solved. In these cases, humans want to be in direct control, without the visible existence of a technique. This is referred to by Norman as 'first–person' interaction. On the other hand, if the task that has to be performed is laborious or repetitive, the visible existence of a technique is preferred. In these cases, humans give commands to the (computerized) technique which then solves the problem. This is referred to by Norman as 'third–person' interaction.

These concepts of transparency and packaging have been applied in a custom decision support system (McKay and Wiers, 2003a) where various groupings of decision rules and functions have been structured. The scheduler can have micro level control when needed, or slightly decomposed or macro-function level decisions selectively applied. The scheduler can also choose to apply more highly-packaged functionality and have many decisions automatically performed. In all cases though, the output of the decisions in the form of reports or task allocation decisions can be manually manipulated. Training was also performed with the schedulers using the system to ensure that transparency of the decision rules existed. This was important for creating the trust level for feasibility. The transparency was also necessary for the three levels of management above the scheduler who were involved in some way with reading and interpreting plans created by the scheduler. The scheduler trusted the system, but the second level manager did not and continually challenged the feasibility of the solutions. To address the trust aspect special reports and the ability to expand or collapse information supporting the decisions was necessary. Based on this experience, a flexible approach using Norman's first and third person concepts might be appropriate if multiple users of the scheduling output exist.

Apart from problems regarding the measurement of performance in production scheduling, there might be another reason against offering certain types of performance feedback to human schedulers. While performance feedback has been found to improve decision rule use, it has also been found to impair effective learning in complex tasks (Johnson et al., 1993). Though feedback about the effectiveness of behavior has long been recognized as essential for learning in tasks, and, as found more recently, stimulating decision rule use, such feedback at least has to be specific and timely to be effective. In complex tasks where the relationship between actions and outcomes is unclear, only offering feedback about performance may be counterproductive. This is because outcome feedback might cue a focus on evaluating one's competence rather than on increasing competence, which could result in a maladaptive behavior pattern (Johnson et al., 1993). Furthermore, because action–effect relations in production systems are very hard to grasp, mental models of schedulers are prone to become inaccurate and variable. This is confirmed by Moray (1995), where a supervisory task controlling a simulated discrete production system was studied. The study of the individuals' behavior showed that there was variability between individual operators in system intervention. Some operators decided to manually schedule parts of the system even when no faults were occurring, possibly to prevent faults from occurring, while others decided to leave the scheduling decisions to the system.

However, there appears to be consensus in the literature that to improve decision behavior in complex tasks, some form of cognitive feedback is required (e.g., Brehmer, 1980; Jacoby et al., 1984; Early et al., 1990; Johnson et al., 1993). In an experiment by DeShon and Alexander (1996) this need for feedback was confirmed for tasks with implicit learning. However, in tasks with explicit learning, they found that setting specific goals appeared to gradually increase performance. Tasks with implicit learning can be characterized by the acquisition of knowledge through repeated exposure to problem exemplars without intention or awareness. In these tasks, it is very difficult for the subject to verbalize the rules used. In tasks with explicit learning, the first step in the solution of any problem is the development of an internal representation of the problem. The internal representation would consist of the perceived initial state of the problem, a goal state, allowable transformations for achieving the goal, and boundary conditions (Newell and Simon, 1972). DeShon and Alexander (1996) state that while explicit learning requires cognitive resources and is sensitive to distraction, implicit learning is relatively resource independent.

3.5 Individual differences

Though believed to be of great importance, there is still insufficient knowledge about the effect of individual differences on the use of computers in general, or on the use of scheduling information systems in particular. According to Wærn (1989), individual differences that influence human–computer interaction from most stable to least stable are: personality factors, cognitive styles, learning styles, and personal knowledge (i.e., user experience). Wærn (1989) argues that user experience is both the most important and the least stable aspect of individual variation. In studies of a supervisory task in a simulated discrete production system, Moray (1995) also found that differences in mental models, which are built by experience, caused differences in decision behavior.

In Levy et al. (1995), a production scheduling task in a laboratory setting was used to study feedback seeking behavior. More specifically, the effect of individual differences and situational characteristics on feedback seeking intent, reconsideration of intent and modifying of intent was studied. The results showed that seeking feedback depends on the perceived privacy of the feedback seeking process and the context in which it is performed. For example, individuals in organizational settings may want feedback but those in public contexts may be very concerned about how they appear to others, especially for individuals with high self–esteem. A finding that relates to individual differences is that people with high public self–consciousness and social anxiety desire feedback more than others.

Self–efficacy, which refers to beliefs in one's capabilities to mobilize the motivation, cognitive recourses, and courses of action needed to meet certain situational demands, is also frequently found to determine computer usage. Individuals who consider computers too complex and believe that they will never be able to control these computers will prefer to avoid them and are less likely to use them. The effect of self–efficacy on computer usage was studied in Igbaria and Iivari (1995) through a survey of 450 microcomputer users in Finland. It was found that self–efficacy influences computer usage through perceived ease of use and perceived usefulness. Also, computer experience and organizational support appeared to increase self–efficacy.

4. CONTEXT OF SCHEDULING IN PRACTICE

4.1 Introduction

While "What is sequencing?" is relatively easy to answer, "What is scheduling?" is not. Dealing with feasibility in sequences and crafting sequences that can be actually executed is part of scheduling, it is not the whole story. In this section, scheduling will be placed in the context of day to day scheduling activities.

When studying human schedulers, it is often difficult to distill the conceptual scheduling and sequencing problem from what human schedulers are actually doing. Because the content of scheduling tasks can vary over organizations, different field studies have consequently focused on different scheduling elements.

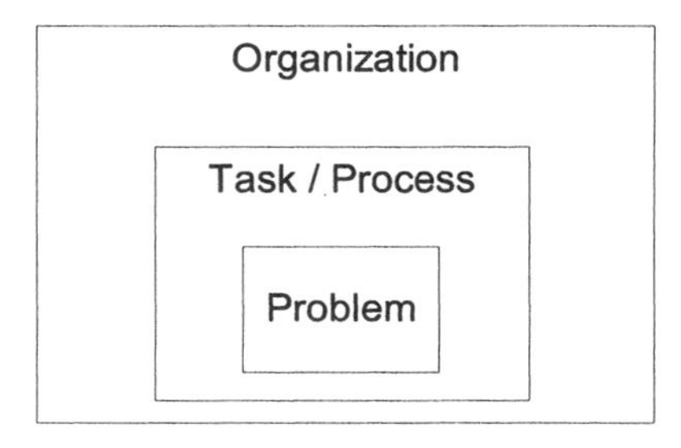

Figure 2-2. Scheduling context

The figure above illustrates three perspectives on scheduling that will be described in this section. In practice, a scheduling *problem* is often tightly coupled with an individual employee: a scheduler who executes the scheduling *process* and hence carries out a scheduling *task*. The human scheduler is part of an *organization* that provides the inputs for, and requires the results from the scheduling process. Unfortunately, the task concept can itself become an issue since different schedulers may draw task boundaries differently and this will vary with experience and skill.

The scheduling *problem* as currently defined in the academic literature highlights a specific part of the scheduling task - the sequencing. Therefore, in the next subsections, the task and organizational perspective will be used to provide an extended definition for scheduling.

4.2 Organization perspective

An operational view of scheduling can be partially derived by studying its context with other organizational production control functions. It is often difficult to make a single schedule for the whole production system of a company. Therefore, production systems are often decomposed into a (more or less) hierarchically organized planning and control structure to reduce the complexity of the scheduling problem. This approach is also known as Hierarchical Production Planning (HPP). For example, Bertrand et al. (1990) distinguish between goods flow control, which concerns planning and control decisions on the factory level, and production unit control, which concerns planning and control decisions on the production unit level. The goods flow control level also coordinates the various underlying production units.

The HPP paradigm is widely used and has become an accepted planning and control strategy for many medium to large manufacturing organizations. The HPP decomposition results in official tasks, task scope, authority, and responsibility. Although the decision model should fit the business requirements of the moment (Robb, 1910), the organizational structures in most firms are reasonably static and inflexible. This results in decision models being used that do not fit the decision problem. Unfortunately, the adoption of the HPP paradigm within a firm usually results in scheduling being done in an hierarchical fashion regardless of the appropriateness of the concept (McKay et al., 1995b).

Wiers (1997) identifies four types of control that are associated with planning, scheduling and dispatching:

- *Detailed control.* Dispatching is seen to be the most detailed control level dealing with the shortest planning horizon in the company. Dispatching answers the questions relating to: What do we do now? What do we do next? How do we fix the mess we are in? Scheduling control refers to work that is planned for the immediate horizon and the scheduler makes the predicted matching of time, resource, and work to be performed. This might also include the release of work to the factory floor. Planning's direct control refers to the ability of the decision maker to accept, interpret, and possibly modify demand. The planner might also be able to orchestrate personnel levels and resource capability - issues usually not possible to manipulate on very short notice.
- *Direct control.* Schedules are transferred to the shop floor without any intermediate control function between scheduling and the shop floor. That is, the scheduler is the person who is turned to for answers and direction. In a similar fashion, the planner transfers to the scheduler a plan for production (without detailed sequencing) and a scheduler to a possible dispatcher (a recommended schedule, but one that can be altered based on the situation).
- *Restricted control.* Short term issues relating to material requirements, material availability and available capacity are usually beyond the direct and immediate influence of the scheduling function and reside at the planning level. The scheduler can request or perform some expediting, additional shifts, and overtime, but in general, they have to live with the situation they have, and deal with the options in front of them for work assignment and operation execution. Each layer in the task structure has some form of direct and restricted control - the planner, the scheduler, the dispatcher.
- *Sustained control.* Each level has a form of sustained control over the level beneath it. For example, scheduling monitors the progress of production and solves problems if the actual situation deviates from the scheduled situation. The scheduler does not generate a release and sequence plan and then check on it the next day. The scheduler, when in

the plant, is typically provided with or seeks out feedback as to schedule execution and fulfillment.

Each of these levels can be observed and documented. The activities and the characteristics of the activities for planning, scheduling, and dispatching can be isolated and studied. However, the titles and official positions of individuals can create difficulties. For the purposes of research, it is necessary to use the traits and types of control suitable for a level to identify the individual and placement of decisions. Someone having the types of control, duties, and interactions with the shop floor typical of what we consider a scheduler to be, is a scheduler, regardless of title or organizational affiliation. For example, the person might be called a planner, but is really a scheduler. The scheduler might be in another department and report to a Materials Manager, but in reality deals directly with the shop floor supervisors or machine operators. If a floor supervisor is making the detailed assignment and sequencing decisions, then the supervisor is a scheduler or dispatcher for the purposes of studying scheduling. If the supervisor is making longer term assignments and time/resource allocations - the supervisor is planning. If the supervisor or operator is deciding what to do next from a set of immediate options, they are dispatching.

Note that the types of control (i.e., detailed, direct, restricted, sustained), specifically help to clarify the distinction between planning, scheduling, and dispatching. The four types and their usage can be considered preliminary and exploratory at this time and further research is required to sharpen this aspect of the scheduling perspective. Clear (or somewhat clear) definitions and distinctions are needed if work is to be compared on equal footing, or if work is to be replicated.

4.3 Task perspective

As noted, a relatively small number of studies have been conducted on real-world scheduling. In McKay (1992), the field studies captured the task specifics associated with what the schedulers were charged to do. The particular decisions associated with detailed, direct, restrictive, and sustained control were analyzed. As a result of looking at the task structure, it was documented that the schedulers' main function was to be a problem anticipator and solver, instead of a simple sequencer or dispatcher. The task analysis provided a clear view of what was being controlled, when it was controlled, and what feedback was used to execute and sustain control. In the field studies, control centered around uncertainty. It is interesting to note that the situation noted by McKay is similar to an early definition of what a scheduler was expected to do:

> "The schedule man must necessarily be thorough, because inaccurate and misleading information is much worse than useless. It seems trite to make that statement but experience makes it seem wise to restate it. He must have imaginative powers to enable him to interpret his charts and foresee trouble. He must have aggressiveness and initiative and perseverance, so that he will get the reasons underlying conditions which point to future difficulties and bring the matter to the attention of the Department Head or Heads involved and keep after them until they take the necessary action. He is in effect required to see to it that future troubles are discounted." (Coburn, circa 1918; pp. 172)

At one of the factories studied, the information and types of information used by the scheduler when dealing with operationally feasible (and desirable) schedules was gathered and analyzed. During the study period approximately 250 non-routine decisions were captured and encoded. These non-routine decisions were those that were not the obvious, straightforward material, job, resource, time decisions. The scheduler at the factory used many types of information for making decisions and was a key information hub - gathering and disseminating. In addition, the information was processed in an active fashion: collected, vetted, augmented, compressed, and reflected upon. Information acting as a cue or signal was used to control secondary information processing activities. Various categories or subject areas of information used by the scheduler are given in Table 2-1.

Table 2-1. Information used by planners, schedulers, and dispatchers

Category	**Examples of information used**
Humans	expertise/skill, motivation, absenteeism, and various other individual characteristics of operators, foremen, management, other schedulers, engineers, salesmen, suppliers, customers, transporters, technicians, subcontractors
Organization	goals, procedures, responsibilities, politics, gossip
Resources	capacity, flexibility, reliability, costs, location, state of maintenance, modes of operation (manual or automatic), age, and sensitivity of: machines, tools, fixtures, personnel, transportation equipment, buffers, pallets, subcontractors
Materials	due–date, required amount, customer, quality, processing time, age (regarding design), specifications, CNC programs, bills–of–material, routings, stability, batch size, stock level, risk

The richness of the information illustrates the challenge made to the ten simplifying assumptions of sequencing theory. It is also important to note that the majority of the scheduler's time and effort is related to anticipatory control based on risk assessment and mitigation - the types of activities difficult to assign to a decision support system.

The five field studies described in Wiers (1997) contained similar findings: the routine tasks could have been allocated to a decision support system, whereas the exceptional situations had to be handled by the human scheduler. Field studies conducted by Crawford (2000) also support the view of the scheduler being an information centre and problem rectifying resource. The two-stage control paradigm presented in McKay et al. (1995c) explicitly discusses a control theoretic role for the human in decision making.

Exceptional situations are those that can make a supposedly feasible plan, infeasible and possibly one not to be trusted. In McKay (1992), it was documented that approximately 10% of all of the scheduling decisions made by the scheduler were "exceptions." The majority of trigger events were considered to be routine by the scheduler – machine failures, wrong parts made, and so on – but the solutions, as represented by the final sequence of work to be done, varied in almost each case. This supports the problem solving view of the scheduler's role and the view of normal/exception information processing. The insights into exceptional decisions made by the scheduler also help to clarify the differences between sequencing and scheduling.

4.4 Daily activities

Previous sections have described the general task of scheduling and a number of different characteristics of the cognitive task. In this section, the daily routine of dispatchers and schedulers will be explored from an integrated perspective and we will use the term scheduler to refer to the combined task. For our purposes, the scheduler's daily routine comprises of what the person does each day at a high level of task description. This daily routine has been described in McKay and Buzacott (2000). The seven steps in a typical daily routine are depicted in Figure 2-3 and will be briefly described.

The decision maker (i.e., scheduler) starts the routine by a situation assessment. For example, the individual might want to see what changes have occurred in the daily shipping requirements, what was built in the last twenty-four hours, and what the current inventory levels are. From this information, the problem definition is refreshed and updated. The decision maker also obtains the net changes in supply and demand and renews the view of what the problem is.

1. Subsequently, special problems or crises are identified. These are likely to be the most constrained or most important activities in the factory. In decomposing the problem, these are the anchor points and will be addressed first.
2. The special problems are sometimes addressed by resequencing and allocating tasks. They might also be addressed by dynamic changes to resources, processes, quantities, materials, dates, crews, operators, and anything else that can get the job done. While a decision support system can help by resequencing, the ability to make and negotiate dynamic changes to the problem definition are currently in the realm of human capabilities.
3. The overall schedule is then updated around the anchors - wanting to see what the updated scenario means when compared to plans already in existence. By making the plan feasible for the hot jobs, the plan might now be infeasible for other work in the immediate future.

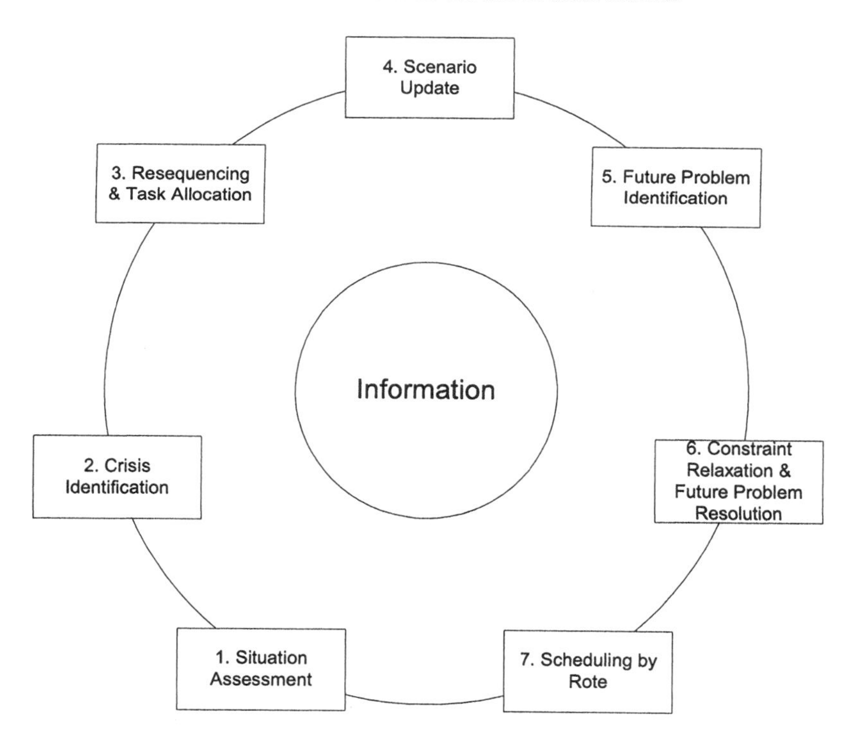

Figure 2-3. Seven subtasks for planning (adapted from McKay and Buzacott, 2000)

4. When the immediate problems have been discounted, the scheduler will identify future problems with the schedule or sequence - problems outside of the immediate dispatching horizon. These are the second order effects or issues that break the feasibility or desirability requirements.
5. The future problems will be attempted to be solved with sequencing or allocation strategies. However, if the problem cannot be addressed, the situation will be dealt with by relaxing constraints.
6. Lastly, the routine work that is not critically constrained is scheduled by rote, i.e., mechanically. This is the area of sequencing rules and systemized methods for coordinating the resources.

The daily routine will change somewhat when an unexpected change in supply, demand, or resource capability is sufficiently large enough to warrant major re-planning and re-analysis, i.e., when a crisis is identified. The daily routine will also vary depending on the day of week, week of month, and month of the year. For example, new short-term forecasts may be updated every Friday and the long term forecast might be updated the last week of the month.

Depending on the factory situation, the tasks might depend on information feeds from the manufacturing software systems (e.g., ERP), or on other activities in the plant. For example, tasks one and two (general situation assessment and crisis control) might be needed to be addressed before 6:30AM each day - before the day shift supervisors arrive. Resequencing and doing a scenario update might be needed to be performed between 6:30AM and 6:50AM - while the supervisors are available and before the day shift workers arrive at 7:00AM and before the 6:50AM production status meeting starts. The remaining tasks might need to be completed before the 8:00AM full production meeting. This is a real example from the field study which inspired the McKay and Buzacott model. It illustrates that the scheduler tasks are not always isolated, can be stressed by time and information requirements, and are affected by other tasks and activities. This is also part of scheduling.

5. INTEGRATING SCHEDULING AND SEQUENCING

The above sections have discussed how scheduling is composed of many factors, one of which is sequencing. While sequencing is mechanistic and algorithmic, scheduling in many real factory situations is not. An operational or practical view of scheduling takes into account operational feasibility and a required trust in the plan. If a scheduler decides to issue a plan that is known to be infeasible, then it should be consciously done, and not done by accident or by blind mandate. The human contribution to scheduling involves dealing with uncertainty, acting as an information hub, and anticipating problems before they occur. The scheduling task has also been discussed as having multiple phases or focus points and these tasks are not isolated or independent of the situation.

This section will take these various concepts and propose how they can be systematically integrated into decision support technology. The discussion focuses on the division of labor between the human and the planning technology and uses four concepts to guide the division and design of functionality: i) how well-defined the problem space is, ii) how much uncertainty exists in the operational environment, iii) problem complexity, and iv) transparency.

5.1 Well–defined vs. ill–defined criteria

To distinguish between formalized data typically found in manufacturing information systems and data used by schedulers, it is possible to consider the following attributes (McKay 1992):

- completeness of information;
- ambiguity in the information;
- error/accuracy in the information;
- presence or existence of the information.

The dimensions can be further discussed as to the implication of high or low completeness, sporadic or pervasive ambiguity, certainty of values and data, and if there is much or only little of this type of information needed for decision making. Where there is incompleteness, ambiguity, errors, inaccuracy, and possibly missing information used in determining sequencing decisions, the decision is ill-defined. A decision that is only partially ill-defined, might still be a candidate for formal techniques, but decisions that exhibit many of these traits will be a challenge for any formalized process. It is also possible that ill-defined decisions can be considered those which contain enriched data – information that is not normally found in manufacturing information systems. Some of the information such as key historical data might be captured and included via enhanced rules, but information such as the current weather conditions at a border crossing or the health of a worker after an evening of partying cannot be practically included.

Table 2-2. The relation between human reasoning level and information used

Human reasoning level	**Information used**
Skill–based	formalized data / well–defined
Rule–based	+ extended formalized data
Knowledge–based	Non–formalized data / ill–defined

The GEMS model presented in Section 0 can be tied to these characteristics by equating *routine*—the GEMS concept—to *formalization*—as shown in the Table 2-2.

The two central characteristics of this table are the amount of *routine* or *formalization* that can be achieved in a particular scheduling task; in other words, it can be used to characterize the information situation as well– or ill–defined. This notion assists in developing a first, general separation in the division of labor: it indicates when people are needed and when they are not. Humans are needed in production scheduling because they can solve ill–defined problems that cannot be modeled by systems designers (Sanderson, 1989). They can provide estimates for incomplete data, provide a judgment on ambiguous data, and correct data (McKay, 1987). A major contribution of humans is that they are social beings; they are continuously gathering information which is not instantaneously relevant to the scheduling task. Consequently, they can fill in blank spots of missing information using this "tacit knowledge." They also provide the interface to the non–formalized information needed when dealing with new situations or changing situations. Furthermore, they can provide information about constraint strengths, constraint relaxation, and penalties for constraint violations (McKay 1992). Thus, humans can outperform systems in problem areas where information is inadequate for any number of reasons. In essence, the humans are the interface to the environment in which the scheduling decisions will be executed and this includes the world at large and the factory itself.

If the situation has relatively few ill-defined aspects, the majority of decision making can rest in the decision support system at the skill-base level. The system should be capable of developing a reasonable starting point requiring few manual modifications. In such a situation, the richness in interpretation, representation, and manipulation functions can be minimized. It should also be possible to encode many of the enhanced or enriched aspects of the problems in rules or decision tables and further reduce the gap between feasible and infeasible schedules – the rule-base level.

If the situation is considered to be largely ill-defined, much of the control has to reside in the knowledge-base level. For example, a reasonable approach might be to use a simple loading heuristic (e.g., forward or backward loading with basic priority) to create a plan. The system should then focus on the added functionality needed to enhance interpretation, feedback, and manipulation. In such cases, the manipulation functions should allow the scheduler to effectively "do what I say" and "not what the database says is possible." If creative problem solving is used routinely to deal with ill-defined aspects of the problem, the scheduler will need complete freedom to say what can be done where, when, and by what – often violating what could be considered hard constraints. For example, the scheduler might assign a task to a machine not specified as being able to perform a task, and assign it at the same time as another job is running on the machine – effectively having the machine do two things at once. This particular example has been observed in the field studies. The machine in question was a two-stage process, normally bolted and welded together. The normally scheduled job was using one of the two stages and with a little bit of work, the single machine was soon two. There was not sufficient time to alter the scheduling and manufacturing database to create this unique solution and the scheduler wanted to make the assignment – immediately and create the necessary paperwork for the factory floor. We call this capability, the ability to lie to the computer. The job shop DSS tool in the longitudinal study by McKay and Wiers (2003b) has this ability and it is used routinely. Approximately 30% of the daily reports are modified daily – some very little, some more so. This particular job shop has many ill-defined problems and the majority of software development has been focused on supporting the knowledge-base of the scheduler and avoiding getting in the way of the scheduler.

5.2 Autonomy and uncertainty

5.2.1 Shop types

The extent to which a certain scheduling task can be supported using decision support systems obviously depends on the characteristics of the underlying production system. As uncertainty in the production system increases, the ability of a 'smart' system to perform decreases. That is, *uncertainty kills smartness*. If a situation exhibits uncertainty, the smart features must be implemented in such a fashion as to effectively and efficiently complement the human tasks and additional functions must exist to support the human tasks. If the functions are not so designed and implemented, there will be a mismatch in requirements/solution – the solutions will not be close to feasible and the human will not be able to do the necessary tasks easily and this will ultimately lead to failure of the system. The smart system ignoring the uncertainty implications will either block the human's tasks, or not provide the information necessary to make a decision, or prevent the human from describing the solution.

One important aspect to consider before implementing any decision support system is the question of where to deal with uncertainty: at the scheduler's level or at the shopfloor. This is the question of what autonomy to allocate to the shop floor regarding production control decisions. In Wiers (1997), a typology of production systems is given that describes the possible strategies in allocating autonomy to the shop floor. The typology is depicted in Table 2-3.

The four categories attempt to capture the dynamic nature of the shop with the corresponding 'general' style of scheduling that can be expected. The uncertainty columns relate primarily to the information base and the rows relate to the ability of the shop to execute the plan – uncertainty in execution. The concept of *human recovery* indicates the positive role that human operators can play in the prevention of disturbances by using flexibility to compensate for uncertainty in the information space. Before linking the two together explicitly, we will briefly describe each of the four categories in the following subsections.

5.2.2 Smooth shop

In the *smooth shop*, there is little uncertainty in the detailed information or execution phases and as a result, there is little need for human intervention and problem solving, i.e., the recovery. Since the shop is stable, optimization can be performed with precise operation timing and sequences. It is likely that the scheduler's life will focus on tactical policies and fine–tuning of the optimization approach – it will not be dominated by exception decisions. A smooth shop is also likely to be considered relatively well-defined and offers the best promise for scheduling systems that provide full, automatic functionality and expect the shop to execute as planned. These types of situations are well suited for the skill-based and rule-based solutions.

Table 2-3. Typology to allocate autonomy

	No uncertainty	Uncertainty
No human recovery	Smooth shop	Stress shop
Human recovery	Social shop	Sociotechnical shop

5.2.3 Social shop

In the *social shop*, there is limited or non–existent uncertainty in the macro or aggregate levels and possibly some minor uncertainty in the detailed situation. In this case, the scheduler can lay out the basic schedule with sequences and timing, but allow for autonomy on the shop floor to tune the final work sequence at any resource. The scheduler provides an optimized recommendation, but acknowledges that some recovery or adjustment will be necessary. Ideally, the schedule identifies the operation sequence, recommended timing, and possible bounds for advancing or delaying the work. Because of their close relation to the production process, human operators are often faster and better able to react to disturbances than the scheduler. Obviously, the social shop is not as well-defined as the smooth shop. As a result, it is possible that if the humans can accept 'close enough' and if the decision support system is robust in terms of re-scheduling and recovery, a computer dominated situation can be achieved. This is possibly the third-best situation in which a scheduling system can be deployed. Depending on how well-defined the actual decisions are, these types of situations might also be well suited for the skill-based and rule-based solutions.

5.2.4 Stress shop

In the *stress shop*, there will be little uncertainty in the planning and state information, but substantial uncertainty in the execution phase – the schedule cannot be carried out as planned. The decisions themselves are not ill–defined and the proper reaction can be determined via rules and static properties. The necessary flexibility needed to resolve the problem can be identified and exploited via pre–specified algorithms and knowledge imbedded in the software. While not as stable as the smooth shop, the scheduling problem can still be considered relatively well-defined in terms of its structure. This is possibly the second-best situation in which a scheduling system can be deployed. It is highly likely that a skill-based and rule-based solution would be sufficient for the purposes of creating a feasible sequence.

5.2.5 Sociotechnical shop

In the *sociotechnical shop*, the worst of all possible worlds exists – substantial uncertainty in information and execution, and the problem is definitely ill-defined. This is the world explicitly studied by McKay (1987, 1992) and provides the most challenging environment for predictive and reactive scheduling. In this situation, it is not possible to *a priori* imbed the necessary flexibility into the system to identify or solve precise problems. It is impossible to know everything in advance when dealing with new inventions or situations and it is best to plan for unknowns and not pretend they will not exist. Appropriately designed decision support systems for supporting the ill-defined problem are required; helping to identify patterns, predicting future problems, and recommending possible solutions. The authors are not aware of any such system which has been developed commercially for these types of manufacturing situations.

5.3 Complexity

Decision support systems have a considerable advantage over humans when the skill-based reasoning is straightforward and the rule–based reasoning task is of sufficient *complexity* to make solutions less than obvious. That is, in order to gain advantage over a human scheduler, the problem domain should be complex in terms of reasoning rules and the number of possible, operationally feasible schedules to consider. A suitable situation for substantial computer assistance in scheduling could be a site such as a process oriented plant where the number of combinations is large; or a group of repetitive flexible manufacturing cells where the number of routings might be large, but where the manufacturing process is well known, stable, and well defined.

5.4 Transparency

In Wiers (1997), transparency of scheduling systems is discussed as an important factor for the amount of confidence a scheduler will have in the system. Especially in situations with much uncertainty, schedulers want to be in direct control, without visible interference of a system. Therefore, the need to be in direct control depends on the amount of re-scheduling required. For information systems to be helpful in re-scheduling actions, they have to be transparent. If many re-scheduling actions have to be carried out, an opaque information system is perceived to get 'in the way' of the human scheduler. The concept of transparency can be considered a research area; in a decision support system, how can the system be designed with variable and controllable degrees of transparency to deal with the routine situations and the exceptional conditions? The schedulers do not want to see the details or be concerned about the parts of shop which are running smoothly, but require complete and comprehensive control for the problem areas. To make a distinction in visualization between aspects that allow routine and exceptions is a major challenge for any kind of scheduling support system.

6. DESIGNING DECISION SUPPORT SYSTEMS

6.1 Design model

How can human schedulers be supported by scheduling information systems? This question is answered by presenting a design model for scheduling decision support systems, which is based on the criteria presented in the previous section and presented in Figure 2-4.

The relationship between these concepts and the required scheduling information system is depicted in the figure above. The use of the system depends on the match between the required system and the actual system. The characteristics on the left side in Figure 2-4 and the scheduler's autonomy are given for a specific situation, although each practitioner probably suggests improvements on aspects such as uncertainty and complexity. Causal relationships are read from left to right in Figure 2-4. From this perspective, it would have been appropriate to place the rectangles related to the scheduler's autonomy to the left of the other characteristics, as the division of autonomy between the scheduler and the shop floor influences these characteristics. Instead, this relationship is represented by the dashed line.

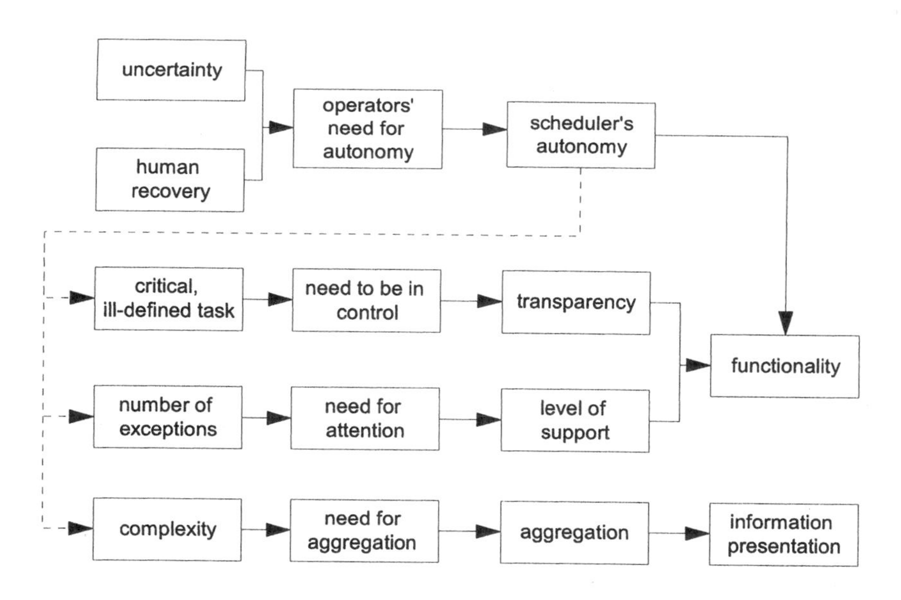

Figure 2-4. Model of designing scheduling decision support (Wiers, 1997)

In the design model, no relationship is assumed to exist between a scheduling information system's functionality and information presentation. However, in information systems, certain functionality might require or impede the presentation of certain information, and vice versa. For example, a low level of support of a scheduling information system's functionality is often combined with information presentation functions. Possible interactions between functionality and information presentation therefore have to be considered during the design of a scheduling information system in practice.

6.2 Semantics and the scheduler's ontology

A decision support system can be quite invasive and disruptive if the system forces the users to adapt to the system – what is done, when it is done, objects used, and the vocabulary employed. In order for the system to be quickly accepted and reduce the disruptive nature, it is possible for the decision support system to adopt the terminology of the users and support their meta-objects. For example, we have observed that the word 'job' has different meanings depending on the factory, as does crew, shift, and any other term possible to use in a scheduling situation.

If production control is viewed as a community and studied via ethnographic methods (McKay and Wiers, 2003a), the communication, interrelationships, and roles of the individuals are highlighted. One of the results of such a study is an ontology of the scheduler's objects and concepts – what they use when they talk about scheduling and what they use to communicate to others. The meta-operations of the scheduler are part of the ontology and are important for understanding the meaning or semantics of the situation. For example, is work effort described in hours, piece rates, shift capacity, or crew capability? It is possible that multiple meta-elements are used simultaneously or in certain situations.

In one field study, we observed and ultimately supported four different ways to describe the load being applied to a resource. Instead of having the human translate the four meta-values to a normalized standard, the system did it internally and preserved the meta-view for the scheduler and other users of the plan. This type of meta-support addresses part of the transparency issue as everyone would realize how the 2,200 for one day was derived – was it from a crew perspective, a rate perspective, etc.

6.3 Task support

Traditional software and systems are designed from a functional perspective – similar to word processing or spreadsheet menus. Schedulers may use their systems continually throughout the day, each and every day. The scheduling systems are not periodic or casual use items, they are mission critical. As noted in the sections on task identification, the schedulers and planners have specific processes they do regularly – each day or each week at roughly the same time, being done in the same way. By packaging functions together, it is possible to create user interfaces based on task allocation (McKay and Wiers, 2003b). In these systems, the interface might have selections for tasks done in the morning (e.g., special optimized interfaces for the 6AM, 6:30AM, 7AM, and 8AM activities), ones in the afternoon and so forth. The job shop system noted earlier has such menus and functions that reduce the user interaction to the bare minimum. Wherever manual, repetitive processes are noted, they are prepackaged. The flow shop system also alluded to earlier in the chapter has menus that change on Fridays (present different tasks and certain functions have enhanced processes). The menus and functions also change on the first of the month and are different if the system was not used on the first of the month and it is now later in the month. The task oriented approach assists when the regular decision maker is ill or on vacation, and when the backup scheduler or manager must use the system.

7. CONCLUSION

In summary, the research on the human scheduler is too meager for any general, quantitatively supported results to be stated in a predictive fashion. At best, the majority of work is descriptive with some insights about what might be reasonable to include in production control practices and decision support systems. At worst, the research is anecdotal without any rigor or scientific value. In this chapter, we have tried to indicate the research which has been done with generally accepted methodologies and from which contributions have been made to the body of knowledge concerning scheduling.

Regardless of methodology, short case studies are always subject to possible bias in what is observed and captured, and what is not. The most rigorous insights have been derived from extensive longitudinal studies which cannot be generalized per se. It is clear that additional, extensive studies are required in this domain. However, the existing studies support each other and general patterns can be observed. The themes of problem solving and task structures are prevalent. The need for transparency and support for dealing with uncertainty are also supported by direct and related research on the scheduling task. Decision support systems can be constructed that support these ideas and working industrial systems have been constructed. The next step is the generalization of the concepts and the inclusion of the ideas in widely available commercial form.

From a theoretical perspective, an enhanced *Theory of Scheduling* could include and embrace these and the other concepts described in this chapter as a starting point. Further research should either dismiss or support them, and possibly add additional concepts to the theory. In any event, the traditional view that the *Theory of Scheduling* = *Theory of Sequencing* is insufficient to bridge the gap between theory and practice.

This chapter has presented a number of guidelines to assist practitioners in designing decision support systems for production scheduling tasks. A design model was presented that is based on four key elements in the scheduler's task support: autonomy, transparency, level of support and presentation of information. Secondly, in the subsection on semantics and the scheduler's ontology, it has been discussed how scheduling decision support systems should speak the language of the scheduler. Lastly, it was emphasized that decision support systems should support a scheduling task, which goes beyond offering a number of functions that are in the eyes of the scheduler structured differently from the daily activities to be supported by them.

ACKNOWLEDGEMENTS

This research has been supported in part by NSERC grant OGP0121274 on Adaptive Production Control.

REFERENCES

Bertrand, J. W. M., Wortmann, J. C. and Wijngaard, J., 1990, *Production Control: A Structural and Design Oriented Approach*, Elsevier, Amsterdam.

Boer, A. A. A. den, 1992, Integration of Information in Logistic Operations, in: *Proceedings of the IFIP WG 5.7 Working Conference: Integration in Production Management Systems*, Eindhoven, August 24-27, pp. 232-246.

Boer, A. A. A. den, 1994, Decision Support for Material Planners, *Production Planning and Control*, **5**(3):253–257

Brehmer, B., 1980, In one word: Not from experience, *Acta Psychologica*, **45**: 223–241.

Buxey, G., 1989, Production Scheduling: Practice and Theory, *European Journal of Operational Research*, **39**:17-31.

Coburn, F. G., circa 1918, Scheduling: The Coordination of Effort, in: *Organizing for Production and Other Papers on Management 1912-1924*, I. Mayer, ed.,Hive Publishing, 1981, pp. 149–172.

Conway, R. N., Maxwell, W. L. and Miller, L. W., 1967, *Theory of Scheduling*, Addison Wesley, Reading MA.

Crawford, S., 2000, *A Field Study of Schedulers in Industry: Understanding their Work, Practices and Performance*, Ph.D. Thesis, University of Nottingham.

Crawford, S., and Wiers, V. C. S., 2001, From anecdotes to theory: reviewing the knowledge of the human factors in planning and scheduling, in: *Human Performance in Planning and Scheduling,* B.L. MacCarthy and J.R. Wilson, eds., Taylor & Francis, London, pp. 15-44.

Davis, F. D., and Kotteman, J. E., 1995, Determinants of decision rule use in a production planning task, *Organizational Behavior and Human Decision Processes*, **63**(2): 145–157.

DeShon, R. P., and Alexander, R. A., 1996, Goal setting effects on implicit and explicit learning of complex tasks, *Organizational Behavior and Human Decision Processes*, **65**(1):18–36.

Dudek, R. A., Panwalkar, S. S., and Smith, M. L., 1992, The Lesson of Flowshop Scheduling Research, *Operations Research*, **40**(1):7-13.

Early, P. C., Northcraft, G. B., Lee, C., and Lituchy, T. R., 1990, Impact of process and outcome feedback on the relation of goal setting to task performance, *Academy of Management Journal*, **33**(1):87–105.

Gary, K., Uzsoy, R., Smith, S. P., and Kempf, K. G., 1995, Measuring the quality of manufacturing schedules, in: D.E. Brown and W.T. Scherer, eds., *Intelligent scheduling systems*, Kluwer Academic Publishers, Boston, pp. 129-154.

Graves, S. C., 1981, A Review of Production Scheduling, *Operations Research*, **29**(4):646-675.

Higgins, P. G., 1999, *Job Shop Scheduling: Hybrid Intelligent Human-Computer Paradigm*, Ph.D. Thesis, University of Melbourne.

Hoffman, P. J., 1960, The paramorphic representation of clinical judgment, *Psychological Bulletin*, **57**(2):116–131.

Igbaria, M., and Iivari, J., 1995, The effects of self–efficacy on computer usage. *OMEGA – International Journal of Management Science*, **23**(6):587–605.

Jacoby, J., Mazursky, D., Troutman, T., and Kuss, A., 1984, When feedback is ignored: Disutility of outcome feedback, *Journal of Applied Psychology*, **69**(3):531–545.

Johnson, D. S., Perlow, R., and Pieper, K.F., 1993, Differences in task performance as a function of type of feedback: Learning–oriented versus performance–oriented feedback, *Journal of Applied Social Psychology*, **23**(4):303–320.

Kleinmuntz, B., 1990, Why we still use our heads instead of formulas: Toward an integrative approach, *Psychological Bulletin*, **107**(3):296–310.

Laforge, R. L., and Craighead, C. W., 2000, Computer-Based Scheduling In Manufacturing Firms: Some Indicators Of Successful Practice, *Production And Inventory Management Journal,* **41**(1):3–24.

Levy, P. E., Albright, M. D., Cawley, B. D., and Williams, J. R., 1995, Situational and individual determinants of feedback seeking: A closer look at the process, *Organizational Behavior and Human Decision Processes,* **62**(1):23–27.

MacCarthy, B. L., and Liu, J., 1993, Addressing the Gap in Scheduling Research: A Review of Optimization and Heuristic Methods in Production Scheduling, *International Journal of Production Research*, **31**(1):59-79.

McKay, K. N., 1987, *Conceptual Framework For Job Shop Scheduling*, MASc Thesis, University of Waterloo.

McKay, K. N., 1992, *Production Planning and Scheduling: A Model for Manufacturing Decisions Requiring Judgement*, Ph.D. Thesis, University of Waterloo.

McKay, K. N., and Buzacott, J. A., 2000, The Application of Computerized Production Control Systems in Job Shop Environments, *Computers In Industry*, **42**(2):79-97.

McKay, K. N., Buzacott, J. A., and Safayeni, F. R., 1989, The scheduler's knowledge of uncertainty: The missing link, in: *Knowledge based production management systems,* J. Browne, ed., Elsevier, New York, pp. 171–189.

McKay, K. N., Safayeni, F. R., and Buzacott, J. A., 1995a, Common Sense Realities in Planning and Scheduling Printed Circuit Board Production, *International Journal of Production Research*, **33**(6):1587-1603.

McKay, K. N., Safayeni, F. R., and Buzacott, J. A., 1995b, A Review of Hierarchical Production Planning And Its Applicability For Modern Manufacturing, *Production Planning and Control*, **6**(5): 384-394.

McKay, K. N., Safayeni, F. R., and Buzacott, J. A., 1995c, An information systems paradigm for decisions in rapidly changing industries, *Control Engineering Practice*, **3**(1):77–88.

McKay, K. N., Safayeni, F. R., and Buzacott, J. A., 1988, Job-shop scheduling theory: what is relevant?, *Interfaces*, **18** (4):84-90.

McKay K. N., and Wiers, V. C. S., 1999, Unifying the Theory and Practice of Production Scheduling, *Journal of Manufacturing Systems,* **18**(4):241-255.

McKay, K. N., and Wiers, V. C. S., 2003a, Planners, Schedulers and Dispatchers: a description of cognitive tasks in production control, *Cognition, Technology & Work,* **25**(2):82-93.

McKay, K. N., and Wiers, V.C.S., 2003b, Integrated Decision Support for Planning, Scheduling, and Dispatching Tasks in a Focused Factory, *Computers in Industry*, **50**(1):5-14.

Moray, N., 1995, Mental models, strategies and operator intervention in supervisory control, in: *Proceedings of the XIV European Annual Conference on Human Decision Making and Manual Control, The Netherlands, June 14–16*, H.G. Stassen and P.A. Wieringa, eds., Delft University of Technology, The Netherlands, pp. 4.3.1–4.3.10.

Muir, B. M., 1994, Trust in automation part I: Theoretical issues in the study of trust and human intervention in automated systems, *Ergonomics*, **37**(11):1905–1922.

Muir, B.M., and Moray, N., 1996, Trust in automation part II: Experimental studies of trust and human intervention in a process control simulation, *Ergonomics*, **39**(3):429–460.

Newell, A., and Simon, H. A., 1972, *Human problem solving*, Prentice-Hall, Englewood Cliffs, NJ.

Norman, D. A., 1988, *The psychology of everyday things*, Basic Books Inc., New York.

Pinedo, M., 1995, *Scheduling Theory, Algorithms and Systems*, Prentice-Hall, New Jersey.

Price, H. E., 1985, The allocation of functions in systems, *Human Factors*, **27**(1):33–45.

Rasmussen, J., 1986, *Information processing and human computer interaction: An approach to cognitive engineering*, North–Holland, Amsterdam.

Reason, J. T., 1990, *Human error*, Cambridge University Press, Cambridge.

Robb, R., 1910, *Lectures On Organization,* private printing, Boston.

Rodammer, F. A., and White, K. P., 1988, A Recent Survey of Production Scheduling, *IEEE Transactions on Systems, Man, and Cybernetics*, **18**(6):841-851.

Sanderson, P. M., 1989, The Human Planning and Scheduling Role in Advanced Manufacturing Systems: An Emerging Human Factors Domain, *Human Factors,* **31**(6):635–666.

Sanderson, P. M., 1991, Towards the model human scheduler, *International Journal of Human Factors in Manufacturing,* **1**(3):195-219.

Sanderson, P. M., and Moray, N., 1990, The Human Factors of Scheduling Behaviour, in: *Ergonomics of Hybrid Automated Systems II*, W. Karwowski and M. Rahimi, eds., Elsevier, Amsterdam, pp. 399-406.

Schaaf, T. W. van der, 1993, Developing and using cognitive task typologies. *Ergonomics*, **7**(4):1439–1444.

Stoop, P. P. M., 1996, *Performance measurement in manufacturing: A method for short term performance evaluation and diagnosis*, Ph.D. Thesis, Eindhoven University of Technology.

Wærn, Y., 1989, *Cognitive aspects of computer supported tasks*, John Wiley & Sons, New York.

Wagenaar, W. A., Hudson, P. T., and Reason, J., 1990, Cognitive failures and accidents, *Applied Cognitive Psychology*, **4**:273–294.

Wiers, V. C. S., 1996, A Quantitative Field Study of the Decision Behavior of four Shop Floor Schedulers, *Production Planning and Control*, **7**(4):381–390.

Wiers, V. C. S., 1997, *Human–computer interaction in production scheduling: Analysis and design of decision support systems for production scheduling tasks*, Ph.D. Thesis, Eindhoven University of Technology.

Chapter 3

ORGANIZATIONAL, SYSTEMS AND HUMAN ISSUES IN PRODUCTION PLANNING, SCHEDULING AND CONTROL

Bart MacCarthy
Nottingham University Business School, UK

Abstract: With global markets and global competition, pressures are placed on manufacturing organizations to compress order fulfillment times, meet delivery commitments consistently and also maintain efficiency in operations to address cost issues. This chapter argues for a process perspective on planning, scheduling and control that integrates organizational planning structures, information systems as well as human decision makers. The chapter begins with a reconsideration of the gap between theory and practice, in particular for classical scheduling theory and hierarchical production planning and control. A number of the key studies of industrial practice are then described and their implications noted. A recent model of scheduling practice derived from a detailed study of real businesses is described. Socio-technical concepts are then introduced and their implications for the design and management of planning, scheduling and control systems are discussed. The implications of adopting a process perspective are noted along with insights from knowledge management. An overview is presented of a methodology for the (re-)design of planning, scheduling and control systems that integrates organizational, system and human perspectives. The most important messages from the chapter are then summarized.

Key words: Production planning, scheduling, organizational structure, human factors

1. INTRODUCTION

Effective planning and scheduling processes are essential for success in manufacturing operations. In today's environments manufacturing operations are typically supported by IT systems that, potentially, provide an abundance of real–time status information. There is a strong inclination to assume that the planning and scheduling process can be 'hard-wired' within the decision structures of the IT system by embedding appropriate models and algorithms. Indeed modern Enterprise Resource Planning (ERP) systems and 'add-ons' such as Advanced Planning and Scheduling (APS) systems try to embrace this philosophy (Padmos et al. 1999). However, the limitations of treating planning and scheduling as essentially mathematical problems capable of being isolated from their environments, fully specified and then solved for feasibility or optimality have been frequently noted (Buxey, 1989; Shobrys and White, 2000; MacCarthy and Wilson, 2001a).

Contemporary ERP systems may bring many benefits to operational control in manufacturing. The benefits are often derived from improvements in data representation, data handling and data integration. Frequently, however, ERP systems come with traditional hierarchical planning and control modules. Although more usable than MRP-based systems from two or three decades ago, they suffer from many of the same issues and limitations – difficulties in supporting responsive planning and control, lack of transparency, limited support for capacity planning and management, poor fit to particular sectors or industrial environments (Davenport 1998, Chen 2001). Many organizations have gone through ERP implementations, often driven by a desire to address operational control, response and order fulfillment problems. Re-engineering of information systems in businesses generally has proved difficult, if not daunting (McAfee, 2003). Many of the 'traditional' planning and control issues may remain after an ERP implementation (Konicki, 2001).

If existing systems, models and algorithms fail to provide full support to planning, scheduling and control functions, then what is missing, what should be included or what should be put in their place? These are difficult questions. This chapter addresses them.

The mathematical approaches to production planning, scheduling and control (PSC) are well known. They are embodied particularly in mathematical programming models that capture decision variables, constraints and objectives (e.g. Hax and Meal, 1975; Shapiro, 1993) and in classical scheduling theory that typically studies algorithms and heuristics to assign jobs to machines to optimize some objective(s) over a time domain (e.g. Baker 1973). Simulation approaches and combined optimization and simulation techniques have also been advocated (Shanthikumar and Sargent 1983).

In this chapter we look at the 'non-mathematical' research in planning, scheduling and control, in particular the key thinking on organizational, systems and human issues and its importance in the context of contemporary manufacturing operations. The emphasis is on identifying key contributions and their relevance to practice rather than a comprehensive review of the literature. Sanderson (1989) presents a review of the literature from a human factors perspective up to the late 1980's. A more recent comprehensive review and analysis of the literature on the human factors of production scheduling is provided by Crawford and Wiers (2001), which is more strongly grounded in operations and manufacturing. MacCarthy and Wilson (2001b) provide a view on the research from the perspective of operations strategy and practice.

In the next section we consider the 'gap' between theory and practice, focusing on two aspects – classical scheduling theory and hierarchical production planning and control. Section 3 begins with a brief description of key studies of industrial practice and also notes some other relevant research. It then presents a description and discusses the implications of a recent model of scheduling derived from detailed studies of practice in real businesses. This is followed by a discussion of socio-technical concepts and their relevance to planning, scheduling and control. Section 4 provides a business process perspective of planning, scheduling and control, discusses insights from knowledge management and presents an overview of a PSC (re)-design methodology that integrates organizational, systems and human perspectives.

2. THE GAP – MODEL 'DEVIANCE'

The gap referred to here concerns the limitations in the applicability and relevance to practice of many of the theoretical models and algorithms. We discuss two areas - classical scheduling theory and hierarchical planning and control.

2.1 Classical scheduling theory

An extensive literature exists on classical scheduling theory. The gap between the theory and real scheduling contexts has been much debated over the last two decades (e.g. Buxey, 1989; MacCarthy and Liu, 1993; Wiers, 1997). The areas of concern relate to the relevance of the models studied and the applicability of the results that have been generated. Here we review, update and rethink some of the ideas.

What is it about actual production environments that classical scheduling models miss? If we view a production system as a transformation of inputs through a conversion process into outputs, we can examine where model deficiencies may occur.

Inputs: In the main, the classical theory assumes a static, finite set of jobs waiting to be scheduled onto a production system. Little consideration is given as to how this set may have arisen, its size, composition or whether it is static. The typical theoretical scheduling problem that is posed in this fashion misses the links with higher level production and capacity planning and with other functions such as materials management. In practice it is likely that the job set will be the output of a planning system, typically operating on a rolling time horizon. In circumstances where the production system is under-utilized, the scheduling problem may be trivial or non-existent. At the other end of the spectrum, a scheduling algorithm may help to extract the most effective utilization from overloaded facilities or reduce the proportion of late jobs but the primary operations management problem is concerned with capacity planning and management, not scheduling.

Ignoring the higher level functions within which lower level allocation decisions are defined and constrained and ignoring the dynamic rolling pipeline of planned production, and the problems and opportunities it provides, are a significant part of the 'disconnect' between scheduling in theory and in practice. The frequency with which plans are updated and the time horizon over which one may assume relatively constant conditions are determined at the higher level. More generally, business strategy and policy have a bearing on how an enterprise positions itself for response – whether it favors stability or reactiveness.

A theoretical scheduling model requires a performance objective to be specified - typically some single regular measure of performance such as minimizing average flow time or maximum completion time. Here the 'gap' issues are a little more subtle. There may be multiple objectives, sometimes competing, conflicting or rapidly changing. In practice the most appropriate objective may not be at all obvious and may only be judged some time in advance. Although some progress has been made in addressing multiple objectives, the precise context in which many theoretical results might be applied is not immediately obvious.

Process: The classical models have considered many different machine configurations from single stage, single machine to complex job shop configurations with multiple potential routes, and parallel non-identical machines at some or all processing stages. The range appears to capture the most likely production system configurations. A reflection may be that the challenging computational complexity of the job shop problem may have received more attention than it deserves from a production perspective, given its relative lack of prevalence in reality.

A more fundamental issue is the widespread development of production systems that 'encourage flow' and that are directly demand driven. In cellular manufacturing, complete products, product modules, sub-assemblies or elements are manufactured in a sequenced and balanced set of operations in a cell. In Just-in-Time approaches, work is pulled through the production system from downstream to upstream often with signaled kanban control. These are systems that are essentially capable of self-regulation. The problems that arise with these systems are in their design and specification - ensuring that there is balanced production flow at the required level. The production planning problems are concerned with facility loading and balancing rather than sequencing and scheduling at the machine or process level.

Outputs: The classical theory assumes that schedule generation is the principal problem and, once generated, is the end of the scheduling problem. In practice (as we shall see), scheduling is strongly about implementation. Any particular scheduled job could require a specific set of resources (materials, personnel, machines) to be available at any or each processing stage. In fact schedule generation may be a small part of a human scheduler's activities in comparison to the effort needed to ensure that the schedule happens i.e. that resources are in place for the desired schedule and that progress against the planned scheduled is acceptable. A schedule that is easy or feasible to resource and implement may often be sought.

Of course the scheduling theorist may respond by saying that all production constraints need to be specified at the outset for any model to generate realistic solutions capable of being implemented. However, as well as potentially magnifying the computational complexity enormously, this goes to the heart of the problem – whether all potential constraints can be adequately specified at the time of schedule generation. It is often unrealistic in dynamic and complex production environments to try to predict likely future conditions and likely resource availability at the level of granularity and time precision that would be needed for such a model. This is an area where human expertise and judgment is most often needed.

One reason cited frequently for the 'gap' is the stochastic nature of disturbances and uncertainties in manufacturing. However, this needs a little more probing. The level and types of uncertainties and disturbances will be context dependent and are often an inherent part of the business environment. Some sectors must live with inherently unreliable processes (e.g. steel rolling mills) whilst others may have inflexible supply sources (e.g. clothing manufacture). Some problems may be self-imposed - unrealistic aggregate plans, poor capacity planning decisions, quality problems, material supply problems, labor absenteeism or poorly maintained plant.

Practical planning and scheduling must however assume deterministic conditions whilst having contingencies in place to address risks that may occur e.g. loading to a limited level of utilization or allowing 'fall-back' schedules to be adopted. A significant contribution of human scheduling is achieving and maintaining the desired level of utilization through the facility or resorting to an appropriate schedule when resource or disturbance problems occur. In both these cases the human scheduler will often strive for minimum deviation from a prescribed schedule – in time, in sequence or in objective. Again the links with higher levels of planning are important in these contexts.

Of course there are many specialized scheduling scenarios that have been researched to address some of the concerns noted above to some degree. Rescheduling and real-time scheduling for instance is an area where some developments have occurred (e.g. Hall and Potts, 2004) and more would be welcome. In addition, classical scheduling theory contributes to our understanding by providing insights into aspects of scheduling problem structure and in helping us to understand the complexity of many manufacturing domains. Given the rich repository of results we now have, it is possible that scheduling theory could contribute more to the development of intelligent and flexible automation (and indeed it does do so in other domains such as computer systems). However the key issues raised here are inherent in managing effective manufacturing enterprises and cannot be addressed by models or algorithms alone.

2.2 Hierarchical planning and control

The need for a decision hierarchy

Manufactured products require materials, components and sub-assemblies to be either sourced from suppliers or produced in-house. Given the cumulative lead-times for sourcing and for manufacturing and assembly operations, and the fact that these are typically longer than customers are prepared to wait, then the necessity to commit to and plan for production over future time periods is clear. In most cases this results in a dynamic pipeline of planned products to meet anticipated demand in future time periods (Vollman et al. 1992). Constraints on the pipeline increase close to actual production with fewer opportunities for changes as planned orders flow from upstream to downstream. With the additional complications of extensive product ranges, complex product structures and various customer stipulations, then planning and control structures are needed to ensure delivery commitments are met whilst utilizing production capacity and resources effectively and efficiently.

A hierarchical approach is intuitive and natural in coping with the complexities of planning, scheduling and control in manufacturing (HPP). Bertrand et al. (1990) describe the hierarchical production planning and control (HPP) paradigm. In the contemporary approach, sales and operations planning take into account different 'interest' groups in the organization e.g. marketing, production and finance, in developing agreed delivery plans, capacity usage, and inventory levels at an aggregate level over a rolling time horizon. Higher level decisions in the hierarchy constrain and drive activities at lower levels such as short term capacity management, materials control and ultimately the release of work orders to the shop floor. Checks on capacity and inventory requirements at each level of the hierarchy are necessary to ensure that realistic and feasible plans are generated and that appropriate constraints are set.

The hierarchical approach separates different kinds of decisions at different levels and over different time periods, enabling a degree of stability in the planning process and allowing complex manufacturing operations to be buffered against too many short-term changes (Bitran et al. 1982). It can also mean that a planning level can have a degree of autonomy within the constraints set. The hierarchical approach is strongly associated with MRP/MRPII control philosophies and the associated computer-based systems. MRP may have been developed originally for complex products in multi-level, batch manufacturing and assembly but it is used much more widely. Even JIT systems may have some level of MRP control (Spencer and Cox, 1995). MRPII type thinking is the control logic of the manufacturing planning and control modules of many of the leading ERP vendors and in that sense may be said to be the dominant planning and control paradigm.

Problems with HPP

The approach may appear to work well in the textbook but the reality is often more problematic – why? Although seemingly natural and intuitive it may be difficult to implement - how many levels are needed for instance? What should constrain what and how tightly? What should we plan in advance? Does a hierarchical approach reduce the speed of decision-making by requiring continual upward referral? How much autonomy and local control should be devolved to lower levels or to distributed production facilities? MRP may bring stability with different aspects of planning being dealt with over different time frames but is stability achieved by rigidity and at the expense of speed and responsiveness?

Accuracy and timeliness of data and appropriate data formats are pre-requisites for effective MRP-based planning and control. Education of managerial personnel is also crucial but probably even more important is the discipline required in running such systems. More fundamental technical issues are the assumptions of fixed lead times, the estimation of lot sizes and safety stock policies and the well-known phenomenon of system nervousness (Koh et al., 2002), resulting in high stock levels and a lack of system predictability and transparency. It is important to realize also that shop floor sequencing is not addressed. All of these considerations make MRP-based planning and control systems challenging to operate and to perform effectively to support business objectives.

Although the problems associated with HPP have been noted, particularly as embodied in MRP-based approaches, in practice such systems must be made to work with structures and solutions that are appropriate to the current environment. Businesses evolve and change in many ways. Technologies and operational resources change, product mixes change, supply chain partners change. In the changing industrial and business landscapes, mergers and acquisitions are ever-present, resulting in significant 'legacy' issues for organizations. Yesterday's formal procedures may no longer work in today's organization. In essence it is the organization, systems and human contributions in combination that make PSC processes work, not just in 'filling in the gaps' but in generating innovative and creative solutions to new problems. It should not be assumed that reliance on human support is an indicator of a poorly performing system; rather that the human contribution may be the essential ingredient. People learn and develop skills to manage and control dynamic systems over time.

There has been an expectation that Information Technology would enable more rapid decision-making and improved responsiveness in industrial organizations without the need for overbearing control structures. Shobrys and White (2000) consider these issues in the context of the process industry sector where levels of the hierarchy are generally more tightly coupled, technically and organizationally, than in the discrete manufacturing sectors. However, even here the anticipated benefits of IT are often difficult to achieve. The principal 'roadblocks' identified in achieving the integration that appears to be technically feasible are concerned with organizational decision-making and human decision-making behavior within organizations. In fact socio-technical theorists and practitioners have long noted the importance of these issues in real systems (see section 3.2).

There have been many interesting developments in the technical side of planning and control in recent years. The movement to pull-type production systems has been noted in the previous section. Also of note are Theory of Constraints (Gupta, 2003), Workload control (Breithaupt, 2002) and POLCA (Suri, 1998). However, regardless of the technical control philosophy, the organizational, systems and human issues need to be addressed in designing and managing effective PSC systems. The remainder of this chapter will discusses these 'inherent' issues.

3. STUDIES OF PRACTICE

A selection of research studies are noted here, focusing on some of the key field studies. The goal of much of this work has been to reduce the gap between theory and practice in planning and scheduling by providing empirical evidence on the factors that influence practice.

Field studies

The early literature stretches back to the seminal work of Dutton (1962, 1964) who attempted to capture scheduling practice in a box manufacturer from a simulation model of scheduler behavior. Dutton and Starbuck (1971) studied a scheduler to develop a model of how he estimated the run-time for two fabricator machines in a textile company, capturing the essence of the estimation process in two non-linear equations. Hurst and McNamara (1967) studied the decision rules of a planner in a textile mill, capturing them in an equation that could be used to decide machine combinations for unscheduled orders. Aspects of planner behavior considered subjective were not included.

There was a considerable gap until the late 1980s before interest in the subject was rekindled. A number of studies in the last two decades have attempted to develop knowledge-based or expert systems to support scheduling processes (Fox and Smith, 1984; Kanet and Adelsberger, 1987). For instance, Duchessi and O'Keefe (1990) studied an experienced planner in a firm producing garden products, eliciting his knowledge to develop a decision aid for production planning. The decision support tool used heuristics to generate realistic and feasible plans.

McKay's work in the field is noteworthy (McKay et al., 1988). McKay et al. (1995a) carried out detailed fieldwork in a Printed Circuit Board (PCB) assembly plant to understand the 'common sense' realities of planning and scheduling. Interesting aspects identified included: the differences between formal and actual scheduling procedures; the importance of the scheduler's information network; the 'political' reality requiring multiple schedules; the impact of performance criteria on scheduler behavior; and the need for schedulers to be proactive. The importance of scheduler expertise and experience is noted. Automating decision-making was problematic because of instabilities in the environment. McKay et al. (1995b) examined how schedulers perform their jobs in a rapidly changing electronics firm. They noted the types of information needed for the design of computer-based decision support systems and presented a framework for studying the scheduling task.

Wiers (1996) conducted a field study of four production schedulers at a truck manufacturing company using a quantitative approach to capture key elements of scheduling decisions – performance variables, action variables and disturbance variables. Observed decision behavior was different between schedulers, which has consequences for studying scheduling in practice. Some schedulers show nervous decision behavior. Some surprising conclusions were reached on scheduler performance and its implications for production unit performance.

Slomp (1997, 2001) has conducted longitudinal case study research on Flexible Manufacturing Systems (FMS). Much of the literature considers the planning and scheduling of FMS as a purely technical problem. Slomp (2001) shows the high level of human involvement necessary to deal with complex FMS planning and control problems. Options are presented on how to allocate planning, scheduling and sequencing tasks in a team responsible for operating the FMS. A human-centered approach is advocated.

Webster (2001) describes a field study of scheduling practice in a cutting tool manufacturer and how an individual scheduler manages to cope with the complexity with a largely manual scheduling process. Those responsible for scheduling may have additional production roles. A major emphasis is on early problem identification to 'nip problems in the bud.' Scheduler knowledge and experience, interpersonal networks, data management, and cognitive factors are significant aspects of the role.

In an empirical study of twelve small and medium-sized Make-To-Order (MTO) companies, Harvey (2001) investigates production supervisors in relation to planning, scheduling, execution and control. Significant diversity is observed in supervisory structures, functions and boundaries, even in ostensibly similar environments. The importance of informal systems in making formal systems work is highlighted. Informal planning includes that which is accepted as being not amenable or not required to be formally planned as well as that which is not officially recognized by the organization. The degree to which the formal system is unable to cope with reality and where the boundary is set between formal and informal planning are highlighted as being crucial for system improvement. Greater improvements in performance may be realized through co-ordination that is specific to a production environment than through the deployment of commercial planning software.

Vernon (2001) describes an observational field study of a production manager responsible for scheduling in a lingerie factory both before and during an MRP implementation. Compiling a schedule is just a part of the manager's role. A significant proportion of his time is spent on information-gathering and trouble-shooting. After MRP implementation, schedule compilation became a more transparent joint activity between the line supervisors and the manager. The need for some activities is questioned as not all add value. In deciding on support for schedulers it needs to be established whether organizational remedies are required to reduce scheduling complexity and improve communication or whether IT support is required to provide better integration of data, better task-technology fit and better scheduling functionality.

It is clear from the above studies that sound methods are needed to study schedulers and scheduling processes in complex industrial settings. Crawford et al. (1999) stress the difficulties and the many practical issues that arise, for instance identifying appropriate personnel to study. They describe a range of methods that have been used successfully to capture key elements of the process, allowing detailed analysis of areas such as decision-making. MacCarthy et al. (2001) broaden this discussion to develop a research framework to conduct and interpret studies of this type in industrial contexts. The framework focuses on understanding the environment, the planning and scheduling processes themselves and related performance issues. It includes a detailed set of generic research questions to underpin field studies.

Related research

A number of studies have investigated planning and scheduling systems through surveys (Barber and Hollier, 1986; Kenworthy et al., 1994; Halsall et al. 1994). Halsall et al. (1994) carried out a survey of planning and scheduling methods and needs of smaller manufacturing companies in the UK. Over 60% of companies had personnel whose principal responsibility was for scheduling of production. Many companies had stability issues deriving from internal and external sources and had to adjust or override schedules frequently. Kenworthy et al. (1994) conducted a survey of 30 companies and found that sophisticated software may not be a prerequisite for best practice and may not be beneficial or cost effective. The main criterion for 'best practice' was the introduction of high caliber, skilled production control personnel in the scheduling environment. The need to reduce scheduling complexity through improved materials and capacity management is noted.

Although somewhat dated, the conclusions from these studies still have validity. It is clear that software for decision support in planning and scheduling has had limited success. Failure to address the reality of scheduling environments, particularly in terms of human-computer interaction, has been a contributory factor. The importance of effective personnel in production control roles, appropriately trained and supported, is stressed. Well-designed software systems should support, not replace, human scheduling activity. Higgins (2001) presents a methodology for designing software tools that support schedulers. The approach focuses mainly on the design of user interfaces and is illustrated with a single case study in a printing firm. A number of analysis approaches are presented to determine the cognitive support needed for scheduling. Allowing the decision maker freedom of action and goal direction are highlighted.

Other studies of note include Haider et al. (1981) who conducted laboratory experiments using a simulator to study interactive job shop scheduling. Nakamura and Salvendy (1988) conducted laboratory experiments on human performance in FMS scheduling. Sanderson and Moray (1990) examined the human factors of scheduling behavior, particularly to understand how time pressure affects tasks. Moray et al. (1991) proposed using scheduling theory as a normative model for strategic behavior for operator tasks. Dessouky et al. (1995) examine the use of classical scheduling theory to develop a conceptual framework for behavior in human-machine settings.

Using perspectives from psychology and the cognitive sciences, Sanderson (1991) proposed the Model Human Scheduler (MHS) to support design decisions on the allocation of functions, decision support needs and optimal information displays in advanced manufacturing technologies. Sanderson notes that the MHS suggests potential areas for future research but the framework has not been validated in practice.

The strongly cognitive perspective on industrial planning and scheduling is necessarily limited in taking account of the organizational factors that influence these activities or the contribution they make to vital business processes. Models more firmly grounded in field studies of practice can capture and explain more of the phenomena seen in industrial situations. In the next sub-section we present a model of human scheduling developed from detailed studies of practice in real businesses.

3.1 A model of human scheduling practice in manufacturing

Jackson et al. (2004) present a new model to describe and understand scheduling in practice in manufacturing industry. It questions the assumptions underlying previous models from the cognitive sciences that view industrial scheduling as purely cognitive or the classical OR models that assume it is primarily concerned with generating job sequences. Instead it focuses on what scheduling consists of and how it is done in practice.

The approach is influenced by Naturalistic Decision Making (NDM) concepts (Lipshitz et al., 2001), which, unlike classical decision theory, attempt to understand and describe the difficulties in decision-making in many real world situations. NDM acknowledges the contextual and environmental factors that affect real world decision-making – uncertainty and limited information, dynamic and sometimes rapidly changing environments. Decisions are made within organizational structures and may involve teams or groups of people that are spatially distributed. In such environments, experience, expertise and judgment play a part in understanding context, in recognizing potential choices and in evaluating trade-offs when adopting courses of action.

The model was developed from extensive field study of individual schedulers in different industries. New investigative tools and methods were developed (Crawford et al., 1999; MacCarthy et al., 2001) to capture scheduling contexts, for observation and interviewing of key personnel and for decision analysis. Feedback to participant schedulers and validation by them were important in obtaining reliable data. Retrospective decision probing in particular was an important part of the research process. Qualitative research methods were used to analyze the large quantities of observational data and to generate findings.

The results showed significant diversity in the sample of schedulers in some respects e.g. in where and how they worked; some were lone practitioners, some operated as part of a team; some were supported by state-of-the-art IT systems whilst others were dealing with legacy systems. Notwithstanding the diversity, there was wide agreement and consistency on the nature of scheduling activities performed by schedulers and on the organizational roles they fulfilled. An immediate finding was clear - little time was spent in actual schedule generation. In fact in most cases the schedule was generated by an information or job release system - 'instead the schedulers managed the scheduling function in order to support the transfer of a virtual production plan into production reality' (Jackson et al., 2004).

Model description

The model is qualitative and explanatory in nature. Figure 3-1 illustrates the overall structure. Here it is described in terms of a single scheduler but it might be a team or more broadly a function involving a group of people. The model distinguishes the *tasks* that human schedulers perform from the *roles* they hold within their organizations.

Tasks are goal-directed activities carried out by schedulers, often dictated by the formal requirements of their job. Three generic task types are identified:

1. Formal tasks: tasks set out explicitly by an organization for a scheduler to carry out, e.g. ensuring a plant is kept loaded, or regular reporting requirements on status.

2. Maintenance tasks: tasks that need to be carried out in order to fulfill the requirements of the job successfully, e.g. information sorting or validation of data from various sources.

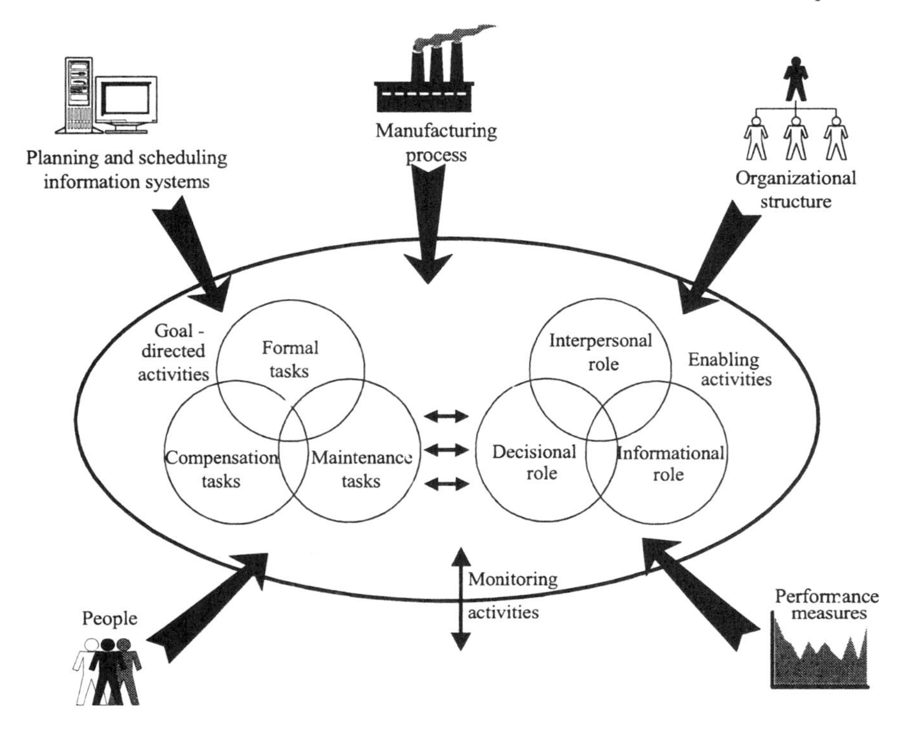

Figure 3-1. Diagrammatic representation of the model of human scheduling practice (from Jackson et al. 2004)

3. Compensation tasks: tasks necessary to overcome limitations or constraints in some aspect of the formal system or processes, e.g. in organizational structures or with respect to technological resources of some form, such as critical shortages resulting from a failure to place an order on time or from an unexpected quality problem.

The categories have some overlap. An activity may fall under more than one heading, e.g. an activity to gather together information to check resource feasibility for a schedule in some future time period may be primarily a maintenance task but in some organizations may have additional complications because of poor information provision, timeliness or accuracy and may thus be judged as compensatory also.

A purely task-based view of scheduling practice provides only a limited perspective on the criticality of the roles fulfilled by schedulers. Roles encompass the wider job requirements – the expectations, obligations and responsibilities associated with the individual holding a particular job. Only so much of a role can be formalized or proceduralized. Roles have strongly social dimensions and are key to ensuring that the formal systems do in fact operate successfully. The roles held by or fulfilled by schedulers were found to be critical in being able to perform the required tasks and add value to the organization in managing the scheduling function. Three generic types of role were identified:

1. ***Interpersonal role*** – These are embodied in the interpersonal networks developed by schedulers over time that complement the formal reporting hierarchies and organizational structures. Informal communication through such networks generates significant contextual data and enables negotiation, bargaining and 'favors' to occur. The perceived status of the scheduler is an important attribute in how effective such interpersonal networks are. The interpersonal role is very significant in 'getting the job done,' particularly in large complex businesses with complex organizational structures and planning levels, multiple supply routes, complex products and manufacturing processes. These networks enable the scheduler to mediate between higher levels of planning and the production function, balancing production concerns such as stability whilst ensuring that high priority work is processed.

2. ***Informational role*** – a key role performed by the scheduler is as an information receiver, processor and transmitter. The model describes this with a 'Hub and Filter' analogy. The 'Hub' captures the concept of an information concentration point for enquiry, receipt and transmission from multiple sources to multiple destinations. The 'Filter' captures the added-value in investigating, interpreting, adjusting, updating and passing on all types of information. Higher levels of planning typically cannot deal with the level of resolution or granularity needed for scheduling close to real time enactment. The scheduler needs 'eyes and ears' in order to obtain accurate information on current or future status, e.g. from the shop floor or from material suppliers.

3. ***Decisional role*** – schedulers are not just problem solvers but problem predictors, taking avoiding or opportunistic action when appropriate. Their decisions extend beyond pure resource allocation to schedule facilitation and schedule management in order to find solutions to cope with, or best exploit, current reality. Plans may be best guesses or just desired future states. Without precise embedding in the current reality, plans may not be realizable to any degree. Information visibility enables interpretation and judgment to be applied and appropriate actions taken. The informational role means that in practice the scheduler receives intelligence that may require action. Information about a possible material shortage for instance may require pre-emptive action. The course of action may be to ensure minimal disruption to a desired schedule, contingency planning to get back on schedule within a limited time span or some opportunistic action that exploits the situation to minimize problems in the long run. The schedule management function and the day-to-day decisions that it requires are key aspects of the decisional role of the scheduler.

Again these roles can overlap and the roles support each other. Thus, an interpersonal network supports information gathering and contingency planning and underpins effective decision actions.

Figure 3-1 also shows two other facets of the model. Firstly, a more general activity (monitoring) can be considered as the 'glue' that supports the totality of scheduling practice. The study showed that monitoring is required to maintain 'situational awareness'. It uses 'hard' manufacturing data such as current stock level or utilization figures, 'soft' manufacturing data such as anticipating jobs not on the current schedule but about to be released and 'exceptional data' such as sudden problems with a catastrophic process failure. It may vary from routine to more concentrated form.

The second additional facet of the model concerns the environmental factors that affect scheduling practice. These comprise the physical technological processes and materials; the organizational structure; the planning and scheduling information systems; the people that the scheduling function interacts with and the performance measures in use. Within any particular environment specific considerations or issues with these factors will influence the nature of scheduling at a detailed level. However, the major types of *tasks* and types of *roles* described in the model can be expected to describe scheduling practice in many manufacturing environments.

Implications for design and management of the scheduling function

The Jackson et al. (2004) model focuses on understanding and improving practice rather than on the development of IT decision support systems. As a model it provides guidance for industrial practice. The model shows the limitations of a purely algorithmic view that concentrates just on the allocation or sequencing decisions. As well as problem solving, it requires scheduling in practice to be seen as an organizational process with a strongly social dimension, typically involving interactions with other key people within an outside the organization. The roles held by schedulers are central to 'getting the job done.' This is an area where businesses need to think carefully in organizational design. Firms need to understand the reality of the job and its contribution – understanding roles in particular. For instance, there may be a gap between the desired and the actual roles held, and between the levels of responsibility and accountability given to the scheduler. The model underscores the importance of the scheduler's authority that enables them to discharge their responsibilities.

Compensatory activities, once identified, may be minimized in a number of ways: by manufacturing improvements that re-design the systems and processes, by information system improvements or by instilling good information practices in an organization.

An important observation is that scheduling in practice centers on facilitation and implementation. This refutes the idea that the only support needed by a scheduler is some form of IT-based decision support system or that in most cases such systems can largely replace the human role. Well-designed decision support systems can indeed aid the *routine task-based aspects* of the job. However, scheduler support needs to concentrate not just on tasks but also on roles, for instance by providing appropriate forums and mechanisms for resolving problems and for conflict resolution. The model gives some hints for job analysis, selection and training. For instance expert schedulers have well-developed networks. They are skilled in problem resolution. How do new recruits develop such attributes? What kind of generic skills are needed to fulfill the interpersonal, informational and decisional roles in a particular organization?

The model also highlights issues in measuring performance of individual schedulers and the scheduling function. The common metrics in use to evaluate operational performance (e.g. resource utilization or proportion of orders completed on time) will dominate in most environments. The model indicates that evaluation of the performance of the function or an individual needs to have a broader base, taken over a relevant and appropriate period of time, and to be cognizant of the factors influencing and affecting performance. Allowing a small number of performance metrics to dominate may lead to inappropriate, sometimes 'pathological' behavior, proving the well-known business aphorism that what gets measured gets managed.

3.2 Socio-technical principles in planning, scheduling and control

Socio-technical concepts are concerned with issues of autonomy and control in how work systems are structured and managed and in how they perform. In a manufacturing context, for instance, how much autonomy should be devolved to local production units? Should production workers be given some control in planning and scheduling to deal with the factors affecting their work such as absenteeism or quality problems? In fact many of the issues noted in this chapter can be viewed from a socio-technical perspective. Socio-technical concepts stretch back to the work of psychologists at the Tavistock Institute in the early 1950's (Trist and Bamforth, 1951). It was stressed that both technical and social systems interact, forming the socio-technical system (STS) that affects work practice and performance.

The ideas have had a resurgence in the last two decades, due in large part to the significant changes in work environments - de-layering, business process reengineering and team-based work practices. Many business processes are now IT-driven. Technology is being used increasingly in tracking and controlling materials, products and people. These changes have been strongly evident in the manufacturing sector with the emphasis on quality initiatives and lean operations. Slomp and Ruel (2001) note the emergence of team and cellular-based manufacturing, and indeed Hyer (1999) has applied STS design concepts in that context, a study that included a planning and control element.

STS principles

A fundamental idea in STS is that the technical and social sub-systems that comprise work systems should complement each other for successful operations. Technology should not dominate how humans work and perform - successful organizations need both to function effectively, in parallel. Socio-technical design recognizes that real systems are open systems, being influenced by, and potentially influencing, entities outside the organization. This is a natural perspective in production systems. The socio-technical approach also advocates a strong focus on organizational choices and self-organization for participants in the system. Slomp and Ruel (2001) note that socio-technical design addresses the problems associated with variety in two ways. First, it attempts to reduce the number of sources of variation in the organizational unit and second, it attempts to add (or decentralize) control tasks to manage variety.

The challenge in using STS concepts in any particular context is how to design and organize the social and technical sub-systems to perform well together. This may mean a degree of sub-optimality in one or other sub-system (Cherns, 1976). It is recognized that full optimization of a complex socio-technical system may not be possible but failure to consider both the social and technical subsystems and their interactions will, according to STS proponents, result in poor performance. Cherns (1976, 1987) developed key principles for socio-technical systems design, covering the design process, the characteristics of an ideal design and the environment for the ultimate design.

Slomp and Ruel (2001) interpret and adapt these guidelines for socio-technical design of planning, scheduling and control systems in manufacturing industry. A production control system comprises its decision hierarchy, organization hierarchy, the information system and the decision support tools. The hierarchical planning and control model of Bertrand (1990) provides a base. The technical sub-system is regarded as comprising the models used for planning, scheduling and control and the software tools and information systems. The social sub-system is represented by the division of decision tasks and the organizational, social and psychological aspects relating to the people responsible for production planning and control. Here the guidelines presented by Slomp and Ruel (2001) for the design of production planning and control systems are briefly re-stated.

Three procedural guidelines are presented: *compatibility* - users of a production control system should participate in its design; *minimal critical specification* - only essential constraints should be specified; *transitional organization* - the designer should be involved in system implementation. Seven design guidelines are presented: *minimal critical specification* - only essential decisions should be taken at each level of the production control hierarchy, objectives rather than detailed procedures should be set for lower levels; *variance control* - decision-making tasks should reflect the variances that may arise at the organizational level; *boundary location* - each level in the decision hierarchy may have its own objectives but co-ordination between levels may be required to avoid sub-optimization; *information flow* - information systems should provide information at the point where action may be needed; *power and authority* - people should only be made responsible for tasks if they have the means, tools and training to deal with them; *multifunctional principle* – more than one employee should be able to deal with each task in the production control hierarchy, which may be helped if the decision complexity at each level is low; *incompletion* - the production control system should be easy to redesign, which may be helped by modularity and simplicity in design. An environmental guideline is also presented: *support congruence* – reward systems, performance systems and support programs should be congruent with the design of the production control system.

Case studies in PSC

Slomp and Ruel (2001) illustrate the guidelines with a case study at a small firm that fabricates a large variety of perforated sheet metal products for other companies. Socio-technical guidelines were followed implicitly rather than explicitly in the re-design of a new production control system but the performance of the new system was not evaluated. The case study is therefore illustrative only and, as the authors' state, 'it does not prove the usefulness of the guidelines.' However, in general they consider that the guidelines fit well in a systematic approach to the design of a production control system, in the allocation and design of production control tasks and responsibilities over the various departments and individuals in a firm. Interestingly, with the approach, production units are defined on the basis of achieving autonomy with respect to production control, and workload control is viewed as an interpretation of the *variance control* design guideline.

Wäfler (2001) gives another perspective on the applicability of socio-technical ideas for the analysis and design of planning and scheduling systems. He agrees with the need for joint optimization of people, organization and technology, but argues that there are difficulties in directly adopting a classical socio-technical approach in the context of planning and scheduling. The classical STS approach focuses on separation of organizational units, whereas planning and scheduling processes aim at coordinating and integrating the activities performed within different organizational units. There are also problems in defining complete tasks that include both planning and execution.

To overcome these deficiencies the concept of a *secondary work system* is discussed that takes an extended view of the scope of planning and control and requires careful definition of work tasks. Three sub-problems need to be considered in this context: design of the organization, design of individual tasks, and design of human-computer function allocation. For the technical aspects of the planning and scheduling system, technology should provide accurate and complete information but human-computer interaction should allow human control over automated planning and scheduling processes, guaranteeing process transparency, human decision authority and flexibility. Possibilities for 'opportunistic planning' and 'situated acting' should be encouraged. Organizational and technical support of planning and scheduling processes should aim at interconnecting people's creativity by facilitating local 'situated acting.' He illustrates the ideas with a case study in a company that produces high-end plumbing items and shows how the existing system is at variance with socio-technical principles, particularly the absence of autonomy-oriented design.

Crawford et al. (2002) question the traditional view of planning, scheduling and control (PSC) as a well-defined, hierarchical set of activities that can be largely automated. Many conventional business process modeling techniques adopt a purely 'technological' view with oversimplified linear representations of work flows that are assumed to be valid for every type of work. Such approaches ignore issues such as power relationships, personal interactions, personalities and motivation. Indeed the 'social' aspects may be wrongly perceived as the problem areas of the organization. Crawford et al. (2002) investigate the relevance of STS principles to PSC through a detailed case study of a consumer products company based on semi-structured interviews, field observations, decision mapping and interaction analysis. They find that PSC processes can be represented more accurately using a socio-technical approach, recording the reality of how people and technology work together. They argue that STS provides a useful framework for analysis and is effective in 'making sense' of holistic business processes but also highlight its limitations. Using an STS approach requires process capture and analysis tools to be developed that are applicable to the specific operational domain. STS also lacks well-developed methods for the definitive re-design of business processes.

Some of the STS ideas may seem like 'common sense,' others may seem to be impractical or 'go against the grain,' appearing to be at variance with traditional views of planning and control, e.g. 'self-organization' and 'local autonomy.' However, developments in manufacturing such as flow-based production systems (Section 2.1) and quality management support STS ideas – simplicity, variance reduction, and local autonomy coupled with responsibility. Many of the field studies of practice discussed in Section 3.1 also highlighted the essential contributions made by human schedulers and the necessity of socio-technical concepts to address current reality, e.g. flexibility in decision-making, opportunistic actions, local autonomy and control.

Production systems, along with their planning, scheduling and control structures evolve over time – as product mixes change, as new technology is acquired, as new information systems and decision support tools are deployed. The reality of the current system may deviate from formal procedures established some time in the past. Some systems may exhibit poor performance and it may be difficult to precisely diagnose the problems. Textbook theory of PSC as well-defined hierarchical activities that can be largely automated may fly in the face of what is apparent in many enterprises. Essentially STS encourages a holistic perspective in systems analysis and design. STS design principles may be less well developed or well tested than we would like, particularly in the production control context, but they do at least acknowledge the importance of the human and social elements of a work system and challenge the primacy of IT. STS concepts may indicate where to look and what to look for in order to address system issues and may indicate potential mechanisms for systems re-design.

4. INTEGRATING ORGANIZATIONAL, SYSTEMS AND HUMAN PERSPECTIVES

Today's industrial environments are characterized by a number of organizational changes that influence the design and management of planning, scheduling and control (PSC) processes. The extended enterprise concept for instance involves strong collaboration with supply chain partners. Collaborative planning, including the sharing of forecasts with supply chain partners, is now more common. Service level agreements stipulate the desired minimum performance level across the supply chain. Part of this trend is to export supply chain complexity to specialists or partners who may be able to handle it more effectively (Frizelle and Efstathiou, 2003). Virtual enterprises involve dynamic and temporary structures for new projects, to exploit new market opportunities or meet changing market demands.

The emergence of flatter organizational structures internally within businesses has been noted. The lean manufacturing model is dominant in much of industry, striving for waste elimination and transparent production flows with an emphasis on self-regulated production systems. Lean initiatives have been challenged somewhat with the rise in product customization and product variety and the ever-present emphasis on time compression and responsiveness.

For some organizations there is an issue in convincing them that their PSC process may merit attention. It may appear as essentially simple - translating customer orders into shop floor schedules. It may be felt that 'off-the-shelf' solutions, usually involving the purchase of a proprietary ERP package, will satisfy their needs and resolve any perceived problems. However, there is now a greater realization in industry of the limitations of 'one size fits all' solutions. The studies discussed in this chapter show the true nature of many PSC processes and that they merit dedicated consideration. MacCarthy et al. (2002), in a practitioner article, outline a 'health check' to evaluate whether a PSC process is in need of attention or re-design. It is clear that effective solutions must address the complexity of the current environment.

PSC processes need to be dynamic, relevant to current business needs and robust enough to absorb the typical shocks associated with the current environment. In highly responsive businesses, decisions in production planning, scheduling and control (PSC) have to be made rapidly and effectively. The human role is, in the broadest sense, to manage these processes. In this section, the design and organization of the planning, scheduling and control function is considered and in particular an approach called PROCHART that develops HPP thinking, taking into account what we know from studies of practice. First we discuss the value of knowledge management thinking in PSC and how it can be embedded in systems design.

4.1 Knowledge management in PSC processes

In the last decade knowledge management has been a burgeoning area for academic research. It has also generated significant interest from business and industry because of the contribution it is perceived to make to effective organizational decision-making (Choo, 1996). Guinery et al. (2001, 2002) have investigated knowledge management concepts in the context of PSC processes and, based on extensive field studies (Guinery and MacCarthy, 2003), examine the types of knowledge used, the preferred forms of knowledge integration and knowledge support mechanisms in the PSC context.

A definition of knowledge as 'a combination of information, ideas, procedures and perceptions that guide actions and decisions' (Bolisani et al., 1999) is used as it relates knowledge specifically to decision-making. A range of knowledge integration practices have been identified and the factors on which they are contingent explored. Significant factors identified for knowledge management include the steadiness of production and of the PSC process, whether knowledge distribution is broad or focused and the decision timescales and timings relative to the planning cycle. A knowledge perspective places the observations from the various studies of scheduling practice in a new light. Just some of the knowledge integration mechanisms are noted here.

Self-forming and self-organizing *informal networks* evolve to communicate, share or transfer information or resources or to solve problems jointly between individuals. The extent of these networks varies depending on the steadiness the PSC decision-making environment and whether rapid decisions, outside of normal planning cycles, are a major feature. *Sense making communities* are groups of individuals within an organization who develop a shared language, objectives and goals to support decision-making. They play a crucial role, particularly at higher levels of planning where the decision-making domain is broad, where there may be a number of interest groups and where objectives and constraints are only loosely defined. They are important in the large, complex and 'messy' manufacturing businesses. *Organizational routines* are the informal ways in which people interact routinely in relatively well-known situations. Established *organizational routines* operating through an appropriate *sense making community* with a shared perspective can support rapid and effective problem solving in more complex PSC situations.

Important concepts are *direction* and *decision frames*. The former relates to the explicit information provided to work units or individuals through procedures, standards, directives or instructions. The latter relates to the explicit objectives and constraints on decision variables transferred from one decision level to another. Where and how *direction* can be effective to inform decision-making in the different businesses was found to vary substantially. The importance of *decision frames* and how they should operate to support the alignment of performance objectives in different environments is highlighted. The work also shows insights on expertise in PSC roles; on cross-functional teams in PSC processes; on co-location and the conduct of meetings in PSC contexts. The study has implications for the design, organization and management of PSC processes, whether in business process improvement initiatives or in fundamental PSC process redesign. Impacting knowledge practices directly may be difficult but the characteristics of the PSC environment can be changed to support effective knowledge integration through reorganization.

4.2 PROCHART – a PSC design methodology

A PSC (re-)design approach called PROCHART is described here briefly. It has been developed from an in-depth study of a number of businesses from different sectors and has been tested in a number of other businesses for refinement and improvements. The research which underpins the work is described in Guinery and MacCarthy (2005), in which a description of the toolkit of methods is given. Only a high level view is given here.

The backdrop to the work is the need to improve responsiveness in many manufacturing businesses (Crawford et al., 2001). The acronym PROCHART was derived from the research project sub-title - from progress chasers to responsive teams. The approach stresses the holistic view of PSC that has been noted throughput the chapter. The knowledge concepts discussed above have contributed to the approach. PROCHART recognizes the complexity inherent in many manufacturing organizations and that competitive advantage may be gained by being able to manage that complexity in a unique way through an effective PSC process.

It is applied in two phases – firstly an audit phase of the process 'as-is' and secondly an analysis phase to determine potential areas for change or improvement. The major areas that the design methodology addresses are illustrated in Figure 3-2. Scoping the study is important in the audit phase to determine the appropriate unit of analysis. This requires a detailed analysis of not just the physical aspects of operations but the business context, the nature of demand and the operations policies that drive the business. Both the audit and analysis phase consider PSC architecture, e.g. planning levels and interfaces, as well as key PSC decision-making activities and the roles, responsibilities and knowledge requirements of the people who make them. The analysis also looks at the most appropriate areas for human and information system decision-making. An important concept is that of a 'hotspot,' a decision centre or interface in which decision-making is complex and in which there is typically a need for significant human input, reasoning and judgment. The analysis helps to identify and categorize hotspots.

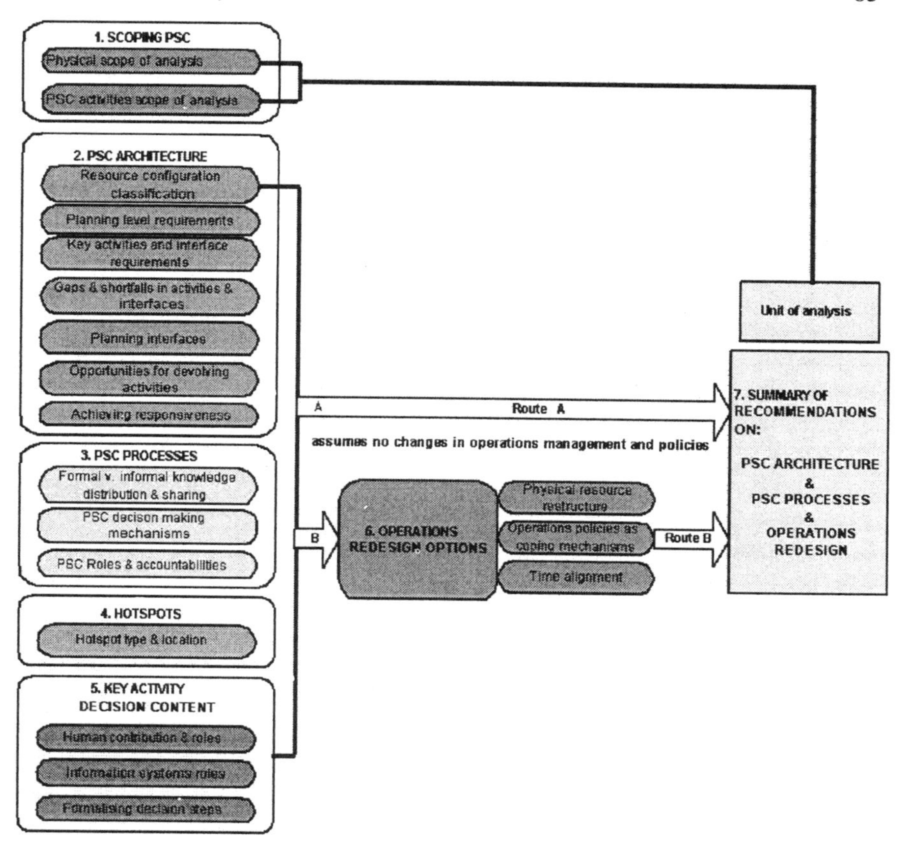

Figure 3-2. The PROCHART process and areas of application

The audit and analysis phases are assisted throughout by a series of interrelated worksheets that operate on relevant data obtained from the unit of analysis. The worksheets play a crucial role in the analysis and re-design process. They incorporate modified GRAI modeling techniques (Doumeingts et al., 1992) for representation of decision centers and for decision analysis but also numerous specialized tools for process visualization, for analyzing interactions and for application of design rules. A number of design rules are incorporated on different aspects of PSC, for instance, on the relationship between the types of production (e.g. flow line, job shop) and the ability to devolve scheduling and resource management responsibilities to production personnel or on the relationship between departmental specialization and the knowledge distribution needs across the organization.

The analysis phase identifies future PSC requirements through a 'gap analysis' that eventually leads to identification of potential opportunities and options for change. Figure 3-2 indicates two routes for application. Route A avoids fundamental changes in operational systems or policies. Route B considers more fundamental re-design options of operational systems or policies. This could entail moving to a different manufacturing system for instance or simplifying PSC processes through facility reorganization or policy changes on how demand is managed.

Typical scenarios where PROCHART can be applied, either in full or in part, include the following: problems in order fulfillment performance; restructuring initiatives; lean manufacturing initiatives; and deployment of new information systems. Two application studies have been published in the practitioner literature. The first is a study of the UK consumer products division of the chemical company Henkel, where PROCHART was used to formalize fast-track fulfillment systems (Hamlin et al. 2005). In the Anglo-Dutch steel producer Corus, PROCHART was used to investigate the impact of restructuring on the PSC process in the central load planning group (Coates et al., 2005). Guinery and MacCarthy (2005) describe a recent successful trial of the approach in a UK bakery company.

It is important to note that PROCHART is a developmental approach. It requires a skilled analyst to conduct a study with relevant business personnel. Both the analyst and the problem owners in the business then work together to identify appropriate and feasible solutions. PROCHART is still under development, and for complex businesses its application is time consuming. Current developments are aimed at making parts of the analysis easier to conduct.

5. IN CONCLUSION

The main thrust of this chapter has focused on the true nature of planning, scheduling and control (PSC) in today's manufacturing organizations. PSC performance is probably of greater significance than in previous years because of the pressures of global competition. This translates into the need to compress order fulfillment times and meet delivery commitments consistently, whilst also maintaining efficient operations to address cost issues. The key messages from this chapter are summarized here:

Models and algorithms can address the task-based aspects of PSC. To improve their applicability and take-up, more attention needs to be given to the overall planning and scheduling environment and the operations management context in which they might apply. Too many studies consider the model and algorithm in isolation from, or with only a very limited perspective on, the application context. The continual emphasis on sequencing algorithms overplays the importance of sequencing in contemporary manufacturing operations.

PSC needs to be seen and treated as *an integrated process* involving an organizational planning structure, information systems and decision makers whose contribution are vital to the process. The PSC process normally operates as a planning and scheduling hierarchy that manages a dynamic pipeline of planned products over a rolling time horizon. A purely task-based view of planning and scheduling provides only a limited perspective on the PSC process.

PSC processes need to be *carefully designed* in an integrated way. Designing integrated processes needs to address organizational structure, planning levels and roles in particular. It needs to address *decision 'hotspots'* where decision activity is intense, outcomes are critical and require significant levels of human judgment. The chapter has stressed the importance of *good practice* in PSC to enable processes that are robust in adjusting to current realities.

Information technology systems, even the latest 'best of breed' ERP systems, are still essentially transaction and execution systems that offer some decision support in some areas and at some levels of planning. They bring enhanced functionality and much greater access to relevant data than previous generations of IT systems. However their decision-making 'intelligence' is still limited and such systems still require human support.

Many roles within PSC involve significant *management and facilitation* to ensure that plans and schedules are realistic and feasible and to ensure that they are enacted i.e. that the agreed or desired plan or schedule is achieved. Such roles often need strong *interpersonal networks* and *sense making communities* with shared perspectives within organizations to utilize organizational resources effectively and efficiently to meet demand. This is where the real human skills lie and where the human contribution is greatest.

A key issue to resolve when reviewing *support for planners and schedulers* is to establish whether it is required in the form of organizational remedies to reduce the complexity, improve co-operation and reduce the effort involved in facilitation or whether improved IT support is required to provide better integration of data, better task-technology fit and better PSC functionality. *Compensatory activity,* which is non-value adding in the long term, needs to be identified, eliminated or reduced.

PSC processes today cannot be divorced from the overall organizational context and changes occurring within and across organizations. They must be cognizant of developments such as lean manufacturing. In particular, the PSC process extends across the supply chain. Of major importance is the international context – today's world is one of global sourcing and global markets. This extends the reach of PSC into *planning and scheduling across international supply networks.*

The changing contexts noted above means that PSC is a continually evolving discipline in theory and in practice and one that is attracting real and renewed interest across industrial and business sectors. There is continuing scope for developments and innovation in all aspects of PSC.

REFERENCES

Baker, K. R., 1974, *An Introduction to Sequencing and Scheduling*, New York, Wiley.

Barber, K.D., Hollier, R.H., 1986, The use of numerical analysis to classify companies according to production control complexity, *International Journal of Production Research*, **24**:203-222.

Bertrand, J.W.M., Wortmann, J.C. and Wijngaard, J., 1990, *Production Control, A Structural and Design Oriented Approach*, Elsevier, Amsterdam.

Bitran, G.R., Tirupati D., 1993, Hierarchical Production Planning, in *Handbooks in Operations Research and Management Science* Vol 4, *Logistics of Production and Inventory* (Eds. Graves S.C., Rinnooy Kan, A.H.G., and Zipkin, P.H.), North-Holland, Amsterdam.

Bolisani, E., Scarso, E., 1999, Information Technology Management: A Knowledge-based Perspective, *Technovation*, **19**:209–217.

Buxey, G., 1989, Production Scheduling: Practice and Theory, *European Journal of Operational Research*, **39**:17-31.

Breithaupt, J-W., Land, M., Nyhuis, P., 2002, The workload control concept: theory and practical extensions of Load Oriented Order Release, *Production Planning and Control*, **13**(7):625 – 638.

Chen I.J., 2001, Planning for ERP systems: analysis and future trends, *Business Process Management Journal*, **7**(5):374-386.

Cherns, A., 1976, The Principles of Socio-technical Design, *Human Relations*, **29**(8):783-792.

Cherns, A., 1987, Principles of Socio-technical Design Revisited, *Human Relations*, **40**(3):153-162.

Choo, C.W, 1996, The Knowing Organization: How Organizations Use Information to Construct Meaning, Create Knowledge and Make Decisions, *International Journal of Information Management*, **16**(5):329-340.

Coates, D., MacCarthy, B.L., Guinery, J., 2005, Coping with restructuring–applying PROCHART principles at Corus, *IEE Manufacturing Engineer*, **84**(1):30-35.

Crawford, S., MacCarthy, B.L., Wilson, J.R., and Vernon, C., 1999, Investigating the Work of Industrial Schedulers through Field Study, *Cognition, Technology and Work*, **1**:63-77.

Crawford, S., Wiers, V.C.S., 2001, From Anecdotes to Theory: A Review of Existing Knowledge on the Human Factors of Production Scheduling, in *Human Performance in Planning and Scheduling* (Eds. MacCarthy, B.L., Wilson, J.R.), Chapter 3, Taylor and Francis, London.

Crawford, S.C., Guinery, J., MacCarthy, B.L., McFarlane, D.C., 2001, Addressing Responsiveness in the planning, scheduling and control Process, Proceedings of 8th International EurOMA Conference, **2**:1082-1094.

Crawford, S., Guinery, J., MacCarthy, B.L., 2002, A sociotechnical approach to the analysis and design of a responsive planning, scheduling and control process. Proceedings of 9th International Annual Conference of the European Operations Management Association: Operations Management & the New Economy, Copenhagen, Denmark, **1**:315-326.

Davenport, T. H., 1998, Putting the enterprise into the enterprise system, *Harvard Business Review*, July – August, 121-131.

Dessouky, M.I, Moray, N. and Kijowski, B., 1995, Taxonomy of Scheduling Systems as a Basis for the Study of Strategic Behaviour, *Human Factors*, **37**(3):443-472.

Doumeingts, G., Chen, D., Marcotte, F., 1992, Concepts, models and methods for the design of production management systems, *Computers in Industry*, **19**(1):89-111.

Duchessi, P., and O'Keefe, R.M., 1990, A Knowledge-based Approach to Production Planning, *Journal of the Operational Research Society*, **41**(5):377-390.

Dutton, J.M., 1962, Simulation of an Actual Production Scheduling and Workflow Control System, *International Journal of Production Research*, **4**:421-441.

Dutton, J.M., 1964, Production Scheduling: a Behaviour Model, *International Journal of Production Research*, **3**:3-27.

Dutton, J.M., and Starbuck, W., 1971, Finding Charlie's Run-time Estimator. In J.M. Dutton and W. Starbuck (eds), *Computer Simulation of Human Behaviour*, John Wiley & Sons, New York, 218-242.

Fox, M.S., and Smith, S.F., 1984, ISIS - A Knowledge-based System for Factory Scheduling, *Expert Systems*, **1**(1):25-49.

Frizelle, G., Efstathiou, J., 2003, The urge to integrate supply chains, *IEE Engineering Management*, **13**(4):14-17.

Guinery, J., Crawford, S., MacCarthy, B.L., 2001, Knowledge in Planning, Scheduling and Control, Proceedings of the Twelfth Annual Conference of the Production and Operations Management Society, POM-2011, Orlando, Florida.

Guinery, J., Crawford, S., MacCarthy, B.L., 2002, Responsive Performance in Production Planning, Scheduling and Control through Effective Knowledge Integration, 2nd International Conference on Responsive Manufacturing, Gaziantep, Turkey, June, 2002.

Guinery, J., MacCarthy, B.L, 2003, A Cross-Sectoral Analysis of Planning and Control Processes in Manufacturing Industry - Evidence from a Study of Practice, in Proceedings of EurOMA 2003 One World, One View of OM, Cernobbio, Italy, June 2003. SGE Ditoriali, Padova.

Guinery, J.E., MacCarthy, B.L., 2005, The Prochart Toolkit for the re-design of production planning, scheduling and control processes, Proceedings of the 18th International conference on Production Research (ICPR18), Salerno Italy, July 2005.

Gupta, M., 2003, Constraints management-recent advances and practices, *International Journal of Production Research*, **41**(4):4647-4659.

Haider, S.W., Moodie, C.L., Buck, J.R., 1981, An investigation of the advantages of using a man- computer interactive scheduling methodology for jobshops, *International Journal of Production Research*, **19**:381-392.

Hall, N.G., Potts, C.N., 2004, Rescheduling for New Orders, *Operations Research*, **52**(3):440–453.

Halsall, D.N, Muhlemann, A.P., and Price, D.H., 1994, A Review of Production Planning and Scheduling in Smaller Manufacturing Companies in the UK, *Production Planning and Control*, **5**(5):485-493.

Hamlin, M., MacCarthy, B.L., Guinery, J., 2005, Fast Track Order Fulfilment at Henkel: the PROCHART approach, *Institute of Operations Management CONTROL* magazine, **31**(4):15- 19.

Harvey,H., 2001, Boundaries of the Supervisory Role and their Impact on Planning and Control, in *Human Performance in Planning and Scheduling* (Eds. MacCarthy, B.L., Wilson, J.R.), Chapter 6, Taylor and Francis, London.

Hax, A.C., Meal, H.C, 1975, Hierarchical integration of production planning and scheduling, In *TIMS Studies in the Management Sciences*, Vol. 1, Logistics (Geisler, M.A., Ed.), North-Holland, Amsterdam.

Higgins, P.G., 2001, Architecture and interface aspects of scheduling decision support, in *Human Performance in Planning and Scheduling* (Eds. MacCarthy, B.L., Wilson, J.R.), Taylor and Francis, London. 245-281.

Hurst, E.G., and McNamara, A.B., 1967, Heuristic Scheduling in a Woollen Mill, *Management Science*, **14**(2):182-203.

Hyer, N.L., Brown, K.A., Zimmerman, S., 1999, A Socio-technical Systems Approach to Cell Design: Case Study and Analysis, *Journal of Operations Management*, **17**:179-203.

Jackson, S., Wilson, J.R., MacCarthy, B.L., 2004, A New Model of Scheduling in Manufacturing: Tasks, Roles, and Monitoring, *Human Factors*, **46**(3):533-550.

Kanet, J.J., and Adelsberger, H.H., 1987, Expert Systems in Production Scheduling, *European Journal of Operational Research*, **29**:51-59.

Kenworthy, J.G., Little, B., Jarvis, P., and Porter, J.K., 1994, Short Term Scheduling and Its Influence on Production Control in the 90s, In K. Case and S. Newman (eds), *Advances in Manufacturing Technology* VIII. Taylor & Francis, London, 436-440.

Koh, S.C.L., Saad, S.M., Jones, M.H., 2002, Uncertainty under MRP-planned manufacture: review and categorization, *International Journal of Production Research*, **40**(10):2399-2421.

Konicki, S., 2001, Nike Just Didn't Do It Right, Says i2 Technologies, Information week.com, March 5, 2001.

Lipshitz, R., Klein, G., Orasanu, J., Salas, E., 2001, Taking stock of naturalistic decision making, *Journal of Behavioral Decision Making*, **14**:331-352.

MacCarthy, B.L., and Liu, J., 1993, Addressing the Gap in Scheduling Research: A Review of Optimization and Heuristic Methods in Production Scheduling, *International Journal of Production Research*, **31**(1):59-79.

MacCarthy, B.L., and Wilson, J.R., 2001a, Eds., *Human Performance in Planning and Scheduling*, Chapter 1, Taylor and Francis, London.

MacCarthy, B.L., and Wilson, J.R., 2001b, Eds., *Human Performance in Planning and Scheduling*, Chapter 20, Taylor and Francis, London.

MacCarthy, B.L., Wilson, J.R., and Crawford, S., 2001, Human Performance in Industrial Scheduling: A Framework for Understanding, *International Journal of Human Factors & Ergonomics in Manufacturing*, **11**(4):1-22.

MacCarthy, B.L., Crawford, S.C., Guinery, J., 2002, Contributing to business improvement, *IEE Manufacturing Engineer*, **81**(3):103-108.

McAfee, A., 2003, When Too Much IT Knowledge Is a Dangerous Thing, *MIT Sloan Management Review*, **44**(2):83–89.

McKay, K.N., Safayeni, F.R., and Buzacott, J. A., 1988, Job-Shop Scheduling Theory: What Is Relevant? *Interfaces*, **18**(4):84-90.

McKay, K.N., Safayeni, F.R., and Buzacott, J.A., 1995a, 'Common Sense' Realities of Planning and Scheduling in Printed Circuit Board Manufacture, *International Journal of Production Research*, **33**(6):1587-1603.

McKay, K.N., Safayeni, F.R., and Buzacott, J.A., 1995b, Schedulers and Planners: What and How Can We Learn from Them? In D. E. Brown and W. T. Scherer (eds), *Intelligent Scheduling Systems*, Kluwer, Boston, 41-62.

Moray, N, Dessouky, M.I., and Kijowski, B.A., 1991, Strategic Behaviour, Workload and Performance in Task Scheduling. *Human Factors*, **33**(6):607-629.

Nakamura, N., and Salvendy, G., 1994, Human Planner and Scheduler, In G. Salvendy and W. Karwowski (eds), *Design of Work and Development of Personnel in Advanced Manufacturing*, Wiley, New York, 331-354.

Padmos, J., Hubbard, B., Duczmal, T., Saidi, S., 1999, How i2 integrates simulation in supply chain optimization, *Proceedings of the Winter Simulation Conference*, 1999, **2**:1350-1355.

Sanderson, P.M., 1989, The Human Planning and Scheduling Role in Advanced Manufacturing Systems: An Emerging Human Factors Domain, *Human Factors*, **31**(6):635-666.

Sanderson, P.M., 1991, Towards the Model Human Scheduler, *International Journal of Human Factors in Manufacturing*, **1**(3):195-219.

Sanderson, P.M., and Moray, N., 1990, The Human Factors of Scheduling Behaviour. In W. Karwowski and M. Rahimi (eds), *Ergonomics of Hybrid Automated Systems* II, Elsevier Science Publishers, New York, 399-406.

Shanthikumar, J.G., and Sargent, R.G., 1983, A unifying view of hybrid simulation/analytical models and modeling, *Operations Research*, **31**:1030-1052.

Shapiro, J.F., 1993, Mathematical Programming Models and Methods for Production Planning and Scheduling, in *Handbooks in Operations Research and Management Science* Vol 4, *Logistics of Production and Inventory* (Eds. Graves, S.C., Rinnooy Kan, A.H.G., and Zipkin, P.H.), North-Holland, Amsterdam.

Shobrys, D.E., White, D.C., 2000, Planning, scheduling and control systems: why can they not work together, *Computers and Chemical Engineering*, **24**(2):163-173.

Slomp, J., 1997, The design and operation of flexible manufacturing shops. In: A. Artiba and S.E. Elmaghraby (eds), *The Planning and Scheduling of Production Systems*, Chapman & Hall, London, 199-226.

Slomp, J., 2001, Human Factors in the Planning and Scheduling of Flexible Manufacturing Systems, in *Human Performance in Planning and Scheduling* (Eds. MacCarthy, B.L., Wilson, J.R.), Chapter 9, Taylor and Francis, London.

Slomp, J., Ruël, G.C., 2001, A socio-technical approach to the design of a production control system: towards controllable production units, in *Human Performance in Planning and Scheduling* (Eds MacCarthy B.L.,Wilson, J.R), Taylor and Francis, London, 383-409.

Spencer, M.S., Cox, J.F., 1995, The role of MRP in repetitive manufacturing, *International Journal of Production Research*, **33**(7):1881-1899.

Suri, R., 1998, *Quick Response Manufacturing: A Companywide Approach to Reducing Lead Times*, Productivity Press.

Trist, E.L., and Bamforth, K.W., 1951, Some Social and Psychological Consequences of the Longwall Method of Coal-Getting, *Human Relations*, **4**(1):3-38.

Vernon, C., 2001, Lingering Amongst the Lingerie: An Observation Based Study into Support for Scheduling at a Garment Manufacturer, in *Human Performance in Planning and Scheduling* (Eds. MacCarthy, B.L., Wilson. J.R.), Chapter 6, Taylor and Francis, London.

Vollman, T.E., Berry, W.L., Whybark, D.C., 1992, *Manufacturing Planning and Control Systems*, 3rd Edition, Dow Jones-Irwin.

Wäfler, T., 2001, Planning and scheduling in secondary work systems, in *Human Performance in Planning and Scheduling* (Eds MacCarthy B.L., Wilson, J.R), Taylor and Francis, London, 383-409, 411-447.

Webster, S., 2001, A Case Study of Scheduling Practice at a Machine Tool Manufacturer, in *Human Performance in Planning and Scheduling* (Eds. MacCarthy, B.L., Wilson, J.R.), Chapter 4, Taylor and Francis, London.

Wiers, V.C.S., 1996, A Quantitative Field Study of the Decision Behaviour of Four Shop Floor Schedulers, *Production Planning and Control*, **7**(4):383-392.

Wiers, V.C.S., 1997, A Review of the Applicability of OR and AI Scheduling Techniques in Practice, *Omega*, **25**(2): 145-153.

Chapter 4

DECISION-MAKING SYSTEMS IN PRODUCTION SCHEDULING

A general approach to understanding, representing, and improving production scheduling systems

Jeffrey W. Herrmann
University of Maryland, College Park

Abstract: In practice, production scheduling is part of the complex flow of information and decision-making that forms the manufacturing planning and control system. This decision-making systems perspective enhances our understanding of production scheduling. The chapter presents a systems methodology for improving production scheduling systems and describes techniques that can be used to represent production scheduling systems.

Key words: Decision-making, information, organization

1. INTRODUCTION

Production scheduling activities are common but complex. This leads to many different views and perspectives. Each perspective has a particular scope and its own set of assumptions. Different perspectives lead naturally to different approaches to improving production scheduling. The following paragraphs will briefly cover three important perspectives: problem-solving, decision-making, and organizational.

The problem-solving perspective is the view that scheduling is a problem to be solved. This approach has dominated the academic literature on scheduling for fifty years. Researchers classify the problems to be solved, invent new algorithms, and develop software. A great deal of effort has been spent trying to generate optimal production schedules, and countless papers discussing this topic have appeared in scholarly journals. Typically, such papers formulate scheduling as a combinatorial optimization problem.

Interested readers should see Pinedo (2005) or similar introductory texts on scheduling.

The decision-making perspective is the view that scheduling is a decision to be made. This perspective is more valid and is the concern of the individual scheduler. Schedulers face many challenges everyday, and they need practical advice on how to schedule a factory (see, for example, McKay and Wiers, 2004). In many cases, those doing planning and scheduling see their job as trying vainly to satisfy requirements in an environment that constantly changes. Managers want to keep resources busy and customers happy, while foremen view these as impossible demands. McKay and Wiers (1999) describe three principles that explain the production scheduling activity. First, it generates partial solutions for partial problems. Second, it anticipates, reacts to, and adjusts for disturbances. Third, it is sensitive to and adjusts the meaning of time in the production situation.

The organizational perspective is a systems-level view that scheduling is part of the complex flow of information and decision-making that forms the manufacturing planning and control system. Such systems are typically divided into modules that perform different functions such as aggregate planning and material requirements planning (Hopp and Spearman, 1996; Vollmann et al. 1997). Production scheduling usually refers to the low-level, shop floor control function. Production scheduling is influenced by the decisions that others have made, and a production scheduling decision influences future decisions in turn. The validity of this view can be seen in the importance of decision-making systems in organizations. Simon (1997) describes decision-making systems in the administration of manufacturing firms and government agencies. Herrmann and Schmidt (2002, 2005) introduce decision-making systems in product development.

This chapter elaborates on this third perspective and uses it as the basis for a general approach to understanding, representing, and improving production scheduling systems. The chapter also discusses ways to represent the decision-making system and gives some examples.

2. THE IMPORTANCE OF DECISION-MAKING

Decision-making is a critical function for any organization, especially manufacturing firms that operate in a hostile, dynamic, and complex environment. This has been known for many years. For instance, Simon (1973) stated that “the decision-making processes, rather than the processes contributing immediately and directly to the production of the organization’s final output, will bulk larger and larger as the central activity in which the organization is engaged." He adds that the central problem is "how to organize to make decisions."

An organization is "the pattern of communications and relations among a group of human beings, including the process for making and implementing decisions" (Simon, 1997). As producing goods becomes a smaller part of the economy and organizations become more aware of the indirect consequences of their activity, the decision-making process in an organization becomes more and more important.

In a decision-making paradigm of organizational design, "effective organizations are those whose decisions are of high quality" (Huber and McDaniel, 1986). This leads, in turn, to the following guidelines for designing organizational structures, processes, and *decision units* (persons or groups that make decisions):

1. Assign decision-making authority to the hierarchical level that minimizes the combined costs that result from a lack of information.
2. Specialization among decision-making units should be commensurate with the complexity of the decision situations that can occur.
3. Create and formalize a structure that has rigid processes for routine decisions and flexible processes for nonroutine decisions.
4. Sensor and message handling units must make appropriate decisions when unusual or unexpected messages arrive.
5. Communication chains should be as short as possible.
6. Message-handling systems should have buffers to protect decision-making units from overload.
7. Maximize the performance of the decision-making systems, not the information processing system.
8. Establish which decisions have priority and which should not be done.
9. Manage decisions as projects.
10. Reward decision units for the quality of their decisions.

The role of decision-making specifically in manufacturing planning and control systems (which form the context for production scheduling systems) has been described by Bertrand et al. (1990). In addition, McKay *et al.* (1995) describe the traditional hierarchical decomposition of production planning and discuss the need for an adaptive framework that can respond to change in the environment. In this framework, each decision level consists of an information filter, a decision controller, and a tactical controller that updates the other two components. This framework is a useful way to describe the overall structure of a manufacturing planning and control system. McKay and Buzacott (1999) also describe the traditional hierarchy of planners and schedulers that works well in a stable situation. However, in a situation of rapid change, a traditional scheme will be ineffective, and new production control organizations will be needed.

3. PRODUCTION SCHEDULING AS A DECISION-MAKING SYSTEM

Manufacturing facilities, which first appeared during the middle of the eighteenth century, are complex, dynamic, and stochastic systems. From the very beginning, workers, supervisors, engineers, and managers have developed many clever and practical methods for controlling production activities. Although dispatching rules, kanban cards, and other decentralized production control policies are in use, many manufacturing facilities generate and update production schedules.

In manufacturing systems with a wide variety of products, processes, and production levels, production schedules can enable better coordination to increase productivity and minimize operating costs. A production schedule can identify resource conflicts, control the release of jobs to the shop, and ensure that required raw materials are ordered in time. A production schedule can determine whether delivery promises can be met and identify time periods available for preventive maintenance. A production schedule gives shop floor personnel an explicit statement of what should be done so that supervisors and managers can measure their performance.

In a manufacturing facility, the *production scheduling system* is a dynamic network of persons who share information about the manufacturing facility and collaborate to make decisions about which jobs should be done when. The information shared includes the status of jobs (also known as work orders), manufacturing resources (people, equipment, and production lines), inventory (raw materials and work-in-process), tooling, and many other concerns.

The persons in the production scheduling system may be managers, production planners, supervisors, operators, engineers, and sales personnel. They will use a variety of forms, reports, databases, and software to gather and distribute information, and they will use tacit knowledge that is stored in their memory.

The following are among the key decisions in a production scheduling system:

- releasing jobs for production,
- assigning resources (people, equipment, or production lines) to tasks,
- reassigning resources from one task to another,
- prioritizing tasks that require the same resources,
- sequencing production tasks,
- determining when tasks should begin and end, and
- interrupting tasks that should be halted.

The production scheduling system is a control system that is part of a larger, more complex manufacturing planning and control system. The production scheduling system includes but is more than a schedule generation process (be it manual or automated). The production scheduling system is not a database or a piece of software. The production scheduling system interacts with but is not the system that collects data about the status of open work orders (often called a manufacturing execution system). The production scheduling system is not an optimization procedure. The production scheduling system provides information that other managers need for other planning and supervisory functions.

Because time estimates are incorrect (indeed, sometimes they are only guesses) and unexpected events occur, precisely following a schedule becomes more difficult as time passes. In some cases, the system may follow the sequence that the schedule specifies even though the planned start and end times are no longer feasible. Eventually, however, a new schedule will be needed.

Therefore, rescheduling is a key concept for understanding production scheduling systems. *Rescheduling* is the process of updating an existing production schedule in response to disruptions or other changes. There are many types of disturbances that can upset a production schedule, including machine failures, processing time delays, rush orders, quality problems, and unavailable material. In practice, rescheduling is done periodically to plan activities for the next time period based on the state of the system. It is also done occasionally in response to significant disruptions.

Vieira et al. (2003) present a rescheduling framework that can be used for classifying and describing rescheduling environments, policies, strategies, and methods. Another chapter of this book reviews this rescheduling framework and discusses considerations involved in choosing between different rescheduling strategies, policies, and methods.

Figure 4-1 shows a conceptual diagram of a typical production scheduling system, modeled as a feedback control system. The input to the production scheduling system is the set of jobs that need to be completed. The order release function checks the status of the jobs and releases those that are ready to begin. The schedule update function takes an existing production schedule, any changes to the state of the jobs, and information about the state of the shop (primarily the jobs and the resources) and creates a new production schedule, which the shop follows as best as possible in the face of disruptions.

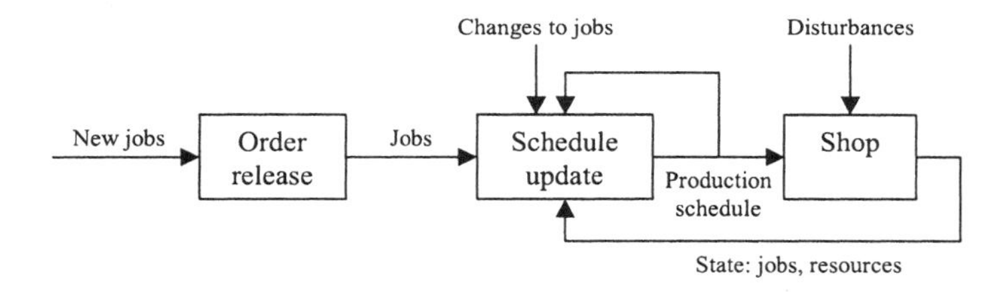

Figure 4-1. Production scheduling as a feedback control system (from Herrmann, 2004)

Though this description is simple, production scheduling systems are complex because the mechanisms for sensing the state of the manufacturing system and generating updated production schedules cannot be expressed (except in very special cases) as mathematical functions. In addition, the randomness of disruptions and other uncertainties make scheduling difficult. The most realistic representation is to view a production scheduling systems as a system of decision-makers that transforms information about the manufacturing system into a plan (the production schedule).

A decision-making systems view of production scheduling does not eliminate the need for better scheduling decisions. Indeed, this perspective is a valuable complement. For instance, McKay *et al.* (2002) discuss the problem of designing the scheduling task. Clearly, this important challenge requires understanding the context in which that task occurs and the information available for the scheduling task. Adopting a decision-making systems view provides an approach to obtaining that understanding and a tool for representing the information flow and decision-making system, as discussed later in this chapter.

The role of scheduling software in the improvement of production scheduling systems is a complex one. Research scientists, software companies, and manufacturing consultants have developed and implemented advanced scheduling systems to reduce the effort of production scheduling and generate better schedules. These scheduling systems include computer algorithms that exploit results from scheduling theory and advanced optimization techniques. However, it is well known that an information system should support the organization's decision-making system (cf. Ackoff, 1967). An improper fit leads to a dysfunctional information system. Thus, it is important, when considering scheduling software, to understand the production scheduling system as a decision-making system and to use this insight to guide the implementation of the scheduling software.

4. IMPROVEMENT METHODOLOGY

Viewing production scheduling as a decision-making system leads to a systems-level approach to improving production scheduling. In particular, this perspective is not concerned primarily with finding an algorithm to solve a combinatorial optimization problem. Moreover, the problem is not viewed only as helping a single production scheduler make better decisions (though this remains important, as discussed above). Instead, the problem is one of organizing the entire system of decision-making and information flow.

As with other efforts to improve manufacturing operations or business processes, improving production scheduling benefits from a systematic improvement methodology. The methodology presented here includes the following steps in a cycle of continuous improvement (as shown in Figure 4-2), which is based in part on ideas from Checkland (1999).

1. Study the production scheduling decision-making system.
2. Build, validate, and analyze one or more models of this decision-making system.
3. Identify feasible, desirable changes.
4. Implement the changes, evaluate them, and return to Step 1.

The important features of the decision-making system are the persons who participate in it, the decisions that are actually made, including the goals, knowledge, skills, and information needed to make those decisions. Also relevant are the processes used to gather and disseminate information. It will also be useful to study other processes that interact with production scheduling, including sales, cost estimation, production planning, and quality assurance.

Many process improvement approaches begin with creating a map or a flow chart that shows the process to be improved. For instance, in organizations adopting lean manufacturing principles, it is common for a team that plans to improve a production line to begin the improvement event with a value stream mapping exercise.

Creating a model of the as-is production scheduling system has many benefits. Though it may be based on pre-existing descriptions of the formal scheduling process, it is not limited to describing the "should be" activities. The process of creating the model begins a conversation among those responsible for improving the production scheduling system. Each person involved has an incomplete view of the system, uses a different terminology, and brings different assumptions to the table. Through the modeling process, these persons develop a common language and a complete picture. Validation activities give other stakeholders an opportunity to give input and also to begin learning more about the system. Even those that are directly involved in production scheduling benefit from the "You are here" information that a model provides.

An especially important part of modeling production scheduling systems is determining the sources that provide information to the scheduler. If they are not documented, changes to the system may eliminate access to these sources, which leads to worse decision-making.

No particular modeling technique is optimal. There are many types of models available, and each one represents a different aspect of the decision-making system. It may be necessary to create multiple models to capture the scope of the system and its essential details. Swimlanes diagrams can be useful, as discussed below. As in other modeling efforts, wasting time on unneeded details or scope is a hazard. The purpose of the model should guide the construction of the model and the selection of the appropriate level of detail.

In general, representing decision-making systems is a difficult task. A decision-making system may involve a complex social network. The information that decision-makers collect, use, and exchange comes in many forms and is not always tangible. Some decisions are routine, while others are unique. The documentation of decision-making systems usually does not exist. If it does, it is typically superficial. (Notable exceptions are the decisions make by government bureaucracies, as when a state highway administration designs a new highway. In such cases, the decision-making process is well documented.)

Analyzing such models quantitatively is usually not possible. A careful review of the model will reveal unnecessary steps or show how one group's activities are forcing others to behave unproductively. The model can show the impact of implementing proposals to change decision-making. For example, if the scheduler works for the plant manager (instead of the machine shop supervisor), how will the scheduler know which machines are down?

Evaluating the feasibility and desirability of potential changes and selecting those to implement requires time and effort to build consensus among the stakeholders. A "to-be" model of the production scheduling system can show how the system will operate after the changes are implemented. If the proposed changes involve different algorithms to generate schedules, discrete-event simulation models can be used to estimate the improvements that will be achieved.

Changes that are implemented should be evaluated to determine if the production scheduling system has improved. That is, is the decision quality increasing? Do decision-makers have better visibility of the production process? Is less time spent creating and updating schedules? Are the schedules more feasible? Are resources used in a more productive manner?

The questions asked depend upon the problems that motivated the improvement effort.

Ideally, production scheduling systems should undergo a continuous cycle of improvement. The organization and its environment are always changing. People come and go. Proposals that were infeasible become possible, and changes that were ignored become desirable. Money becomes available for software, or the scheduling software vendor goes out of business. Better information is appearing, and decisions that were easy become hard.

5. REPRESENTING PRODUCTION SCHEDULING SYSTEMS

Creating a model that represents the production scheduling system is a key step in the process of improving production scheduling, as discussed in Section 4. Because there are many different potential representations, this section will review and discuss the most relevant.

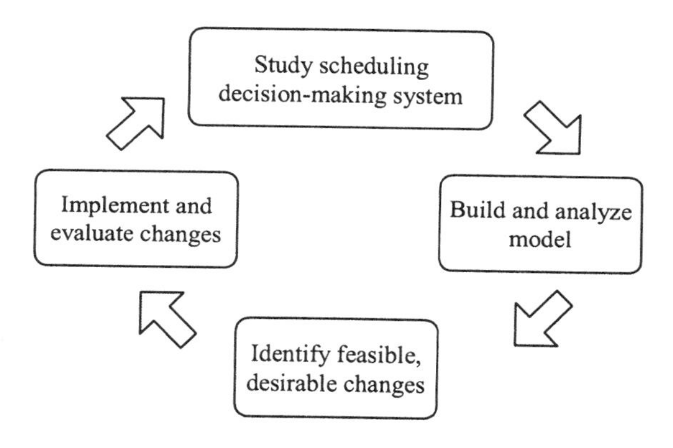

Figure 4-2. The methodology for improving production scheduling systems.

5.1 Organization charts, flow charts, and IDEF

Organizations are decision-making systems. The most typical representation of an organization is the organization chart, which lists the employees of a firm, their positions, and the reporting relationships. However, this chart does not explicitly describe the decisions that these persons are making or the information that they are sharing.

Another common representation is a flowchart that describes the lifecycle of an entity by diagramming how some information (such as a customer order, for example) is transformed via a sequence of activities into some other information or entity (such as a shipment of finished goods). This is a particularly useful representation for an organization designed around a workflow.

A suite of IDEF techniques exists for modeling an organization's functions, information systems, and processes. The "box and arrow" graphics of IDEF0 diagrams show the inputs, outputs, controls, and mechanisms that are relevant to each function. Comprehensive documentation of these techniques is now available online at www.idef.com.

More generally, the Viable System Model (Beer, 1972) is a cybernetics model that decomposes an organization into predefined subsystems. In an interpretive systems approach, a rich picture can indicate the many components of a complex system and encourage system-level thinking (Checkland, 1999). Control systems theory provides another way to represent a decision-making system (as shown in Figure 4-1).

5.2 Swimlanes

Swimlanes are a special type of flowchart that adds more detail about who does which activities, a key component of a decision-making system. Sharp and McDermott (2001) provide a good introduction to the use of swimlanes.

Our research has used swimlanes to represent a decision-making system since the swimlanes model yields a structured model that describes the decision-making and information flow most efficiently and clearly shows the actions and decisions that each participant performs. One limitation is that the model does not show the structure of the organization. Also, representing a larger, more complex system would require swimlanes models at different levels of abstraction to avoid confusion.

A swimlane diagram highlights the who, what, and when in a straightforward, easy-to-understand format. Unlike other forms, they identify the actors in the system. There are other names used to describe this type of diagram, including process map, line of visibility chart, process responsibility diagram, and activity diagram (the name used in the Unified Modeling Language).

A swimlane diagram includes the following components:

- *Roles* that identify the persons who participate in the process.
- *Responsibilities* that identify the individual tasks each person performs.
- *Routes* that connect the tasks through information flow.

Sharp and McDermott (2001) present techniques for modeling branching, optional steps, the role played by information systems, steps that iterate, steps that are triggered by the clock, and other details. The following summarizes some key points.

A single diagram is the path of a single item (e.g., form or schedule) as it goes through a process. Each person gets a row from left to right. An organization, a team, an information system, or a machine can have a row. In the row go boxes, one box for each task that the person performs. Arrows show the flow of work from one task to another and also indicate precedence constraints (what has to be done before another task can start).

Tasks can involve multiple actors, so the task should span the different actors' rows. While there are multiple flowchart symbols available, Sharp and McDermott recommend a simple box with occasional icons to represent a inbox or a clock. Boxes should be labeled with verb-noun pairs (e.g., "create schedule" rather than "new schedule"). Transportation steps and other delays should be included.

Flow should go generally from left to right, with backward arrows for iteration. A conditional flow should have one line that leaves an activity and then splits into two lines. Flow from an activity to two parallel steps should have two lines.

Managing detail requires multiple diagrams. The highest level shows one task per person per handoff. This clarifies the relationships and flow of information between persons. Another diagram can show the tasks that are key milestones that change the status of something, decisions, communication activities (passing and receiving information), and iteration. An even more detailed diagram can describe the specific ways in which the tasks are done (via fax or email, using specific tools or other special resources).

6. EXAMPLES

This section presents examples of representing production scheduling systems with swimlanes models.

6.1 CAD/PAD production scheduling

The Naval Surface Warfare Center, Indian Head Division (NSWC/IHD) serves the armed forces by developing, manufacturing, and supporting energetics products, including cartridge-actuated devices (CADs) and propellant-actuated devices (PADs) that are typically found in aircrew escape systems and in other aircraft systems. The CAD/PAD assembly facility assembles devices using cartridges, primers, and other hardware that are made in other NSWC/IHD facilities and at contractors.

Workorders arrive from the acquisitions organization that is responsible for purchasing devices for the armed services. The branch manager logs the workorder. The production controller adds it to the long-range schedule. The production engineer determines if the key hardware will be ready on time and informs the branch manager if the required delivery date is feasible. The branch manager accepts the workorder and informs the acquisitions organization.

The production system operates with two schedules: a long-range schedule (discussed below) and a weekly schedule. The weekly schedule records the status of about 24 operations (for 13 workorders) that are currently in process or ready to start.

At the end of each week, the shop foreman tells the production controller how many hours were worked on which workorders. The production controller updates the weekly schedule (the one created at the beginning of the week) with this information and brings this interim schedule to the weekly meeting.

The primary communication mechanism in the production scheduling system is a weekly meeting (first thing Monday morning) of all the participants. The primary objective of the meeting is to create an accurate picture of which workorders are ready for production and which have priority so that the shop foreman can determine what the shop will do. The participants discuss the workorders scheduled for that week, the work performed the previous week, new workorders that are ready for production, and any other updates. For example, a workorder may be ready to be shipped to the X-ray facility, hardware may have been moved from storage to the production building, or a piece of necessary equipment may be unavailable. Based on information from a monthly meeting with the acquisitions organization, the branch manager identifies the workorders that have priority that week. After the meeting, the production controller updates the schedule accordingly, signs it, and distributes it to all personnel that day.

The shop foreman makes decisions about which operators will work on which activities, and when during the week tasks will be done. The shop foreman records the hours worked. When changes occur during the week (to the status of equipment, hardware, or workorders), the production engineers, production controller, and shop foreman react appropriately without changing the weekly schedule. These events are discussed at the next weekly meeting, and the schedule is updated accordingly then.

The long-range schedule lists approximately 80 workorders and, for each one, the number of production labor hours scheduled in each of the next 12 months. Once a month the weekly meeting also discusses the long-range schedule. The group discusses each workorder on the long-range schedule and its status. The production controller updates the long-range schedule accordingly and distributes this to personnel in the branch and elsewhere.

Figures 4-3 and 4-4 illustrate the production scheduling process using swimlanes. Each horizontal bar corresponds to a particular person and shows the activities in which that person participates. The links between the activities show the flow of information. Figure 4-3 represents the activities that receive workorders and update the long-range schedule. Figure 4-4 represents the activities that update the weekly schedule.

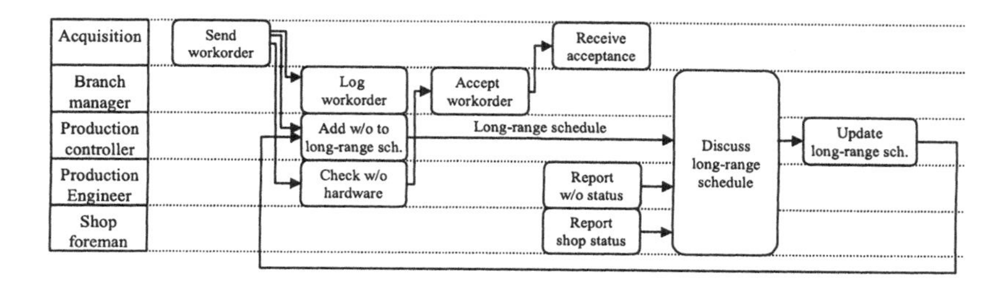

Figure 4-3. Long-range production scheduling.

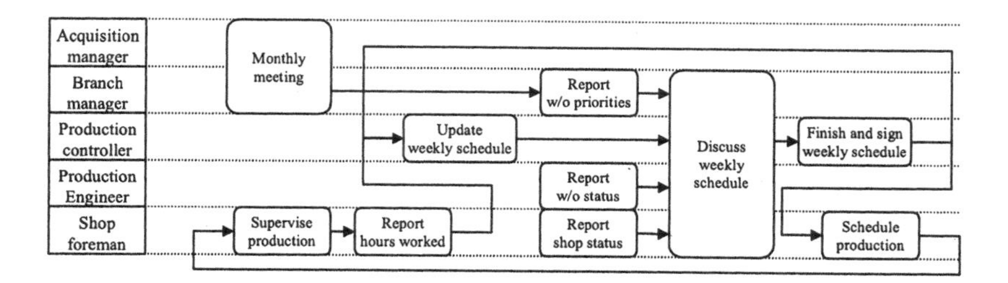

Figure 4-4. Weekly scheduling of CAD/PAD production.

6.2 Conservatory assembly

Tanglewood Conservatories is a conservatory builder that designs, fabricates, and constructs custom conservatories (rooms with wood frames and glass panels typically used for dining, entertaining, or recreation). The firm completely constructs the conservatory at their facility and then (usually) disassembles the structure, ships the components to the site, and re-assembles the conservatory. Each conservatory project involves nearly 300 tasks and takes months to complete. The firm usually has four to six projects in process at one time.

The firm has many types of resources, including engineers, sales personnel, shop associates, woodworkers, a glass shop, and painters.

The firm has recently implemented a new scheduling system to give the owner better information about the status of each customer order and to create better schedules. The key personnel in the scheduling system are the owner and the administrative assistant. The foreman and the lead engineer also participate. The scheduling system uses standard office applications enhanced with customized data processing and scheduling macros (see Figure 4-5 for a representation of this system).

When a customer agrees to order a conservatory, the owner creates a new project using a template in his project management software. He then emails the new project file to the assistant, who saves it in the scheduling system folder on her computer.

The production scheduling routine occurs weekly. On Wednesday the assistant walks around the shops and talks to the foreman about the status of tasks that were supposed to be done that week. The assistant also talks to vendors who are supposed to deliver material that week and to the lead engineer about his tasks.

Based on this information, on Thursday the assistant updates the project files for each customer order, which specify the task information. The assistant also maintains a file that specifies the resources available (the "resource pool"). Each one of the customer order files is linked to the resource pool file. The assistant runs resource leveling routines that resolve scheduling conflicts and produce a detailed production schedule. The schedule is exported from the project scheduling software and into an electronic spreadsheet that lists the tasks that need to be done. If a customer order is projected to be tardy, the assistant reviews the tasks on the critical path to locate the problems. The spreadsheet, which contains all of the unfinished tasks that have already begun and the tasks that need to start in the next two weeks, is the production schedule. For each task, it specifies when the task should begin and which resource group should perform it.

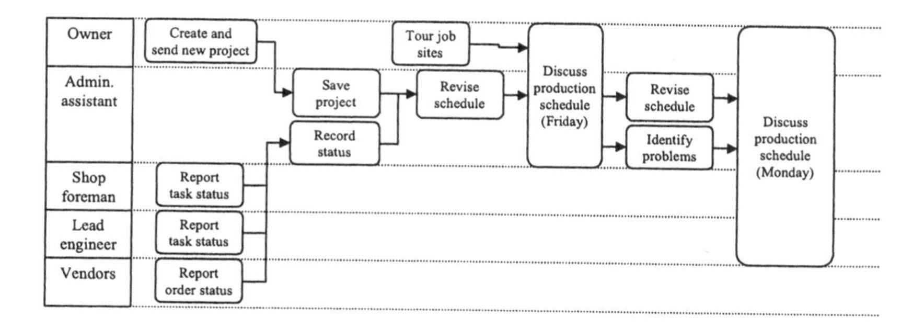

Figure 4-5. Swimlanes diagram for production scheduling at Tanglewood Conservatories.

On Friday morning the owner and the assistant meet to discuss this schedule. The owner brings additional information about new orders and other updates from his visits to job sites. If necessary, the assistant updates the project files and creates a new schedule later that day. If the schedule shows that a customer order will be tardy, the assistant reviews the tasks on the critical path to identify the problem.

On Monday morning, the owner, the assistant, the foreman, and the lead engineer meet to discuss the schedule and any issues that have not been considered. The foreman and the lead engineer then know what they and their staff need to do that week. The foreman has the freedom to make small adjustments and to expedite tasks when possible to work ahead and improve resource utilization. (Note that, because installation operations at the job sites are independent of the fabrication and assembly operations at the shop, they are managed locally by a supervisor at each site.)

7. SUMMARY AND CONCLUSIONS

This chapter has described a decision-making systems perspective on production scheduling. This view leads to a particular methodology for improving a production scheduling system. A key feature of this methodology is the role of creating a model to represent the production scheduling system. There are many types of models available. This chapter discussed swimlane diagrams and used real-world production scheduling systems to give examples of the swimlane diagrams.

The objective has been to show how, in practice, production scheduling is part of a complex flow of information and decision-making. It is not an isolated optimization problem. It is hoped that this material will help engineers, analysts, and managers improve their production scheduling systems by considering the structure and behavior of the system. This will encourage researchers to develop new representations and to employ innovative methodologies for studying and improving production scheduling.

The study of production scheduling as a system of decision-makers is not complete. It would be useful to determine how representations such as swimlanes can be used most effectively to improve production scheduling systems. At what point in the process of improving production scheduling are such representations most appropriate? What level of detail is best? Are there good ways to prototype production scheduling systems?

For those who are trying to improve production scheduling in their factory, the ideas presented in this chapter should help them to take a more holistic approach to the problem. It can be tempting to think that the solution is to invent a better scheduling algorithm or to purchase and install new software. However, such solutions will fail unless one has thought carefully about the persons who do the production scheduling, the tasks that they perform, the decisions that they make, and the information that they share. Redesigning this part of the system will be a significant chore. Software does not complain when it is uninstalled, but people complain when their routines are changed.

Improving production scheduling requires investments of funding, people, and time. The material presented here is no substitute for those resources, but it is hoped that these ideas will encourage and assist those who seek success.

ACKNOWLEDGEMENTS

The author would like to thank Charles Carr for providing information about the production scheduling system at Tanglewood Conservatories and NSWC Indian Head Division for their support.

REFERENCES

Ackoff, R.L., 1967, Management misinformation systems, *Management Science*, **14**(4):147-156.

Beer, S., 1972, *Brain of the Firm,* Allen Lane, London.

Bertrand, J.W.M., Wortmann, J.C., and Wijngaard, J., 1990, *Production Control: a Structural and Design Oriented Approach*, Elsevier, Amsterdam.

Chacon, G.R., 1998, Using simulation to integrate scheduling with the manufacturing execution system, *Future Fab International*, 63-66.

Checkland, P., 1999, *Systems Thinking, Systems Practice*, John Wiley & Sons, Ltd., West Sussex.

Flanders, S.W., and Davis, W.J., 1995, Scheduling a flexible manufacturing system with tooling constraints: an actual case study, *INTERFACES*, **25**(2):42-54.

Graham, R.L., Lawler, E.L., Lenstra, J.K., and Rinnooy Kan, A.H.G., 1979, Optimization and approximation in deterministic machine scheduling: a survey, *Annals of Discrete Mathematics*, **5**:287-326.

Hasenbeign, J., Sigireddy, S., and Wright, R., 2004, Taking a queue from simulation, *Industrial Engineer*, 39-43.

Herrmann, J.W., 2004, Information flow and decision-making in production scheduling, 2004 Industrial Engineering Research Conference, Houston, Texas, May 15-19, 2004.

Herrmann, J.W., and Schmidt, L.C., 2002, Viewing product development as a decision production system, DETC2002/DTM-34030, Proceedings of the ASME 2002 Design Engineering Technical Conferences and Computers and Information in Engineering Conference, Montreal, Canada, September 29 - October 2, 2002.

Herrmann, J.W., and Schmidt, L.C., 2005, Product development and decision production systems, in *Decision Making in Engineering Design*, W. Chen, K. Lewis, and L.C. Schmidt, editors, ASME Press, New York.

Hopp, W.J., and Spearman, M.L., 1996, *Factory Physics*, Irwin/McGraw-Hill, Boston.

Huber, G.P., and McDaniel, R.R., 1986, The decision-making paradigm of organizational design, *Management Science*, **32**(5):572-589.

Katok, E., and Ott, D., 2000, Using mixed-integer programming to reduce label changes in the Coors aluminum can plant, *INTERFACES*, **30**(2):1-12.

Leachman, R.C., Kang, J., and Lin, V., 2002, SLIM: short cycle time and low inventory manufacturing at Samsung Electronics, *INTERFACES*, **32**(1):61-77.

McKay, K.N., and Buzacott, J.A., 1999, "Adaptive production control in modern industries," in *Modeling Manufacturing Systems: From Aggregate Planning to Real-Time Control*, P. Brandimarte and A. Villa, editors, Springer, Berlin.

McKay, K.N., Pinedo, M., and Webster, S., 2002, "Practice-focused research issues for scheduling systems," *Production and Operations Management*, **11**(2):249-258.

McKay, K.N., Safayeni, F.R., and Buzacott, J.A., 1995, "An information systems based paradigm for decisions in rapidly changing industries," *Control Engineering Practice*, **3**(1):77-88.

McKay, K.N., and Wiers, V.C.S., 1999, "Unifying the theory and practice of production scheduling," *Journal of Manufacturing Systems*, **18**(4):241-255.

Moss, S., Dale, C., and Brame, G., 2000, Sequence-dependent scheduling at Baxter International, *INTERFACES*, **30**(2):70-80.

Pinedo, M., 2005, *Planning and Scheduling in Manufacturing and Services*, Springer, New York.

Sharp, A., and McDermott, P., 2001, *Workflow Modeling*, Artech House, Boston.

Simon, H.A., 1973, Applying information technology to organization design, *Public Administration Review*, **33**(3):268-278.

Simon, H.A., 1997, *Administrative Behavior*, 4th edition, The Free Press, New York.

Skinner, W., 1985, The taming of the lions: how manufacturing leadership evolved, 1780-1984, in *The Uneasy Alliance*, Clark, K.B., Hayes, R.H., and Lorenz, C., eds., Harvard Business School Press, Boston.

Vieira, G.E., Herrmann, J.W., and Lin, E., 2003, Rescheduling manufacturing systems: a framework of strategies, policies, and methods, *Journal of Scheduling*, **6**(1):35-58.

Vollmann, T.E., Berry, W.L., and Whybark, D.C., 1997, *Manufacturing Planning and Control Systems*, fourth edition, Irwin/McGraw-Hill, New York.

Chapter 5

SCHEDULING AND SIMULATION

The Role of Simulation in Scheduling

John W. Fowler, Lars Mönch, Oliver Rose

Arizona State University, Technical University of Ilmenau, Technical University of Dresden

Abstract: This chapter discusses how simulation can be used when scheduling manufacturing systems. While deterministic scheduling and simulation have often been seen as competing approaches for improving these systems, we will discuss four important roles for simulation when developing deterministic scheduling approaches. After an overview of the roles, we will use a case study to highlight two of the roles.

Key words: Scheduling, discrete event simulation, performance evaluation, semiconductor manufacturing

1. INTRODUCTION AND DEFINITIONS

Simulation and scheduling are well-established techniques in operations research and industrial engineering. Many papers discuss either simulation or scheduling by itself. However, much less literature exists on the interface between the two different approaches because usually researchers from different communities consider scheduling and simulation problems. Traditionally, scheduling approaches are more mathematical in nature. Simulation techniques are used to solve a problem that cannot be tackled by analytical methods because either it is too complex or an analytical description itself is impossible.

Scheduling is defined as the process of allocation of scarce resources over time (Pinedo, 2002; Brucker, 2004). The goal of scheduling is to optimize one or more objectives in a decision-making process. The two major categories in scheduling are deterministic and stochastic scheduling. Deterministic scheduling is characterized by processing times, set-up times, and job priorities that are known in advance. They are not influenced by uncertainty. In contrast, stochastic scheduling problems do not assume the existence of deterministic values for processing times, set-up times, or other quantities that are used within the scheduling model. The deterministic values are replaced by corresponding distributions. Deterministic scheduling problems can be further differentiated into static problems where all jobs to be scheduled are available at time $t=0$. Dynamic scheduling problems relax this condition. In this situation, jobs are ready at different points in time $t \geq 0$.

Simulation is used to describe a certain process in a time-dependent manner (Law & Kelton, 2000; Banks *et al.*, 2005). Depending on time progression, we differentiate between discrete-event and continuous simulation. In the first case, future events are determined, and the simulation clock jumps to the next future event. In some cases, we consider an equidistant time progress; however, because of computational reasons, time progress is very often made in a non-equidistant manner. Continuous simulation means that infinitesimally small time steps are considered. Hence, continuous simulation is basically the (numerical) treatment of differential equations. Difference equations are used in case of a discrete time steps.

In this chapter we consider scheduling and simulation approaches in the manufacturing domain. Based on our research interests and practical needs, we mainly focus on complex manufacturing systems such as semiconductor wafer fabrication facilities. However, a large majority of our statements can be extended to the service sector in a straightforward manner.

Simulation of manufacturing systems is still challenging. Fowler & Rose (2004) summarize the grand challenges of manufacturing simulation. The challenges are to reduce modeling effort, to increase the acceptance of modeling and simulation within industry, to establish real plug-and-play capabilities of simulation software, and to develop advanced interoperability capabilities of simulation tools. Modeling and simulation is therefore still an area of ongoing research that attracts many researchers.

The remainder of this chapter is organized as follows. We describe some application areas of scheduling and simulation based on system theory. Then, we focus on the use of simulation techniques as part of scheduling approaches. Finally, we present a case study where we focus on the simulation-based emulation and evaluation of a modified shifting bottleneck heuristic for semiconductor wafer fabrication systems (wafer fabs).

2. OVERALL SETTING

The overall setting for simulation and scheduling in manufacturing systems can be derived from general considerations of production control. A manufacturing system consists of a base system that contains all the resources, i.e., machines and operators. The corresponding base process is given by jobs that consume capacities of the resources during processing. The resource allocation process of the jobs is influenced by the production control process that is performed by using the production control system (PCS). The PCS consists of the computers and the software used to determine production control instructions m, i.e., software with production scheduling capabilities. The production control process determines when and under which circumstances a certain control algorithm is used to determine production control instructions. We present the overall setting in Figure 5-1.

Now we can use Figure 5-1 to come up with application areas for simulation and scheduling. Scheduling capabilities are provided by the production control system and process. At this stage, simulation can be used as part of the PCS; i.e., simulation is used directly or indirectly in order to derive production control instructions.

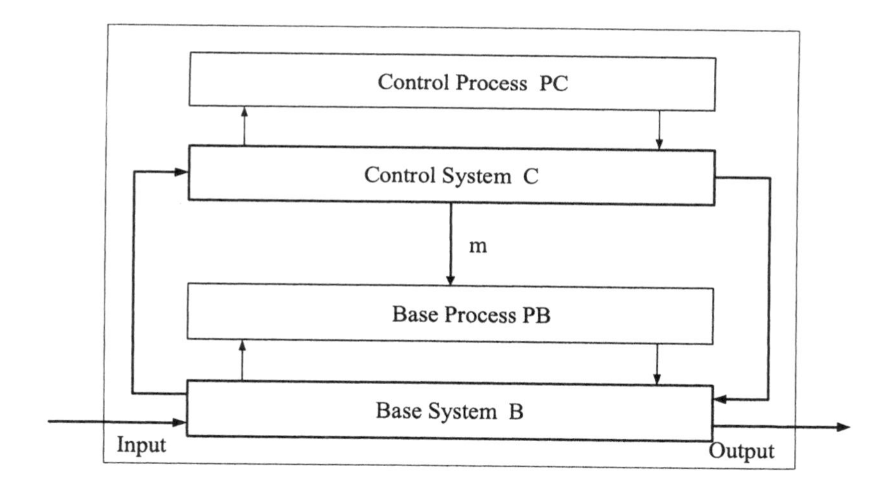

Figure 5-1. Overall setting of scheduling and simulation within the production control context.

The second opportunity of using simulation is given by the emulation of the base process. In this situation, the scheduling approach produces control instructions whereas the simulation model is used to represent the base system and the base process. Simulation can be used to evaluate the performance of a certain production scheduling approach.

In the remainder of the chapter we want to describe the two principle possibilities in more detail. Furthermore, we present a case study where we apply the second type of using simulation.

3. SIMULATION TECHNIQUES WITHIN SCHEDULING

We begin with describing the use of simulation as part of the production control approach. We identify four different uses for simulation in this context. First, as described in Section 3.1, simulation-based schedule generation and refinement use simulation directly in order to determine initial schedules (generation) and to improve existing schedules (refinement), generally for short periods of time. In addition, simulation-based optimization applies simulation in a repeated fashion in order to estimate a certain objective function value and to improve it. The second use, as described in Section 3.2, is given by deterministic forward simulation in order to evaluate certain parameter setting strategies for a given production control scheme. The third use of simulation is in emulating a scheduling system. The final use is to evaluate deterministic scheduling approaches. Section 3.3 describes these last two uses of simulation.

3.1 Simulation-based schedule generation, refinement, and optimization

3.1.1 Simulation-based schedule generation

Simulation-based scheduling means that simulation is used to determine schedules with a scheduling horizon ranging from several hours to a day. Dispatching rules that are already part of the simulation engine are used to allocate machines to jobs. The assignment and the sequencing of jobs observed in the simulation are used to produce a control instruction in the original production control system that is used to influence the base system. The major parts of a simulation-based scheduling system are:

1. a simulation engine that contains several dispatching rules for next job selection within an appropriate (up-to-date) simulation model,

2. a graphical user interface that produces Gantt charts based on the results of a simulation run,
3. an interface to the information systems on the shop-floor that develops a dispatch list from the Gantt chart.

Simulation-based scheduling relies to a large extent on the capability to produce simulation models that represent the base process and the base system in an appropriate manner. Automated or semi-automated simulation model generation abilities based on data in systems like a manufacturing execution system (MES) are necessary in order to run a simulation-based scheduling system. The selection of a final schedule as a production control instruction set can be based on several criteria (see Sivakumar, 2001, for a more detailed description of this approach). There are several recent simulation-based scheduling systems described in the literature. Among them we refer to Potoradi *et al.* (2002).

All stochastic effects (for example, machine breakdowns) are generally turned off because of the short horizon. Appropriate model initialization is a non-trivial issue in simulation-based scheduling.

3.1.2 Simulation-based schedule refinement

Once a schedule has been generated, simulation can be used to further refine the schedule in order to get better estimates of system performance, such as the staring and completion times of jobs. This is generally done by using the sequence of operations from the initial schedule and then providing additional detail to the model used to generate the initial schedule. The additional detail may involve including resources not in the original model (such as jigs, fixtures, or operators) or may involve sampling from distributions for processing times instead of using deterministic values.

3.1.3 Simulation-based schedule optimization

Simulation-based optimization starts from the idea that it is difficult in some situations difficult to evaluate an objection function (Fu *et al.*, 2001; Fu, 2002). In some situations, a simulation model implicitly gives the objective function. Optimization is typically performed using local search methods (meta-heuristics) like genetic algorithms, tabu search or simulated annealing. These methods tend to be computationally costly. Therefore, the level of detail for the simulation model is important in simulation-based optimization applications. Simulation-based optimization therefore also requires a simulation model. However, in contrast to simulation-based scheduling, very often stochastic effects are not neglected. They have to be taken into account both for the meta-heuristic and the simulation model.

The overall scheme for scheduling applications can be described as follows:

1. Start from a given initial solution,
2. Use the simulation model in order to calculate the objective function,
3. Modify the given schedule by local changes,
4. Repeat the algorithm from Step 2 onwards until a given stopping criterion is met.

Simulation-based optimization can be used either on the entire manufacturing system level or for parts of the manufacturing system. The latter case is, for example, important for cluster tool scheduling. In Dümmler (2002) a genetic algorithm is used for scheduling jobs on parallel cluster tools. The genetic algorithm assigns jobs to single cluster tools and determines simultaneously the sequence of the jobs on each cluster tool. Because of the complex control algorithms of cluster tools, which cannot be modeled analytically, simulation is used to determine the objective function value for a given sequence of jobs on a cluster tool.

3.2 Simulation used for parameter setting and test instance generation for scheduling approaches

Many scheduling problems are NP hard. Therefore, efficient heuristics have to be applied for their solution. Some heuristics need to be parameterized in order to adapt them to a large range of different situations. A scheduling heuristic often contains certain parameters that influence the performance of the heuristic. The general framework of parameter selection can be described as follows.

1. Find quantities that describe a certain situation in a manufacturing system. High workload or tight due dates are examples for such situations.
2. Find a mapping that assigns to each situation description the appropriate parameters of the heuristic.
3. Use the parameters selected in Step 2 to determine production control instructions.

Deterministic forward simulation can be used to construct the mapping between situation description and appropriate parameters of the heuristic. Usually, discrete points are chosen from a range of parameters. The objective function value obtained by the heuristic using these parameters is determined by simulation. The parameters that lead to the smallest objective function value for a given situation are stored for later usage. The constructed mapping can be explicitly given by a formula (Lee and Pinedo, 1997) or implicitly by a knowledge representation structure like a neural network (Kim *et al.*, 1995; Park *et al.*, 2000).

A second application closely related to parameter setting is the construction of test problem instances and benchmarks for scheduling approaches. Hall and Posner (2001) describe techniques for generating random problem instances of scheduling problems. Deterministic forward simulation techniques can be used in order to determine objective function values for benchmark purposes. This approach is followed, for example, by Biskup & Feldmann (2001).

3.3 Simulation for emulation and evaluation of scheduling approaches

Given a production scheduling approach, we are interested in the behavior of the base process and base system under the influence of this production scheduling approach. In contrast to the approaches in Sections 3.1 and 3.2, we are not interested in improving the performance of the production control system by changing the production scheduling algorithm via simulation.

Before actually implementing a production scheduling algorithm, it is useful to simulate the use of the algorithm. Simulation has two important roles to play in this context. First, it acts as an emulator for the real system and second it can be used to evaluate the performance of the system. In the emulation role, simulation is used to specify the data requirements for extracting the current system state (both job and machine states) that is fed to the algorithm and for feeding the resulting schedule information back to the production system. It can also be used to determine what decisions will have to be made when things happen in the production system that lead to non-conformance to the schedule.

In the evaluation role, simulation is used to compare the performance of alternate production scheduling approaches. While this role is easy for simulation when the scheduling approach is to simply apply dispatching rules, it is considerably more difficult when a deterministic scheduling approach is applied. This role requires the simulation to also emulate the information flows described in the previous paragraph. The overall approach of simulation-based performance assessment of production control approaches can be summarized as follows:

1. Represent the base system and the base process by a discrete-event simulation model,
2. Build an interface between the production scheduling approach, i.e., the production control system and process, and the simulation model, i.e., the base system and process,
3. Describe the production control process, i.e., determine when the scheduling approach should be called, and

4. Implement the schedules obtained via the interface within the simulation model, i.e., within the base system and process.

Many researchers use the described evaluation approach (see, for example, Toba, 2000; Horiguchi, 2001). However, most of the applications are proprietary with respect to a specific simulation tool. True plug-and-play interoperability is not always ensured.

The use of simulation for performance assessment has the main drawback that the execution of the simulation is often computationally expensive.

4. CASE STUDY IN SEMICONDUCTOR MANUFACTURING

In this section, we describe a case study from the semiconductor manufacturing domain. We begin with a short description of the semiconductor manufacturing process. Then we recall the main ingredients of a modified shifting bottleneck heuristic that takes the main process characteristics of wafer fabs into account. We describe an efficient software implementation and a simulation framework for emulation and performance assessment of this scheduling heuristic. Then we continue with the description of simulation experiments for the performance assessment of the shifting bottleneck heuristic.

4.1 Manufacturing environment

The manufacturing of integrated circuits (ICs) on silicon wafers is an example of a manufacturing process that is interesting from a complexity point of view. There are multiple products, routes with several hundred process steps, and a large number of machines (tools). Semiconductor manufacturing is characterized by the following process conditions:

- a mix of different process types, for example, batch processes, i.e., several lots can be processed simultaneously at the same time on the same machine versus single wafer processes on the other hand,
- unrelated parallel machines,
- sequence-dependent setup times that are very often a multiple of the raw processing time,
- a changing product mix, and
- a customer due date related type of manufacturing.

The machines used for processing lots are extremely expensive. Therefore, they are often scarce resources. This is the main reason for a reentrant flow of the lots through the wafer fab. This type of flow causes problems related to production control of wafer fabs that are different than production control problems in classical job shops (for example, the occurrence of dynamic bottlenecks).

Complicated machines are used to manufacture ICs. Therefore, the manufacturing system wafer fab is influenced to a large extent by stochastic behavior like machine breakdowns. Very often preventive maintenance tasks are necessary because of the difficulty of the technological processes. These tasks also reduce machine capacities. There is a competition between the production lots and the prototype lots for processing times on the machines.

The starting point for the research described in this chapter is a modification of the shifting bottleneck heuristic for complex job shops by Mason *et al.* (2002). Several assessment efforts for static test instances are described in the literature (Mason *et al.,* 2002; Pinedo and Singer, 1999; Ovacik and Uzsoy, 1997; Demirkol *et al.*, 1997). However, the goal of our research was the development of a scheduler prototype that allows for an assessment in a dynamic environment. The problems discussed in previous research were significantly smaller with respect to the number of machines, number of lots, and number of process steps of the routes. It was unknown whether we could use the shifting bottleneck heuristic for scheduling wafer fabs or not. Furthermore, we did not know which modifications of the original heuristic were required in order to come up with a scheduling heuristic that provides high quality results with acceptable computational efforts.

4.2 Shifting bottleneck heuristic for complex job shops

4.2.1 Statement of the problem

We use in this chapter the (α,β,γ) notation for deterministic scheduling problems due to Graham *et al.* (1979). The α field describes the manufacturing environment from a machine point of view. For example, single machine, parallel machine, and job shop type of problems can be represented. The β field indicates process restrictions like sequence dependent setup times or batching restrictions. The γ field specifies the objective function. Using this notation scheme, the scheduling problem to be solved can be described as follows:

$$Jm \mid batch, incompatible, s_{jk}, r_j, recrc \mid \sum w_j \left(c_j - d_j\right)^+ . \tag{1}$$

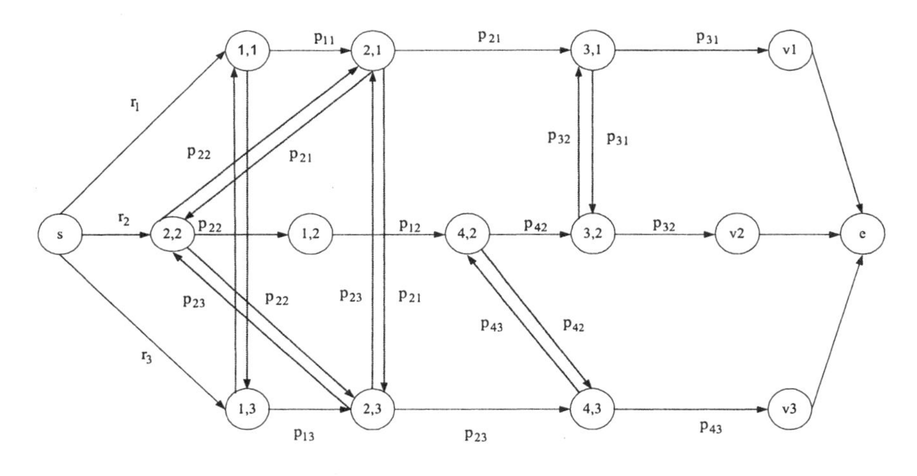

Figure 5-2. Disjunctive graph for representing a job shop.

The notation *Jm* is used for a job shop. Batch processing with incompatible families is denoted by *batch, incompatible*. All lots of one family can be used for batch formation. We denote sequence dependent setup times by s_{jk}. Dynamic job arrivals are indicated by r_j and reentrant process flows by *recrc*. The performance measure of interest is given by the total weighted tardiness (TWT) of all lots. It is defined by

$$TWT := \sum w_j \left(c_j - d_j\right)^+, \tag{2}$$

where $x^+ := max(x,0)$ and w_j denotes the weight, c_j the actual completion time, and d_j the due date of lot j.

The scheduling problem (1) is NP hard by reduction to the simpler single machine scheduling problem $1//\Sigma w_j(c_j - d_j)^+$, which is known to be NP hard (Lenstra *et al.*, 1977). Hence, we need to find efficient heuristics.

4.2.2 Disjunctive graph representation for job shop problems

Job shop scheduling problems can be represented by disjunctive graphs (see Adams *et al.*, 1988, for more details on this concept). A disjunctive graph is given by the triple

$$G := \left(V, E_c, E_d\right). \tag{3}$$

We denote by V the set of nodes of the graph. E_c is used for the set of conjunctive arcs, whereas E_d denotes the set of disjunctive arcs. The node set contains an artificial starting node s and an artificial end node e. We need an end note V_j for each lot. The end nodes represent the due dates d_j of each lot. The process step of lot j that has to be performed on machine i is modeled by a node $[i,j]$ in G. Conjunctive arcs are used to link consecutive process steps of a lot. Conjunctive arcs are directed. The processing time of the lot that is associated with node $[i,j]$ on machine i is denoted by p_{ij}. The arrival dates of lots in the manufacturing system are represented by arcs between the start node s and the nodes that are used to model initial processing steps on the machines. The weight of this node for lot j is chosen as the ready time r_j of lot j. Arcs with weight 0 are used to link the nodes V_j and e. Disjunctive arcs are used to model scheduling decisions. Disjunctive arcs are undirected. They are introduced between nodes that represent process steps of lots with respect to a singe tool group. The disjunctive arc between node $[i,j]$ and node $[i,k]$ is transformed into a conjunctive one if lot j is processed immediately before lot k on tool group i. We remove the disjunctive arc if we know after a scheduling decision that lot j and lot k are not processed in a consecutive manner on a certain machine of the tool group. Figure 5-2 shows a disjunctive graph for a job shop that consists of four machines and three jobs.

The disjunctive graph allows us to evaluate the influence of single scheduling decisions on the entire manufacturing system. In order to do so, we calculate ready times for the lots on each single tool group by applying longest path calculations between the start node s and the corresponding node. The ready time with respect to node $[i,j]$ is denoted by r_{ij}. The planned completion time of the process step that is represented by the node $[i,j]$ is calculated as the longest path between node $[i,j]$ and the artificial end node e. We denote the due date by d_{ij}. Sophisticated production conditions like parallel machines, sequence dependent setup times, batch tools, and reentrant flows are important in complex manufacturing systems. Therefore, we describe how these characteristics can be modeled by the disjunctive graph G.

Parallel machines are included in a natural way in the scheduling graph. Only those nodes are connected after a scheduling decision that represent lots to be processed on a specific parallel machine.

Sequence dependent setup times can be integrated into the scheduling graph only after the scheduling decisions for the tool group are made. Therefore, we increase the weight of the corresponding arc by the setup time. It holds

$$p_{ij} := p_{ij} + s_{kj}, \quad (4)$$

if for the processing of lot j on machine i the setup time s_{kj} is necessary. An appropriate modeling of batching tool groups is more sophisticated (Mason *et al.*, 2002). A scheduling decision for a batch machine includes three types of decisions:

1. Batching decisions: which lots should form a certain batch,
2. Assignment decisions: which batch should be assigned to a certain machine, and
3. Sequencing decisions: in which sequence should batches be processed on a given machine.

After solving the scheduling problem for a batch tool group, artificial batch nodes are added to the scheduling graph G. The predecessors of the artificial batch nodes are those nodes that represent the lots that form the batch. We consider batch processes, which are characterized by the same processing time of all lots that form the batch. Therefore, it makes sense to use the weight 0 for the incoming arcs. The outgoing arcs of the artificial batching node connect the batch node with the successor nodes according to the routes of the lots that form the batch. The weight of each outgoing arc is set to be the processing time of the batch.

In Figure 5-3, we depict a disjunctive graph where the lots represented by node *[1,1]* and node *[1,2]* form a batch. Furthermore, we consider batches consisting of one lot for the nodes *[1,3]*, *[4,1]*, and *[4,2]*.

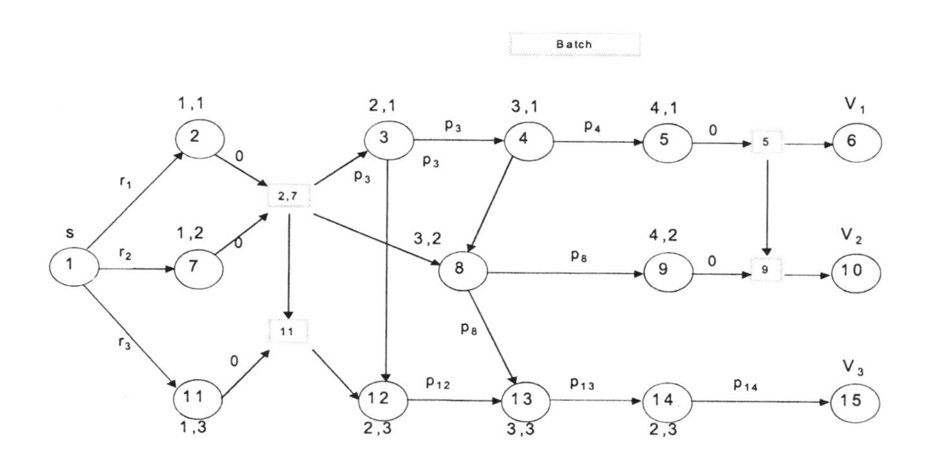

Figure 5-3. Disjunctive graph for representing batch machines.

Reentrant flows, i.e., the multiple usages of a machine by the same lot, are modelled in such a way that the corresponding nodes are not connected by disjunctive arcs. In this situation, the nodes are already connected by conjunctive arcs according to the routes of the products.

4.2.3 Decomposition approach

Information on earliest start dates r_j and planned completion times d_j are provided. We use these data items to formulate subproblems. Because of the production conditions in semiconductor manufacturing we have to solve problems of type

$$Pm \mid batch, incompatible, s_{jk}, r_j, prec \mid TWT . \qquad (5)$$

Here, P_m is used for parallel machines. We denote with *prec* precedence constraints in the disjunctive graph that have to be considered during the scheduling decisions. We have to follow these precedence constraints in order to make sure that no cycles are created in the disjunctive graph. We refer to the work of Pabst (2003) for more details on this topic.

The performance of the shifting bottleneck heuristic is influenced by the sequence of solved subproblems and also by the choice of the solution approach, called *subproblem solution procedure* (SSP), for the scheduling problem in Equation (5).

4.2.4 Algorithm

The shifting bottleneck heuristic can be formulated as follows (Mason *et al.*, 2002):

1. We denote by M the set of tool groups that have to be scheduled. Furthermore, we choose the notation M_0 for the set of tool groups that are already scheduled. Initially, $M_0 = \varnothing$ holds.
2. Determine and solve subproblems for each tool group $i \in M - M_0$.
3. Determine the most critical tool group $k \in M - M_0$.
4. Implement the schedule determined in Step 2 for the tool group $k \in M - M_0$ into the scheduling graph. Set $M_0 = M_0 \cup \{k\}$.
5. (Optional) reoptimize the determined schedule for each tool group $m \in M_0$ by considering the newly added disjunctive arcs from Step 4 for tool group $k \in M - M_0$.
6. The algorithm terminates if $M = M_0$. Otherwise, go to Step 2.

For determining the most critical tool group in Step 3, we may use different approaches. For example, it is possible to consider the total weighted tardiness as an appropriate measure. In Pinedo (2002), it is suggested to use the change of the completion times after the implementation of the schedule for a tool group relative to the graph in the previous iteration as a measure for machine criticality. Furthermore, it is possible to exploit more global information like, for example, knowledge of planned and dynamic bottlenecks during decision-making.

4.3 Software implementation of the shifting bottleneck heuristic

4.3.1 Implementation approach

It was the goal of our research to develop a scheduler prototype to test the applicability of the shifting bottleneck heuristic for realistic scenarios in a dynamic production environment. The anticipated problems in runtime performance and memory usage of the application resulted in a number of requirements for the selection of the programming language and the efficiency of the data structures used. Since we did not intend to shift the focus of our research work to implementation issues, we tried to apply mainly commercial or semi-commercial software libraries.

We implemented the scheduler prototype in the programming language C++ because it generates fast and efficient code. In addition, there are many software libraries developed in this programming language. The third reason for selecting C++ was the opportunity of designing our application according to the object-oriented paradigm from which we expected several advantages concerning the complexity of the prototype implementation.

4.3.2 Efficient data structures

The execution speed of the prototype is mainly determined by the applied data structures. In our case, the software implementation of a disjunctive graph plays an important role.

The reason is that, during the execution of the scheduling algorithm, such a graph has to be traversed completely from the source to the destination nodes and back repeatedly to compute the start and end times for processing the jobs. As a consequence, we have to select a data structure that facilitates that in a fast manner. Apart from that, we modified the scheduling algorithm in order to minimize the number of traversals. In general, this was possible only by increasing the memory usage of the algorithm by storing intermediate results in an additional data structure.

Another problem that had to be solved by means of special data structures was avoiding cycles in the schedule. Cycles are not allowed in a feasible schedule. During schedule computation it is possible that the algorithm generates cycles because the exact cycle times of the tool groups are available only at the end of the algorithm. If no additional precautions are taken the final schedules may have cycles. We avoid the generation of cycles in our prototype implementation by means of additional data structures that store precedence constraints.

Apart from the representation of disjunctive graphs, lists for collecting pointers to business objects are the most important data structures. AVL trees are used to facilitate efficient access to these lists. AVL trees are binary trees that are balanced with respect to the height of their subtrees (Ottmann and Widmayer, 1996).

4.3.3 Generic approach for integrating subproblem solution procedures

We chose an object-oriented approach for developing subproblem solution procedures (SSPs). The basic idea was to implement an abstract class named "Subproblem Solution Procedure" that encapsulates all fundamental characteristics of a SSP for the shifting bottleneck heuristic. We integrated the following abstract concepts into the class "Subproblem Solution Procedure."

- Management of schedules for parallel tools; the schedules contain the combination of lots to adequately represent scheduling entities, i.e., batches can be represented in the schedules,
- Management of reference schedules,
- Avoidance of cycles in the disjunctive graph due to scheduling decisions on a tool group level,
- Transformation of a node set from the disjunctive graph into a lot set for which a tool group schedule has to be computed,
- Communication of the schedules to the shifting bottleneck heuristic, and
- Selection of alternative performance measures.

On the second level of the heritage hierarchy we derive further relatively general classes from the abstract class "Subproblem Solution Procedure," e.g., "Dispatched Subproblem Solution Procedure." The basic characteristic of these derived classes is to consider always only jobs that are actually available at the instant of the scheduling decision.

In concrete classes that provide a certain scheduling solution approach, we only have to implement the solution algorithm in its "Solve_Problem" method. As a consequence, we are able to integrate C++-based optimization libraries in a straightforward manner. For instance, we integrated the GaLib library for Genetic Algorithm methods (Wall, 1999; Voss and Woodruff, 2002) into our prototype.

4.3.4 Simulation-based approach for performance evaluation

We use a generic architecture for performance evaluation as described in Mönch *et al.* (2003). The core of this architecture is a blackboard-type data layer that contains all objects that are relevant for the scheduling decision. We consider the following types of information:

- Dynamic data
  - Lot release information for new orders,
  - Status information of the tools,
  - Status information of the lots (each order is split into a set of lots), and
  - Setup conditions of the tools.
- Static data
  - Order-related process plans,
  - Operation-related resource lists,
  - Setup times, and
  - Information required for the computation of processing times for the operations.
- Statistical data
  - Actual start time of an operation,
  - Actual tool used for an operation,
  - Order completion times, and
  - Utilization of the tools.
- Control data for the production system
  - Planned start time for an operation,
  - Planned tool for an operation, and
  - Tool schedules.

The data blackboard is implemented in C++ and kept in the main memory of the computer. At periodic intervals the data objects are saved into an object-oriented database. The objects are updated by simulator events. In return, we transfer schedules to the data blackboard. They are provided as tool dispatch lists that are used by the simulator. The architecture has the advantage that we need little effort to integrate the prototype into an existing shop floor control system. We need only to replace the interfaces between the blackboard and the simulator by interfaces to a real factory information system such as a MES or the ERP system. Figure 5-4 depicts the developed architecture.

In our prototype, we use the discrete event factory simulator AutoSched AP 7.2 and the object-oriented database POET. We note that the development of the prototype took over five person-years of effort.

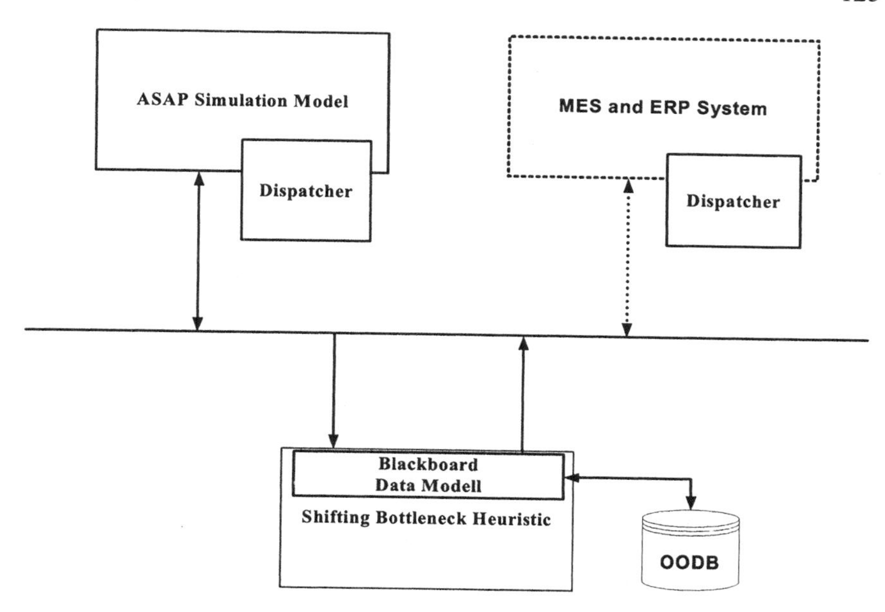

Figure 5-4. Performance assessment architecture for the Shifting Bottleneck heuristic.

4.4 Performance evaluation of the scheduling approach

We assess the performance of our approach based on the architecture described in Section 4.3. We started the prototype testing with a problem of moderate complexity. We considered the MiniFab model that is often used as a reference factory model for the evaluation of shop floor control mechanisms in the semiconductor industry (El Adl *et al.*, 1996). The model, however, consists only of five tools and is not well suited as a realistic testing environment. Nevertheless, we found out that even for this small model our scheduling prototype clearly outperforms traditional dispatching with local priority rules such as EDD or Critical Ratio. An overview of dispatch rules typically applied in semiconductor industry can be found in Atherton and Atherton (1995). We denote this simulation model by Model A.

Next, we present a selection of results for a large factory model. We considered the MIMAC Testbed Data Set 1 that is publicly available and is used in a number of simulation studies (MIMAC, 2003; Fowler and Robinson, 1995). It is the model of a semiconductor fabrication facility with 286 tools in 84 tool groups. The factory manufactures two products with 210 and 245 processing steps, respectively. The model is called Model C.

Model B is between Model A and Model C. We consider a simulation model where we included the eleven highest utilized tools from Model C. The model contains 45 machines and the adapted routes from Model C.

4.5 Design of experiments

It is important to note that the prototype has to be tested in a realistic, i.e., dynamic, environment. In our case, this means that we started new lots into the factory according to a target factory utilization during the whole evaluation period. In contrast to other publications, we do not compute a static schedule for a given set of lots but calculated a number of schedules under different factory conditions.

We simulated the factory for 70 days including a warm-up phase of 30 days at a utilization of 95% (with respect to the bottleneck tool group). The evaluation period of 40 days is sufficient because we deactivated all stochastic effects on the factory, in particular, tool breakdowns. The due dates were tight, i.e., the target cycle time was typically less than twice the raw processing time. We used the sum of the total weighted tardiness over all lots as performance measure.

We examined the following parameters of the scheduler prototype:

- Interval between schedule computations (scheduling interval): every 2 hours to 16 hours,
- Additional horizon: 0 hours to 16 hours (the effective scheduler horizon equals the interval between schedule computations plus the additional horizon),
- The subproblem solution procedure (SSP): FIFO, Earliest Due Date (EDD), Critical Ratio (CR), Operation Due Date (ODD), BATCS (for batch tools) +ATCS (for regular tools), BATCS+CR, BATCS+ODD, and a genetic algorithm SSP for tool groups. Note that ATCS is the Apparent Tardiness Cost with Setups (Pinedo, 2002), and BATCS is Batch ATCS.
- BATCS parameters,
- Number of reoptimization steps of the Shifting Bottleneck Heuristic.

First, we examined the parameters separately. Then, we simulated a full factorial design to determine the interactions between the parameters. In our case, there were no significant interactions.

4.6 Computational results

In total, we made about 1500 simulation runs that lead to significant results for all tested parameters.

With respect to the horizon, the results show that the shifting bottleneck heuristic (SBH) scheduler works best with a small time interval between schedule computations and no additional horizon. Applying the BATCS+ATCS SSP and the BATCS+ODD SSP leads to the smallest deviations from the due dates. Appropriate BATCS parameters (Pinedo, 2002) have to be determined by a series of simulation runs. They depend both on the factory model and the product mix. With respect to the number of reoptimization steps it turns out that two steps are generally sufficient.

Table 5-1 shows the results for Model C. Critical Ratio was the best dispatching rule. For the SBH scheduler, the best parameter settings were used. The flow factor (FF) was used to determine due dates. A lot's due date was set as its release time plus its lead time. The lead time was the flow factor multiplied by the total raw processing time. The values given in Table 5-1 are the flow factors for Products 1 and 2, respectively.

Figure 5-5 shows the how the scheduling interval and the additional scheduling horizon influence the results for Model B when the SSP was BATCS-ODD, which uses the ODD dispatching rule as the SSP for tool groups without batching and uses BATCS for tools with batching.

Table 5-1. Performance comparison between the Shifting Bottleneck Heuristic (SBH) and the Critical Ratio (CR) dispatching rule for Model C

FF	Rule	Lots Completed	Tardy Lots	TWT (hours)
1.4, 1.5	CR	537	356	2265.5
	SBH with BATCS	561	52	790.3
	SBH with BATCS-OOD	567	56	792.9
1.47, 1.66	CR	536	0	0.0
	SBH with BATCS	561	1	6.1
	SBH with BATCS-OOD	567	2	7.8

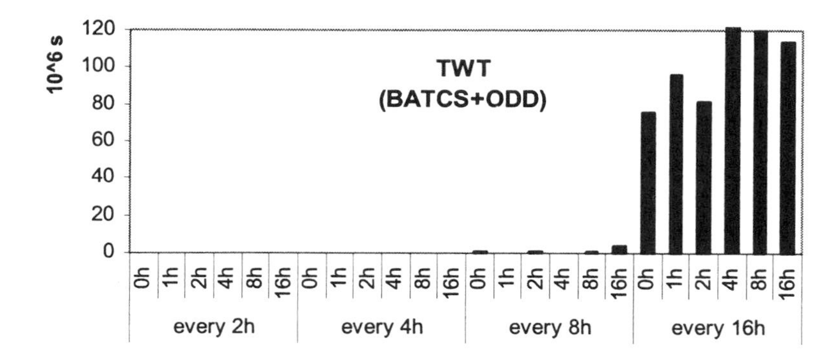

Figure 5-5. Influence of the scheduling interval and additional horizon on TWT performance of the SBH with BATCS+ODD SSP for Model B.

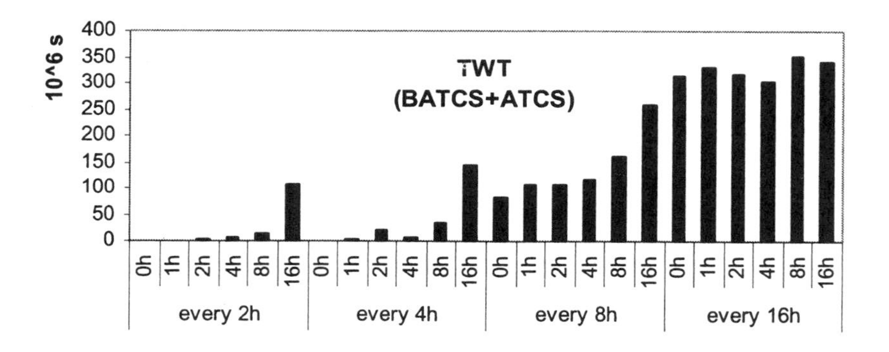

Figure 5-6. Influence of the scheduling interval and additional horizon on TWT performance of the SBH with BATCS+ATCS SSP for Model B.

Figure 5-6 shows the how the scheduling interval and the additional scheduling horizon influence the results for Model B when the SSP was BATCS-ATCS, which uses the ATCS dispatching rule as the SSP for tool groups without batching and uses BATCS for tools with batching.

We conclude from both Figure 5-5 and Figure 5-6 that it is useful to consider a small scheduling interval and a small additional scheduling horizon.

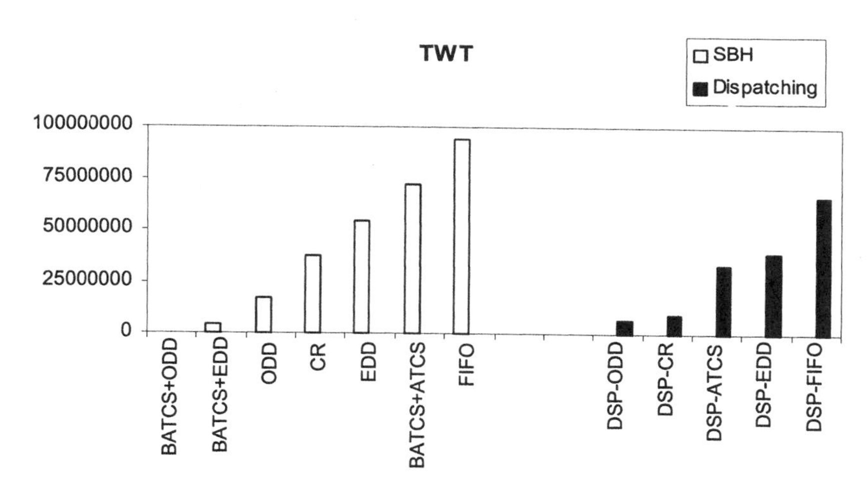

Figure 5-7. TWT performance of SBH with different SSPs and different dispatching rules for Model B.

Table 5-2. Performance of the SBH with a genetic algorithm SSP for Model A. TWT performance is relative to that of the BATCS SSP (2+0).

Scheduling interval + additional horizon	Tight Due Dates FF=0.75 FIFO-FF	Moderate Due Dates FF=0.9 FIFO-FF	Wide Due Dates FF=1.2 FIFO-FF
2+0	0.9960	0.8968	0.0000
2+2	0.7057	0.6259	0.0000
2+4	0.7460	0.4794	0.0081
4+0	0.9252	0.7147	0.0161
4+2	0.7656	0.2355	0.0027
4+4	0.7228	0.3464	0.3880
6+0	0.9826	0.8269	0.0408
6+2	0.6327	0.3466	0.2392
6+4	0.7662	0.3091	0.4614
8+0	0.7595	0.6269	0.3249
8+2	0.7875	0.3371	0.1007
8+4	0.8322	0.7405	0.2639

Figure 5-7 presents, for Model B, the TWT results for different SSPs (used with the SBH scheduler) and different dispatching rules. It turns out that the use of BATCS for batching tools and ODD for non-batching tools performs best. Note that the use of certain dispatching rule SSPs may lead to TWT values that are worse than TWT values of schedules generated using only dispatching rules.

The results of some experiments with a more advanced SSP and Model A are described in Table 5-2. Here we use a SSP that is based on genetic algorithms. The overall scheme first forms batches by using variants of BATCS. Then, in a second step, a genetic algorithm is used to assign the batches to the different machines. Several variants of the BATCS dispatching rule are used to determine sequences of batches on each single machine. The algorithm is described in more detail in Mönch *et al.* (2005).

We apply the genetic algorithm SSP only for the three most critical tool groups (using a fixed criticality measure). We perform simulation experiments with tight, moderate, and wide due dates. In order to set these due dates, we first measure the flow factor *FF* for a FIFO controlled system. Then, in a second step, we multiply this factor (called FIFO-FF) with the scaling factors presented in Table 5-2. For tight due dates, a scheduling interval of 6 hours and an additional horizon of 2 hours leads to TWT improvements of 36 percent compared to the TWT values that are obtained by using the SBH schedulers with BATCS SSPs, a scheduling interval of 2 hours, and no additional horizon.

For moderate due dates a scheduling interval of 8 hours and an additional horizon of 2 hours leads to improvement rates of approximately 66 percent. In the case of small scheduling horizons the genetic algorithm SSP performs similar to BATCS SSPs. It is interesting to note that, for the genetic algorithm SSP, the scheduling interval and the additional horizon should be chosen larger than those for dispatching rule SSPs. The larger horizon leads to more room for improvement.

4.7 Discussion

In summary, we conclude that it was possible to find scheduler prototype parameters that lead to considerable tardiness reductions for all tested scenarios with tight due dates. As shown in Table 5-1, for Model C with tight due dates (FF equal to 1.4 and 1.5), the TWT of the best schedules found using the shifting bottleneck heuristic and a SSP was about 34% of the TWT of the best schedule found using a dispatching rule. With respect to the number of tardy lots, the improvement is even more substantial.

In terms of performance the scheduler is able to beat traditional shop floor control methods. So far, the advantage of a rule-based control was its small computation time. The generation of a schedule for a model with 286 tools, 95% utilization, and a planning horizon of 2 hours took at most 13 seconds on a 2 GHz Pentium 4 computer. As a consequence the application of the scheduler is suitable for real-world problems.

From Table 5-2, we conclude that the use of more advanced, i.e., optimization-based, SSPs offers some advantage compared to pure dispatching rule SSPs.

5. CONCLUSIONS

In this chapter, we described the interface between scheduling and simulation research. Simulation techniques are used in simulation-based schedule generation, refinement, and optimization, and for the situation-dependent parameterization of scheduling approaches. Furthermore, discrete-event simulation techniques can be applied in order to emulate and evaluate the performance of production scheduling approaches.

We also presented the details for a rather complex case study. Here, we used simulation in order to assess the on-time delivery performance of a modified shifting bottleneck heuristic for scheduling wafer fabs. We described our simulation framework and presented the results of several computational experiments based on it.

REFERENCES

Adams, J., Balas, E., and Zawack, D. "The Shifting Bottleneck Procedure for Job Shop Scheduling." *Management Science*, **34**, 391-401, 1988.

Atherton, L.F., and Atherton, R. W. *Wafer Fabrication: Factory Performance and Analysis* (Kluwer Academic Publishers, Boston, Dordrecht, London, 1995).

Banks, J., Carson, J. S., Nelson, B. L., and Nicol, D. *Discrete-Event System Simulation*, (Fourth Edition. Prentice Hall, Upper saddle River, (2005).

Biskup, D., and Feldmann, M. "Benchmarks for Scheduling on a Single Machine Against Restrictive and Unrestrictive Common Due Dates." *Computers & Operations Research*, **28**, 781-801, (2001).

Brucker, P., *Scheduling Algorithms*. (Fourth Edition Springer, Berlin, 2004).

Demirkol, E., Mehta, S., and Uzsoy, R. "A Computational Study of Shifting Bottleneck Procedures for Shop Scheduling Problems." *Journal of Heuristics*, **3**(2), 111-137, (1997).

Dümmler, M.A., "Using Simulation and Genetic Algorithms to Improve Cluster Tool Performance." In: Proceedings of the 1999 Winter Simulation Conference, P. A. Farrington, H. B. Nemhard, D. T. Sturrock, G. W. Evans (Eds.), 875-879, (2001).

El Adl, M.K., Rodriguez, A. A., and Tsakalis, K. S. "Hierarchical Modelling and Control of Re-entrant Semiconductor Manufacturing Facilities." In: Proceedings of the 35th Conference on Decision and Control. Kobe, Japan, (1996).

Fowler, J.W., and Robinson, J. "Measurement and improvement of manufacturing capacities (MIMAC): Final report." Technical Report 95062861A-TR, SEMATECH, Austin, TX, (1995).

Fowler, J.W., and Rose, O. "Grand Challenges in Modeling and Simulation of Complex Manufacturing Systems." *Simulation – Transactions of the Society for Modeling and Simulation International*, **80**, 469-476, (2004).

Fu, M.C., "Optimization for Simulation: Theory vs. Practice." *INFORMS Journal on Computing*, **14**, 192-215, (2002).

Fu, M.C., Antradóttir, S., Carson, J. S., Glover, F., Harrell, C. R., Ho, Y.-C., Kelly, J. P., and Robinson, S. M. "Integrating Optimization and Simulation: Research and Practice." In:

Proceedings of the 2000 Winter Simulation Conference, J. A. Joines, R. R. Barton, K. Kang, P. A. Fishwick (Eds.), 610-616, (2001).

Graham, R.L., Lawler, E. L., Lenstra, J. K., and Rinnooy Kann, A. H. G. "Optimization and Approximation in Deterministic Sequencing and Scheduling: a Survey." *Annals of Discrete Mathematics*, **5**, 287–326, (1979).

Hall, N.G., and Posner, M.E., "Generating experimental data for computational testing with machine scheduling applications." *Operations Research*, **49**, 854-865, (2001).

Horiguchi, K., Raghavani, N., Uzsoy, R., and Venkateswaran, V. "Finite Capacity Production Planning Algorithms for a Semiconductor Wafer Fabrication Facility." *International Journal of Production Research*, **39**(5), 825-842, (2001).

Kim, S.Y., Lee, Y. H., and Agnihotri, D. "A Hybrid Approach to Sequencing Jobs Using Heuristic Rules and Neural Networks." *Production Planning and Control*, **6**(5), 445-454, (1995).

Law, A.M., and Kelton, D. W. *Simulation Modeling and Analysis*. (Third Edition. McGraw-Hill, New York, 2000).

Lee, Y.-H., and Pinedo, M. "Scheduling jobs on parallel machines with sequence-dependent setup times." *European Journal of Operational Research*, **100**, 446 – 474, (1997).

Lenstra, J.K., Rinnooy K. A. H. G., and Brucker, P. "Complexity of Machine Scheduling Problems." In: *Annals of Discrete Mathematics* **1**, 343-362, (1977).

Mason, S., Fowler, J. W., and Carlyle, M. W. "A Modified Shifting Bottleneck Heuristic for Minimizing the Total Weighted Tardiness in Complex Job Shops." *Journal of Scheduling*, **5**, 247-262, (2002).

MIMAC (2003) Testbed Data Sets der Arizona State University, Tempe, Arizona, USA: http://www.eas.asu.edu/~masmlab/ftp.htm.

Mönch, L., Balasubramanian, H., Fowler, J. W., and Pfund, M. "Scheduling Heuristics for Jobs with Unequal Ready Times and Incompatible Families on Parallel Batch Processing Machines." *Computers & Operations Research*, **32**(11), 2731-2750, (2005).

Mönch, L., Rose, O., and Sturm, R. "Simulation Framework for Performance Assessment of Shop-Floor Control Systems." *Simulation – Transactions of the Society for Modeling and Simulation International*, **79**(3), 60-67, (2003).

Ottmann, T., and Widmayer, P. *Algorithmen und Datenstrukturen*. (Spektrum Akademischer Verlag, Heidelberg, 1996).

Ovacik, I.M., and Uzsoy, R. *Decomposition Methods for Complex Factory Scheduling Problems* (Kluwer Academic Publishers, Dordrecht, MA, 1997).

Pabst, D., "Handling Precedence Constraints for the Shifting Bottleneck Heuristic Applied in a Dynamic Semiconductor Manufacturing Environment." Master Thesis. Institut für Informatik, Universität Würzburg, (2003).

Park, Y., Kim, S., and Lee, Y.-H. "Scheduling jobs on parallel machines applying neural Network and heuristic Rules." *Computers & Industrial Engineering*, **38**, 189-202, (2000).

Pinedo, M.L., *Scheduling: Theory, Algorithms, and Systems*. (Second Edition. Prentice-Hall, Englewood Cliffs, NJ, 2002).

Pinedo, M.L., and Singer, M. "A Shifting Bottleneck Heuristic for Minimizing the Total Weighted Tardiness in a Job Shop." *Naval Research Logistics*, **46**, 94-109, (1999).

Potoradi, J., Boon, O. S., Mason, S. J., Fowler, J. W., and Pfund, M. E. "Using Simulation-Based Scheduling to Maximize Demand Fulfillment in a Semiconductor Assembly Facility." In: Proceedings of the 2002 Winter Simulation Conference, E. Yücesan, C.-H. Chen, J. L. Snowdon, J. M. Charnes (eds.), 1857-1861, (2002).

Sivakumar, A.I., "Multiobjective Dynamic Scheduling Using Discrete Event Simulation." *International Journal of Computer Integrated Manufacturing*, **14**(2), 154-167, (2001).

Toba, H., "Segment-Based Approach for Real-Time Reactive Rescheduling for Automatic Manufacturing Control." *IEEE Transactions on Semiconductor Manufacturing*, **13**(3), 264-272, (2000).

Voss, S., and Woodruff, D. L. *Optimization Software Class Libraries*, Kluwer Academic Press, 1-23, (2002).

Wall, M., GaLib. A C++ Library of Genetic Algorithm Components. http://lancet.mit.edu/ga/, (1995).

Chapter 6

RESCHEDULING STRATEGIES, POLICIES, AND METHODS

Using the rescheduling framework to improve production scheduling

Jeffrey W. Herrmann
University of Maryland

Abstract: This chapter reviews basic concepts about rescheduling and briefly reviews the rescheduling framework. Then the chapter discusses considerations involved in choosing between different rescheduling strategies, policies, and methods.

Key words: Rescheduling, schedule repair, predictive-reactive rescheduling

1. INTRODUCTION

Many manufacturing facilities generate and update production schedules, which are plans that state when certain controllable activities (e.g., processing of jobs by resources) should take place. In dynamic, stochastic manufacturing environments, managers, production planners, and supervisors must not only generate high-quality schedules but also react quickly to unexpected events and revise schedules in a cost-effective manner. These events, generally difficult to take into consideration while generating a schedule, disturb the system, generating considerable differences between the predetermined schedule and its actual realization on the shop floor. Rescheduling is then practically mandatory in order to minimize the effect of such disturbances in the performance of the system. There are many types of disturbances that can upset the plan, including machine failures, processing time delays, rush orders, quality problems, and unavailable material.

This chapter reviews basic concepts about rescheduling and briefly reviews the rescheduling framework presented by Vieira et al. (2003). Then the chapter discusses considerations involved in choosing between different rescheduling strategies, policies, and methods.

Note that Vieira et al. (2003) reviewed numerous papers that describe specific approaches in each area of the rescheduling framework. Interested readers should refer to that paper for those details. This chapter focuses on using the rescheduling framework to help one improve a production scheduling system.

McKay *et al.* (2002) identify some key research opportunities for conducting high-impact research in scheduling. Rescheduling is one of these topics, along with flexible algorithms, adaptation and learning, multiple objectives, user interfaces, and task design. The framework that this chapter discusses will be useful for guiding research on rescheduling as well as for helping organizations understand and improve their production scheduling systems.

2. RESCHEDULING BASICS

Although dispatching rules, kanban cards, and other decentralized production control policies are in use, many manufacturing facilities generate and update production schedules. In manufacturing systems with a wide variety of products, processes, and production levels, production schedules can enable better coordination to increase productivity and minimize operating costs. A production schedule can identify resource conflicts, control the release of jobs to the shop, and ensure that required raw materials are ordered in time. A production schedule can determine whether delivery promises can be met and identify time periods available for preventive maintenance. A production schedule gives shop floor personnel an explicit statement of what should be done so that supervisors and managers can measure their performance.

Note that, after a schedule is generated, manufacturing operations begin. Managers and supervisors want the shop floor to follow the schedule. In practice, operators may deviate from the schedule. Ideally, the schedule is followed as closely as possible. Small deviations from scheduled start times and end times are expected and usually ignored. (The definition of small depends on the facility in question.) Larger deviations or changes to the sequence occur when unexpected events disrupt the initial schedule. Even if the managers and supervisors do not explicitly update the schedule, schedule repair occurs as the operators react to the disruptions, delaying tasks or performing tasks out of order.

Rescheduling is the process of updating an existing production schedule in response to disruptions or other changes. The following is a partial list of possible disruptions:

- Arrival of a new, urgent job,
- Cancellation of a job, change to a job's due date, or other change in job specification,
- Machine failure, repair, or other change in status,
- Delay in the arrival of required material or other problem with material delivery,
- Poor quality parts that require rework or making new parts,
- Incorrect predictions of setup time, processing time, or other actions, and
- Absentee workers or changes to worker assignments.

Possible responses include changing when activities will occur, using a substitute machine or person, authorizing overtime, subcontracting work to another firm, and changing the manufacturing process.

3. RESCHEDULING FRAMEWORK

Unlike the scholastic study of scheduling, which has been organized around a particular classification scheme (Graham et al., 1979), the area of rescheduling has a much greater variety. Like a great swamp, it resists an orderly classification into mathematical abstractions.

Traditionally there was no standard way of describing rescheduling approaches, which includes methods for repairing a schedule that has been disrupted, methods for creating a schedule that is robust with respect to disruptions, and studies of how rescheduling policies affect the performance of the dynamic manufacturing system. To understand this work, Vieira et al. (2003) presented a framework for understanding rescheduling not only as a collection of techniques for generating and updating production schedules but also as a control strategy that has an impact on manufacturing system performance in a variety of environments.

This rescheduling framework includes rescheduling environments, rescheduling strategies, rescheduling policies, and rescheduling methods. The rescheduling environment identifies the set of jobs that need to be scheduled. A rescheduling strategy describes whether or not production schedules are generated. A rescheduling policy specifies when rescheduling should occur. Rescheduling methods describe how schedules are generated and updated.

Rescheduling Environments				
Static (finite set of jobs)		Dynamic (infinite set of jobs)		
Deterministic (all information given)	Stochastic (some information uncertain)	No arrival variability (cyclic production)	Arrival variability (flow shop)	Process flow variability (job shop)

Rescheduling Strategies				
Dynamic (no schedule)		Predictive-reactive (generate and update)		
Dispatching rules	Control-theoretic	Rescheduling Policies		
		Periodic	Event-driven	Hybrid

Rescheduling Methods				
Schedule generation		Schedule repair		
Nominal schedules	Robust schedules	Right-shift rescheduling	Partial rescheduling	Complete regeneration

Figure 6-1. Rescheduling framework (from Vieira et al. 2003)

Figure 6-1 presents the framework for understanding rescheduling. The framework includes rescheduling environments, rescheduling strategies, rescheduling policies, and rescheduling methods. Either rescheduling strategy (dynamic scheduling or predictive-reactive scheduling) can be used in any rescheduling environment with uncertainty or variability. However, dynamic rescheduling environments are the ones most relevant to manufacturing systems, and the predictive-reactive rescheduling strategy is the approach most commonly used in practice. Still, there are a great variety of rescheduling policies used in predictive-reactive scheduling.

3.1 Rescheduling environments

The rescheduling environment, which identifies the set of jobs that need to be scheduled, is an important component of the rescheduling framework, as Figure 6-1 illustrates. Static rescheduling environments have a finite set of jobs. Dynamic rescheduling environments have an infinite set of jobs (that is, jobs continue to arrive over an infinite time horizon).

Deterministic, static scheduling problems can be viewed as a special case of rescheduling, where there is a finite set of jobs and no uncertainty about the future. The specified schedule can be followed without any modifications.

Stochastic, static rescheduling environments are an important special case of rescheduling. Again, there is a finite set of jobs, but some variables are uncertain. For instance, the exact task processing times may be unknown. A production schedule may specify resource assignments and task sequences, but the actual task start times and completion times will not match the expected ones. At the minimum, executing the schedule requires some rule or policy for reconciling the error in the schedule. However, other policies exist. One can modify the schedule at some point during execution to react to additional information, or one can construct a solution that only partially specifies the schedule, leaving details unspecified until the appropriate time comes.

A dynamic rescheduling environment has an infinite stream of jobs. Each job requires scheduling before it can be processed. As shown in the framework, three important cases exist.

First, when a manufacturing system produces a specific set of jobs repeatedly (and there is no uncertainty or variability in the arrival process), then the system can follow the same production schedule repeatedly. The production schedule specifies a sequence of operations that are done. This is sometimes called a "cycle." After the system finishes one cycle, it starts the next cycle using the same sequence. The cycle might be a daily schedule that is repeated every day, or it could be a relatively short interval that is repeatedly many times in a day. The scheduling decision requires solving a cyclic scheduling problem. A flexible manufacturing system that performs milling and drilling operations to produce the same set of aluminum and steel transmission components each day is an example of a cyclic environment (Flanders and Davis, 1995).

Second, there may exist some uncertainty in job arrivals, but all jobs follow the same route through the manufacturing system, and the arrival rate is steady. When there exist significant setups between different classes of jobs or reentrant flow, scheduling is necessary to determine when a resource should switch from processing one type of job to another.

Third, there may exist process flow variability along with the variability in job arrivals. Job shops often have this characteristic, since there are many products, but a limited subset of them are being produced at any given time. Thus, a specific product's arrival process has great variability. In some situations, no advance information is available about jobs before they arrive. Otherwise, some information about future arrivals may be known, but the information is subject to change as new jobs are added and existing jobs are delayed or deleted.

Another aspect that characterizes rescheduling environments is the presence of potential additional capacity using subcontracting or overtime. A facility that works 24 hours every day is not the same as a facility that work five eight-hour shifts a week. The presence of capacity buffers (in the form of potential overtime) affects the rescheduling environment, since it relaxes a set of constraints (capacity) but adds a set of costs (overtime).

3.2 Rescheduling strategies

As shown in the rescheduling framework there are two basic rescheduling strategies: dynamic scheduling and predictive-reactive scheduling.

Dynamic scheduling does not create production schedules. Instead, decentralized production control methods dispatch jobs when necessary and use information available at the moment of dispatching. Sometimes dynamic scheduling schemes are called on-line scheduling or reactive scheduling. Such schemes use dispatching rules or other heuristics to prioritize jobs waiting for processing at a workstation or waiting to be moved by a material handling vehicle. In some facilities, pull production control schemes such as kanban or constant work-in-process (CONWIP) are used instead of scheduling.

Predictive-reactive scheduling is a common strategy for rescheduling dynamic manufacturing systems. Predictive-reactive scheduling has two primary steps. The first step generates a production schedule. The second step updates the schedule in response to a disruption or other event to minimize its impact on system performance. Rescheduling can occur frequently in a dynamic rescheduling environment, or it can simply be a single revision of the schedule of a stochastic, static rescheduling environment.

Within this rescheduling strategy, there are various rescheduling policies that govern when rescheduling occurs.

1. A *periodic* rescheduling policy reschedules the facility periodically and implements the schedules on a rolling time horizon basis.
2. In an *event-driven* rescheduling policy, events in a certain class trigger rescheduling. In the extreme, a new schedule is created (or revised) every time an event that alters system status occurs. Triggering events may be machine breakdowns, the arrival of urgent jobs, job cancellation, or job priority changes.
3. A *hybrid* rescheduling policy reschedules the system periodically and also when special (or major) events take place.

In practice, periodic and hybrid policies appear to be the most common. Katok and Ott (2000) describe a weekly scheduling policy for an aluminum can plant, and Moss et al. (2000) describe a weekly scheduling policy for a health care products factory. Leachman et al. (2002) describe a rescheduling policy that creates a new schedule each shift (which is eight hours long) and updates it every ten minutes.

Chacon (1998) describes a hybrid policy for a semiconductor wafer fabrication facility operated by Sony. The production scheduling system periodically downloads data from a manufacturing execution system and generates a schedule. According to Chacon, the users determine which events trigger rescheduling: "if an unscheduled event makes the schedule significantly obsolete, the fab supervisor has an option to recreate the schedule manually."

3.3 Rescheduling methods

Rescheduling methods are used to create or update schedules as part of a predictive-reactive scheduling strategy. Schedule generation methods include most of the literature in the area of scheduling. For a general introduction to the topic, see, for instance, Pinedo (2002) and Pinedo (2005).

Traditional approaches to creating schedules do not consider the possible need to adjust the schedule by making small changes when disruptions occur. Thus, manufacturing system performance may suffer significantly when a change is necessary.

However, robust scheduling is an attempt to create a schedule that achieves good performance even though changes occur. This is similar to the concept of robust design, which creates products that can perform well in a variety of environments.

When rescheduling becomes necessary, there are various methods available for changing the old schedule to create a feasible plan for future production. Such methods are called *schedule repair* methods, and there are three basic types:

1. *Right shift rescheduling* postpones each remaining operation (shifting it to the right on a Gantt chart) by the amount of time needed to make the schedule feasible.
2. More generally, *partial rescheduling* reschedules only the operations that were affected directly or indirectly by the disruption. This preserves the original schedule as much as possible. Right-shift rescheduling is a special case of this method.
3. *Complete regeneration* reschedules everything not processed before the rescheduling point, including those operations (jobs) not affected by the disruption.

To illustrate, consider the following example of a simple three-machine flowshop (Figure 6-2). Each job is processed on the three machines M1, M2, and M3. The current production schedule has a number of jobs already scheduled. As job J1 finishes on M1 and job J4 finishes on M2, M2 fails and will be down as shown by the thick line.

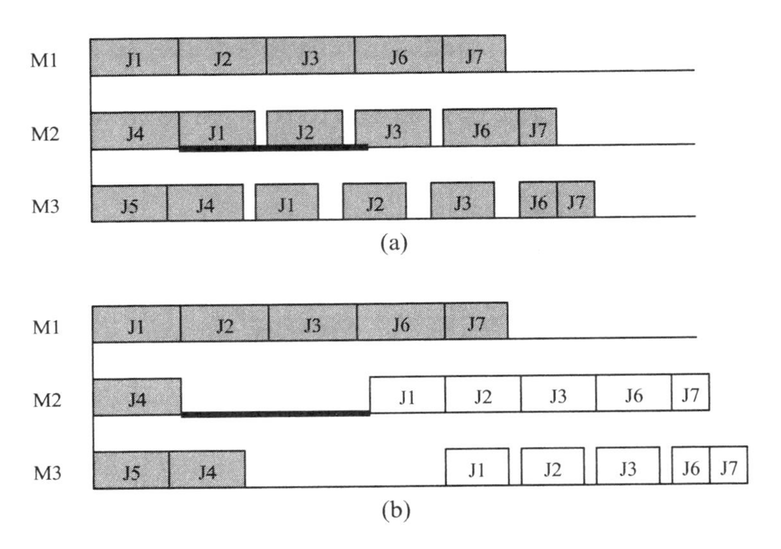

Figure 6-2. (a) Original schedule. The thick horizontal line shows the time that machine M2 will be down. (b) The schedule after using right-shift rescheduling to repair it. (The white tasks are those that have been rescheduled.)

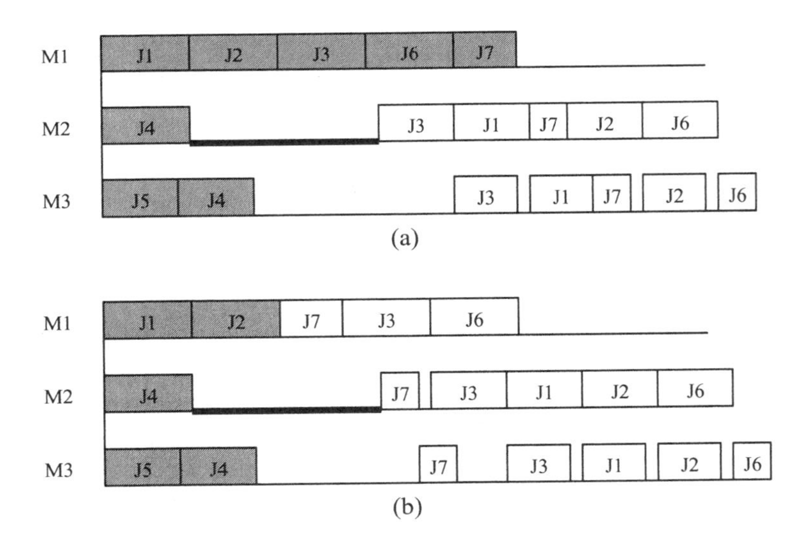

Figure 6-3. (a) The schedule after partial rescheduling. (b) The schedule after complete regeneration. (The white tasks are those that have been rescheduled.)

A right-shift rescheduling repair method simply delays jobs J1, J2, J3, J6, and J7 on both machines M2 and M3. The sequence is not changed, and the schedule on M1 is not changed. A partial rescheduling method (Figure 6-3) rearranges jobs J1, J2, J3, J6, and J7 on machines M2 and M3, in this case scheduling job J3 before job J1 and job J7 before J2 to get these critical jobs done earlier. A complete regeneration method reschedules not only machines M2 and M3 but also M1, though the operations on M1 are not affected by the failure of M2.

4. RESCHEDULING STRATEGIES

This section discusses the factors that influence the desirability of the two basic rescheduling strategies, dynamic scheduling and predictive-reactive scheduling.

4.1 Dynamic scheduling

Dynamic scheduling is an appropriate strategy in factories that have relatively high-volume, low-mix production and for automated systems that need to react quickly to uncertain events.

Factories that have relatively high-volume production of a small number of products tend to have manufacturing cells and assembly lines that feed other cells or lines. In this case, the system lends itself to a pull production control strategy that uses kanban signals (or another similar system) both within the manufacturing cell and to link production of different cells or lines. Black and Hunter (2003) call such systems *linked-cell manufacturing systems*. McKay and Wiers (2004) call controlling these types of systems *minimal production control*.

Dynamic scheduling is also useful for automated systems that need to react quickly to uncertain events. Unlike humans, who can easily adapt a production schedule when large disruptions occur, an automated system needs rules to react. For instance, Hasenbein et al. (2004) describe the use of dispatching rules for the automated material handling system in a semiconductor wafer fabrication facility. Because the wafers are large (300mm in diameter), semiconductor manufacturing firms are using automated material handling. However, there is a large amount of variability in semiconductor manufacturing, so creating a material handling schedule is not practical. Instead the control system uses a dispatching rule to determine which set of wafers an automated guided vehicle should move next.

Determining which dispatching rule to use is a difficult decision. There are many different types of rules available. Common rules include first-in-first-out (FIFO), which gives priority to the job that entered the queue first. In unique settings, sometimes unique rules perform better. To avoid trial-

and-error, some firms have used discrete-event simulation to evaluate how different dispatching rules affect manufacturing system performance. Hasenbein et al. (2004) used simulation to evaluate FIFO and other dispatching rules for an automated material handling system in a semiconductor wafer fabrication facility. They determined that the best rule gives priority to the job whose destination queue has the largest total number of jobs currently waiting to be moved. The benefit of this rule is that it moves the automated guided vehicle to the longest queue, and the vehicle then takes a job from this queue. This scheme should keep all of the queues short.

4.2 Predictive-reactive scheduling

Predictive-reactive scheduling, unlike dynamic scheduling, requires generating and updating production schedules. Predictive-reactive scheduling is common in discrete-parts manufacturing facilities that make a wide variety of products. The success of this strategy depends upon the ability of the scheduler (enhanced by the scheduling system in place) to predict what is going to happen, to prepare for problems that are likely to occur, and to react quickly and appropriately when they do.

The schedules that are created specify which orders or operations should be produced when. Schedules can be more or less detailed. It is common for factories to use schedules that specify only which week an order or operation should be completed. On the other extreme, some scheduling systems can generate very precise production schedules that specify the day, hour, and minute that each operation should begin (and the time at which it should end). Clearly, schedules that are very precise are highly likely to be "wrong" (wrong because there may be many small discrepancies), whereas schedules that are less precise make it easy to meet the schedule. More precise schedules require and enable better coordination and thus can improve system performance.

In a middle-of-the-road approach, the production schedule will specify the set of jobs that each resource should perform each shift. Katok and Ott (2000) developed a scheduling system for an aluminum can plant, and the weekly schedule specified the particular product that each can line should make each eight-hour shift that week.

Mathematical and simulation studies of rescheduling (see Vieira et al., 2003, for a more complete discussion) show that increasing rescheduling frequency (which reduces the rescheduling period) improves manufacturing system performance. Rescheduling frequency

In practice factories often follow a formal rescheduling policy that is periodic (but not very frequent) and an informal rescheduling policy that is event-driven. A typical shop will have a weekly scheduling meeting where the manager, scheduler, support staff, and foremen (with possibly the operators) review the orders that need to be done, discuss the state of the shop, and decide which orders have priority and should be done this week (this is the schedule). In addition, the participants determine what other activities need to be done to expedite orders, fix problems, and pacify customers.

During the days after the meeting, the scheduler receives information about new orders that need to be done. If they can wait until the next scheduling meeting, the scheduler simple record the order. Otherwise, someone must make preparations for completing the order and inform the foreman of the change. Meanwhile, other disruptions occur (see the set listed above) and force the foreman and operators to deviate from the schedule in other ways.

This type of situation may be reasonable and tolerable if the scheduler has tactics in places for reacting to these disruptions and minimizing the overall impact on performance. However, when scheduling is done poorly, these disruptions lead to ad-hoc, shortsighted, and panicky reactions that lead to serious performance problems. McKay and Wiers (2004) discuss in great detail the need for schedulers to anticipate and prepare for trouble.

5. RESCHEDULING METHODS

The distinction between generating a schedule and repairing a schedule can seem a bit arbitrary since, in practice, schedulers may not start from scratch when generating a schedule. They often update an existing schedule, removing the operations and orders that are complete, adding new orders and operations, and resequencing operations as needed.

5.1 Generating robust schedules

Generating robust schedules is an important part of rescheduling. As discussed before, a robust schedule achieves good performance even though changes occur. This involves two key objectives:

1. Reduce the probability that something bad will occur.
2. Reduce the impact if something bad does occur.

The first aspect has not been studied in the traditional scheduling literature, which treats disruptions as exogenous random variables. However, it is clear that scheduling can have affect the chance of a disruption happening. A robust schedule will assign sensitive operations to machines or workers that are more reliable.

From a traditional scheduling perspective, the only way to achieve the second objective is to include idle time in the schedule. This idle time serves as a buffer that can absorb delays due to longer-than-expected setup or processing times or machine failures. The idle time also provides openings that a new, urgent job can use if one arrives. However, when a disruption occurs, one can react by bringing additional resources into service. These may be backup equipment or personnel who were assigned to other tasks. Some factories have highly-skilled and versatile personnel who can be sent to the rescue when a serious disruption occurs. McKay and Wiers (2004) list other practical actions that schedulers can take to generate robust schedules. They suggest that schedulers identify potential problems as early as possible, regularly meet with individuals who may have information about future changes, schedule a number of small batches (instead of one large one), expedite learning by scheduling difficult operations first, avoid troublesome operations during a period with little slack, do simple operations after a process change, do work earlier than scheduled, keep the most powerful and flexible resources as free as possible (so that they're free if needed), and maintain backup plans for the most critical problems that could occur.

5.2 Repairing schedules

The methods available to repair a schedule that is no longer feasible depend upon the scheduling system in place. If the scheduler generates schedules manually, it is unlikely that he will have time for complete regeneration when reacting to a disruption. Delaying the affected operations (while pulling ahead work where possible) may be the only choice. On the other hand, a computer-based scheduling system that is closely integrated with the shop floor and manufacturing execution system is likely to have up-to-date information about the state of the machines, operators, and jobs. Thus, it can use its scheduling engine and completely regenerate the schedule.

Of course, as mentioned before, informal schedule repair occurs automatically as operators and foremen deal with disruptions on their own. Operators and foremen typically respond to a disruption by finding something else to do. There is usually some order that is waiting for processing and can be done today. If the original schedule is not precise, such actions can be reasonable and don't even affect the schedule. However, this type of reaction makes a more precise schedule even more infeasible and leads to more confusion and miscommunication.

6. SUMMARY AND CONCLUSIONS

The rescheduling framework presented by Vieira et al. (2003) provides a way to understand the topic of rescheduling. Unlike that paper, which reviewed numerous papers that describe specific approaches in each area of the rescheduling framework, this chapter addresses more practical issues associated with rescheduling and gives some indication of how different rescheduling strategies and methods can be used in practice.

The objective of this chapter is to help those who are trying to improve production scheduling systems by highlighting the importance of rescheduling and the key concepts associated with rescheduling. Rescheduling is done both formally and informally in practice. Informal rescheduling is a good thing if it reflects the ability of the shop floor personnel to meet production goals even when disruptions occur. However, if it reflects the fact that the formal scheduling system is unable to predict problems and plan responses to them, informal rescheduling is a symptom of a dysfunctional production scheduling system.

REFERENCES

Black, J.T., and Hunter, S.L., 2003, *Lean Manufacturing Systems and Cell Design*, Society of Manufacturing Engineers, Dearborn, Michigan.

Chacon, G.R., 1998, Using simulation to integrate scheduling with the manufacturing execution system, *Future Fab International*, 63-66.

Flanders, S.W., and Davis, W.J., 1995, Scheduling a flexible manufacturing system with tooling constraints: an actual case study, *INTERFACES*, **25**(2):42-54.

Graham, R.L., Lawler, E.L., Lenstra, J.K., and Rinnooy Kan, A.H.G., 1979, Optimization and approximation in deterministic machine scheduling: a survey, *Annals of Discrete Mathematics*, **5**:287-326.

Hasenbeign, J., Sigireddy, S., and Wright, R., 2004, Taking a queue from simulation, *Industrial Engineer*, 39-43.

Katok, E., and Ott, D., 2000, Using mixed-integer programming to reduce label changes in the Coors aluminum can plant, *INTERFACES*, **30**(2):1-12.

Leachman, R.C., Kang, J., and Lin, V., 2002, SLIM: short cycle time and low inventory manufacturing at Samsung Electronics, *INTERFACES*, **32**(1):61-77.

McKay, K.N., Pinedo, M., and Webster, S., 2002, Practice-focused research issues for scheduling systems, *Production and Operations Management*, **11**(2):249-258.

McKay, K.N., and Wiers, V.C.S., 2004, *Practical Production Control: a Survival Guide for Planners and Schedulers*, J. Ross Publishing, Boca Raton, Florida. Co-published with APICS.

Moss, S., Dale, C., and Brame, G., 2000, Sequence-dependent scheduling at Baxter International, *INTERFACES*, **30**(2):70-80.

Pinedo, M., 2002, *Scheduling: Theory, Algorithms, and Systems*, second edition, Prentice Hall, Upper Saddle River, New Jersey.

Pinedo, M.L. 2005, *Planning and Scheduling in Manufacturing and Services*, Springer, New York.

Vieira, G.E., Herrmann, J.W., and Lin, E., 2003, Rescheduling manufacturing systems: a framework of strategies, policies, and methods," *Journal of Scheduling*, **6**(1):35-58.

Chapter 7

A PRACTICAL VIEW OF THE COMPLEXITY IN DEVELOPING MASTER PRODUCTION SCHEDULES: FUNDAMENTALS, EXAMPLES, AND IMPLEMENTATION

Guilherme Ernani Vieira
Pontifical Catholic University of Parana

Abstract: Although Master Production Scheduling (MPS) has been studied and used by both academia and industries for quite a long time, the real complexity involved in making a master plan when capacity is limited, when products have the flexibility of being made at different production lines, and when performance goals are tight and conflicting, has not yet been presented in the literature in a simple and practical way. In this context, one should consider how to attain a given performance by balancing different objectives, such as maximizing service level, and minimizing inventory levels, risk of stockouts, overtime, and setup time. Many decisions need to be made during the development of an MPS, such as: Which product should be scheduled, in what quantity, and to which resource? Is overtime needed? Should inventory be built for future periods? Should backlogging be considered? Clearly, an MPS process depends on the combination of many different parameters. For this type of problem, it is extremely difficult to find a solution that satisfies all objectives involved simultaneously, mainly because of the great number of variables involved. It is known that finding an optimal MPS solution for industrial scheduling scenarios is time consuming - despite nowadays computers being extremely fast. It is common, therefore, to use heuristics (or meta-heuristics) to find good plans in reasonable computer time. Using a plain language, this chapter describes some of the complexity involved in the MPS creation without, however, paying too much attention to mathematical formalisms and definitions, using mostly the author's industry experience and practical examples faced during research in the production scheduling area.

Key words: Master production planning, optimization, complexity, information systems, and implementation.

1. INTRODUCTION

Although master production scheduling (MPS) has been used and studied for several years, there is still a wide unexplored field to work on. McClelland (1988) mentioned that "although folklore exists concerning effective MPS methods, research is still needed to scientifically investigate the MPS practices advocated by conventional wisdom." Higgins & Browne (1992), Raffish (1981), and Moran (1986) stated that "despite the recognition of the importance of the MPS, researchers have published very little on the development of an effective MPS in either a make-to-stock or a make-to-order environment." These statements, made several years ago, still hold true.

Development of an MPS is a very complex task, especially in production environments with limited capacity, a common situation in industrial environments. It is possible that, in specific scenarios, with little capacity restrictions or with low use of production resources, the creation of a master plan may not be cumbersome, however this is not a rule but exception. Usually, resources should be well utilized, workers correctly assigned to working stations, tools are limited, and when a production line unexpectedly stops, it is a cry-for-help situation.

Within this scenario of restrictions, based on demand forecasts and customer orders, the company should accurately define what to make, when, where, and how much product quantity should be assigned to the different workstations. Moreover, when a quantity can be scheduled to more than one production resource (*flexible routing*), planning becomes even harder. What is the most adequate resource to use when more than one can be picked? What is the best load balancing among resources and workers? What is the best assignment of product quantities to resources so that changeover can be minimal? Therefore, it is clear that only those intimately involved with actual MPS creation know exactly how complex the task of making a master plan really is, even though this is not a new subject.

It is not enough to work with simply good production plans. To be and stay competitive, one should strive for optimal inventory levels, resource utilization, production costs, changeover times, and service levels. Computer systems can tremendously aid and accelerate the production planning and scheduling process considering such optimization objectives. The objective function drives the logic behind the execution of these systems. This type of function is usually related to cost reduction and productivity increases. For this, several techniques have appeared proposing to implement the creation of very good, perhaps optimal in some cases, production plans and schedules. This chapter mentions some of them, like linear programming, tabu search, neural networks, simulated annealing, and genetic algorithms. These last two have been implemented and tested and are detailed in this chapter.

This chapter informally shows what usually does not appear in the literature: the complexity inherent in search for an optimal master production schedule. The ideas presented are taken mainly from the author's industry and research experience, both in Brazil, Canada and United States, and, therefore, do not follow any mathematical or theoretical formalism that this subject in fact deserves. The chapter is organized as follows: Next section reviews some of the fundamentals of master production scheduling. Section 3 describes a (computer) system for MPS creation. Section 4 illustrates MPS complexity through a heuristic running on an example. Section 5 details two artificial intelligence techniques applied to the MPS process: genetic algorithms and simulated annealing. Conclusions and ideas for future studies are given on section 6.

2. FUNDAMENTALS OF MASTER PRODUCTION SCHEDULING

MPS is at the interface between strategic planning and tactical planning in an integrated production planning and control system. As such, it is a key decision-making activity, in which strategic goals from business planning are translated into an anticipated statement of production, from which all other schedules at lower levels are derived. The MPS function is an essential part of the production management architecture, and, as such, it should be given high priority when developing an integrated manufacturing system (Higgins & Browne, 1992).

According to Slack *et al.* (2001), the master production schedule is the most important planning and control schedule in a business, and it forms the main input to materials requirements planning (see Figure 7-1). It contains a statement of the volume and timing of the end products to be made; this schedule drives the whole operation in terms of what is assembled, what is manufactured, and what is bought. It is the basis of planning the utilization of labor and equipment and it determines the provisioning of materials and cash.

The master production schedule also provides the information to the sales function on what can be promised to customers and when delivery can be made. Therefore, sales function can load known sales orders against the MPS and keep track of what is available to promise – ATP (Slack *et al.*, 2001).

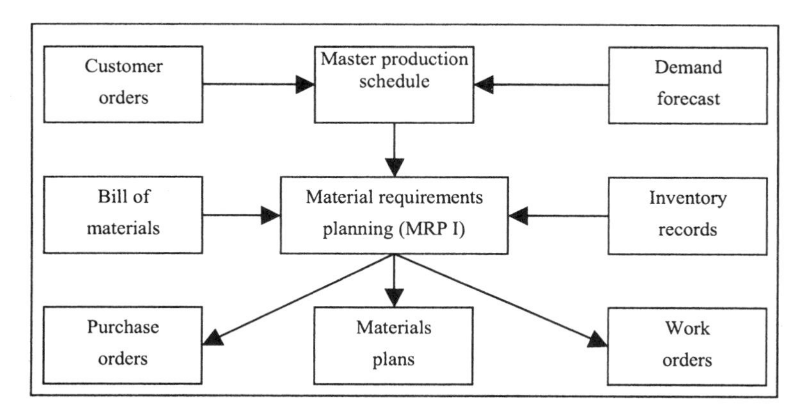

Figure 7-1. The master production schedule in the MRP I schematic (Slack *et al.*, 2001).

The American Production and Inventory Control Society (APICS) defines master production schedule as:

> *1) The anticipated build schedule for those items assigned to the master scheduler. The master scheduler maintains this schedule, and in turn, it becomes a set of planning numbers that drives material requirements planning. It represents what the company plans to produce expressed in specific configurations, quantities, and dates. The master production schedule is not a sales forecast that represents a statement of demand. The master production schedule must take into account the forecast, the production plan, and other important considerations such as backlog, availability of material, availability of capacity, and management policies and goals. 2) The master schedule is a presentation of demand, forecast, backlog, the MPS, the projected on-hand inventory, and the available-to-promise quantity. (Cox III & Blackstone Jr., 1998)*

From the production, sales and/or operations plans, which consider products organized in families or product lines and a long time horizon, the MPS transforms general information into detailed, disaggregating such plans into detailed programs, individually defined for each end product, usually written in weekly and/or monthly time periods. In other words, tactical production planning processes a decomposition of the goals established by the aggregate planning (Fernandes *et al.*, 2000). In case of manufacturing, tactical planning disaggregates groups of resources into machines or production lines, years into months and months into weeks, if applicable.

In general, manufacturing enterprises must have these objectives in mind: maximize customer service and resource utilization and minimize inventory levels. Ideally, this means operating the plant on levels next to production available capacity during all the time, with inventory levels next to zero, and maximum service level. This would imply that when a customer places an order, that product would, at that moment, be leaving the production line towards the dispatching area. The challenge is to plan production to operate it in a comfortable steady pace, building minimum inventory, and taking into consideration costs caused by changing production rates and carrying inventory (Bonomi & Lutton, 1984). But one knows that these are conflicting objective measures. If one tries to minimize inventory level, for instance, not having enough products to meet unexpected orders may result in degradation of service levels. The contrary is true; having inventory is acceptable in order to meet customer demand, however too much of it will increase costs. Production planning, MPS especially, must take all these matters into consideration and also that production is generally a multi-task procedure (different operations), distributed in a multi-period discrete horizon.

The objective of the planning process is to plan all production activities necessary to meet demand forecasts and, secondly, to meet immediate requirements and promised orders. The production planning form most used seems to be hierarchical (see Figure 7-2), proposed by Vollmann *et al.* (1992), although similar hierarchies have also been presented by others (see Figure 7-3). Following Vollmann *et al.* (1992), initially an aggregate plan is established, considering aggregation of end-items into classes of families and covering a long term horizon. Decreasing the planning horizon and considering end-items (or stock keeping units – SKUs), the production planning becomes what is known as a master production schedule. The following hierarchical level comprises the materials requirements planning (MRP or MRP I), which will define when and how much should be ordered, mainly in terms of raw materials, components and, if appropriate, sub-assemblies. The last hierarchical level relates to scheduling tasks and operations needed to accomplish the master plan – it is called production scheduling. The MPS, therefore, is the crucial input information for production scheduling, and this chapter focuses on it.

Cavalcanti & Moraes (1998) show an approach to the master production scheduling process with the intent to cover a gap existing in scientific publications in the field, which only superficially consider the real complexity involved in such a process. In this sense, it is easily seen that the literature, in order to introduce the actual MPS process complexity to the novice, ends up making too many simplifications that hide the real difficulty inherent in the MPS creation. Some of these simplifications are:

- Not considering that production capacity (work centers, production lines or cell, machines, workers and tools) is limited;

- Often, a changeover time will incur every time a new product is to be made at a production line or work center. These changeover or setup times usually demand a non-negligible time, which varies from product to product and their production sequence. Sequence dependent setup times are then to be considered since a different production sequence can yield dramatic savings in the use of limited resources. In such cases, the MPS process should consider a changeover matrix.
- Avoidance of routing flexibility, that is, there is not only one production resource that can produce the product. This routing flexibility increases the complexity in the MPS process.

Even authors of renowned books in this area simplify the explanation of how complex the MPS process really is (Vollmann *et al.*, 1992; Slack *et al.*, 2001; and Gaither & Frazier, 2002).

The MPS constitutes one of the modules part of the production planning and control structure. There are not, however, a commonly adopted form for this structure. It is the result of several factors like promised delivery dates from suppliers, production capacity, strategies and objectives (e.g., minimum inventory levels), and considers information exchange between departments, such as between manufacturing and marketing – for the production and sales forecasting.

Master production scheduling becomes a very complex problem as the number of products, number of periods, and number of resources (production lines assembly lines, machines, production cells) increase. In fact, Garey & Johnson (1979) proved that production planning problems are NP-hard. Yet, setup times and overtime can make this problem even more complex. Moreover, as seen previously, production planning problems usually involve conflicting objectives, like minimizing inventory and maximizing service levels. Because of all this, use of heuristics or meta-heuristics is suggested for the resolution of these types of problems. Since absolute optimal solution finding might be extremely time consuming, a good, perhaps close to optimal, in reasonable computer time is preferred. Several artificial intelligence meta-heuristics have been applied to optimization, among them, genetic algorithms, tabu search, ant colony, beam search and simulated annealing. Some of these techniques are explained in the following sections.

3. A SYSTEM FOR MPS CREATION

A system (heuristics or algorithm) for the creation of master production schedules needs to include:

- A clear definition of objectives and respective performance measure indicators (see Section 3.4). Multiple objectives may exist in an optimization approach, such as minimization of ending inventory levels and maximization of service levels. Coefficients specifying the importance – weight – of each one of the performance measure considered should also be defined.
- Parameters to be used, like initial inventories, gross requirements, standard lot sizes, minimum or safety inventory levels (Sections 0 and 3.2).
- Final adjustments by the master scheduler. No planning (scheduling) information system will generate a plan (schedule) that will perfectly fit the expectation of those responsible for this task in the industry. There are always annoyances that an experience person considers which are often not considered by standard software packages. There are also specificities and cultural aspects particular of an industry that such packages were not intended to deal with. For these reasons, the planner will generally have to make final adjustments to master plans generated by computer systems.

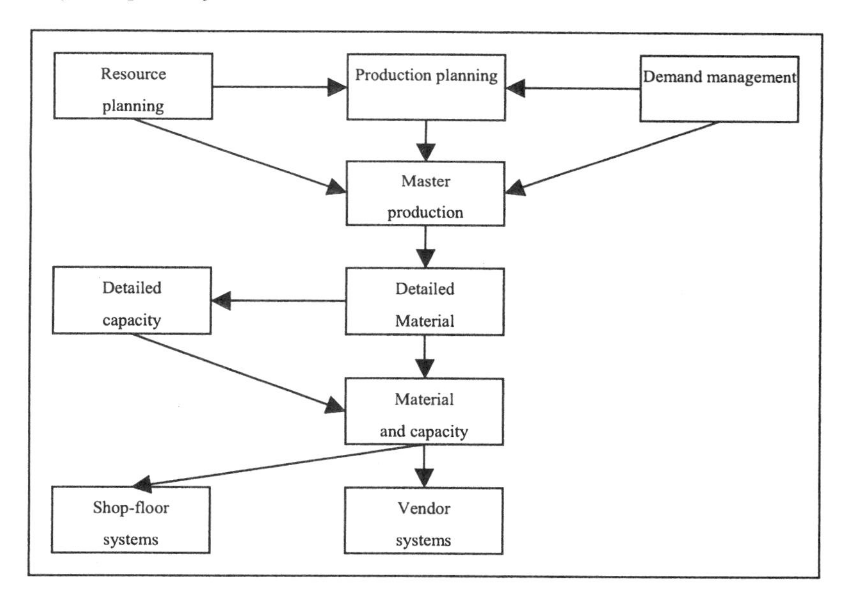

Figure 7-2. Manufacturing Planning and Control System – simplified (Vollmann *et al.*, 1992)

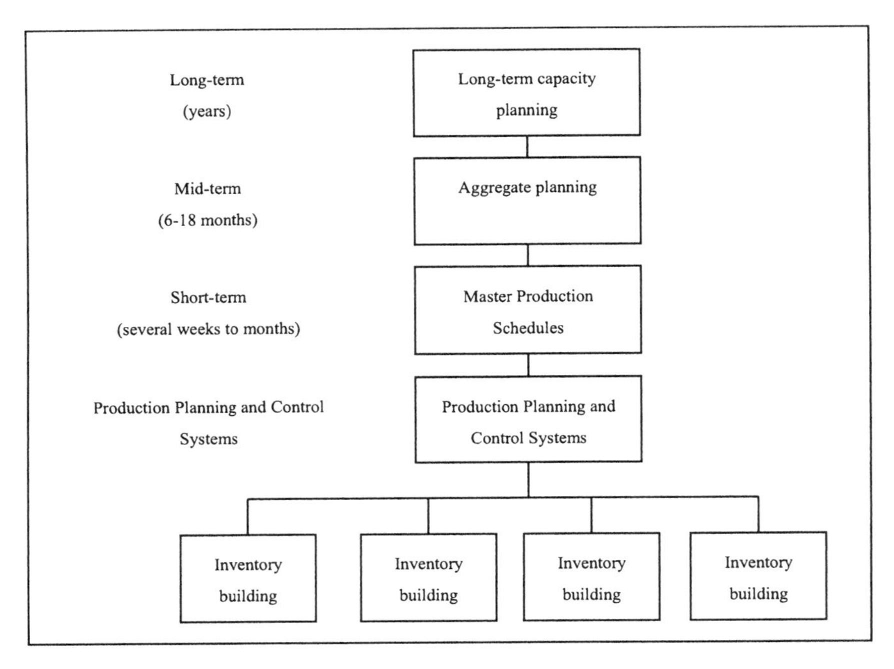

Figure 7-3. Manufacturing Production Planning (Gaither & Frazier, 2002)

Three categories of parameters exist in a master production scheduling system, especially one based on an optimization (pseudo-optimization, heuristics, or meta-heuristics) approach. These are input parameters, output parameters, and objective function parameters. MPS parameters can also be categorized into *updatable* and *non-updatable* parameters. Updatable parameters can be manually altered by the scheduler while, as its name suggests, the non-updatable cannot. Examples for these parameters are given in the following sections.

3.1 Main input parameters

The following are some of the main input parameters often considered in the MPS process:

- Planning horizon: Usually weeks to a couple of months.
- Resources and products (SKUs).
- Gross requirements: mainly demand or production forecasts and customers orders. Usually, initial periods (time buckets) rely more on orders while further periods more on forecasts.

- Subcontracted: Quantity to be manufactured by third-party companies at a certain period of time.
- Standard lot size: The quantity to be manufactured should be a multiple of the standard lot size. It can be estimated based on costs, pallet sizes, minimum raw material (or components) purchase order size, number of parts per box.
- Minimum lot size: the minimum quantity to be scheduled.
- On-hand inventory: Represents the SKU inventory at the beginning of the planning horizon. Sometimes it is confused with beginning inventory (output parameter explained later).
- Safety (or minimum) inventory: Quantity of inventory kept to deal with uncertainties, usually when demand surpasses forecast.
- Maximum inventory: Maximum quantity the company can carry in a time period. This is particularly important for perishable products, where maximum shelf-life is an important issue. It can also be given in terms of maximum inventory coverage, meaning that inventory should not cover more than a given period.
- Production rate: How much a resource can manufacture of a product per time unit. The reciprocal would be how much time the production of one unit consumes of capacity. (A value equal to zero means that a product cannot be made at the resource.)
- Changeover (or setup) time: Time needed to prepare a production resource (it is usually assumed that this time consumes capacity from the resource, also known as internal setup). It can depend on product type sequence (sequence dependent setup) or not (sequence independent setup).
- Backlogging: Maximum quantity of a product (derived from customer orders) that can not be made at the desired time bucket but can be manufactured in future periods.
- Capacity: Regular capacity available from a resource. Usually number of hours or days.
- Overtime: Maximum number of hours (or days) that can be used as overtime per time period per resource. (Usually, government rules specify a maximum number of hours a worker can do in overtime.)

3.2 Main output parameters

The following are some of the main output parameters:

- Beginning inventory: Quantity of a product available at the beginning of a time period. In the first time bucket, it equals the on-hand inventory; for the remaining periods, it equals the ending inventory of the previous period.

- Ending inventory: Quantity of products available at the end of a time period.
- Net requirements: Represents what should be manufactured. It will be in fact manufactured if there is enough available capacity. It is directly calculated from gross requirements, initial inventory levels, maximum inventory, subcontracted, minimum lot size, and standard lot size. This can be approximated with the following expression:

Minimum {Multiple {Maximum {Maximum {Gross Requirements – Initial Inventory - Subcontracted; 0}; Minimum Lot Size}; Standard Lot Size}; Maximum Lot Size}.

- Master production schedule (MPS) row: Contains the final result with the product quantities to be manufactured, by which resources, through the planning horizon. When more than one resource is to be used, there will be several MPS rows, one for each resource used, and a "Total MPS" row.
- Used capacity: For each resource, it shows the capacity used by the MPS at each period of time. It can also be given in relation to the total available capacity (% used capacity).
- Requirements met: Shows in absolute terms how much of the gross requirements will be met by the master plan.
- Requirements not met: Shows in absolute terms how much of the gross requirements will not be met by the master plan. This quantity can become backlogging and be transferred to future periods, if allowed by the scheduler.
- Service level: A percentage representing how much of the gross requirements (demand and orders) will be met by the MPS. In other words, it is the ratio between requirements met and gross requirements.

One can see that planning horizon, resources and products, gross requirements, subcontracted, standard lot size, minimum lot size, safety inventory, maximum inventory, on-hand inventory, production rate, changeover time, backlogging, capacity, overtime, and the MPS row are updatable parameters, while beginning inventory, ending inventory, net requirements, total MPS row, used capacity, % used capacity, requirements met, requirements not met, and service level are non-updatable parameters.

Most output parameters result from the calculation of input parameters and, therefore, are non-updatable. Other parameters, especially those related to costs (holding inventory, production and backlogging costs), also are part of an industrial master production scheduling system.

Table 7-1. Example of item and resource tables.

Item-table: Product AAA

		Week 2	Week 3	Week 4	Month 2	Month 3
On-hand						
Beginning inventory						
Gross requirements						
Standard lot size						
Safety stock						
Net requirements						
MPS						
Service level						
Ending inventory						
Production rate	Line XYZ					

Resource-table: Line XYZ

	Week 2	Week 3	Week 4	Month 2	Month 3
Regular capacity					
Used capacity					
% Used capacity					
Overtime allowed					

3.3 Item and resource tables

In a master production scheduling system, parameters can be grouped in two types of tables. Here, they are called “item-table” and “resource-table,” as illustrated for a fictitious product and production resource (*Product AAA* and *Production Line XYZ*, respectively) shown at Table 7-1.

All products and resources are grouped in these tables, as shown on later examples.

3.4 The multi-objective function

In production planning and scheduling, optimization regards maximizing profits, e.g., “maximize { revenues – costs }.” Since usually maximizing revenues is left to the sales and marketing personnel, the manufacturing focuses more on the costs side of the function, that is to minimize costs: "minimize { costs }." However, the approach described here considers both of them. Costs are easily understood by the parameters shown below, however, maximizing revenues is indirectly attained, by "minimizing requirements not met" as explained below.

Therefore, for the search of a good master production schedule, an objective function should consider the minimization of, at least, the following aspects:

- Ending inventory;
- Requirements not met (remember that demand is mainly given by customer orders and forecasts and service level is directly related to demand met);
- Overtime;
- Setup time;
- Risk of not meeting requirements when operating under safety inventory levels (in other words, risk of stockouts).

Since these variables operate in quite different ranges, the objective function should consider them in a common scale, or, in other words, they need to be normalized. There exist some standard normalization procedures on the literature. This study considers the following approach. Consider three variables A, B, and C that operate under minimum and maximum values given by (MinA, MaxA), (MinB, MaxB), (MinC, MaxC), respectively.

Other parameters that need to be considered regard the importance or weighting of these performance measures used in the objective function. This work considers five weighting coefficients, one for each performance measure that can be used (C1, C2, C3, C4, and C5). These coefficients set the importance of each factor to the MPS quality to be created. Their appropriate definition is fundamental and depends on each company's own interests.

The multi-objective function can then be generically stated as:

Minimize { C_1[inventory] + C_2[requirements not met] + C_3[overtime] + C_4[setup time] + C_5[operating below safety levels]}

With this objective function, there are only five adjustable parameters (C_1, C_2, C_3, C_4, and C_5), which facilitate its use and, at the same time, allow one to use different policies by varying the weighting combination.

3.5 A generic MPS process

Gaither & Frazier (2002) mention that a master production schedule horizon has four sections. The first section includes the first few planning weeks and is referred to as the *frozen* section. The following section, also having a few weeks, is called the *firm* section. Third section, having weeks to a few months, is the *full* section. The last section, also lasting weeks to months, is called the *open* section. These sections can actually be simplified into only two: the *frozen* and *open* sections. Basically, the planner can not, except on very rare and extraordinary situations, change the contents on the *frozen* part of the planning horizon – since resources have already been allocated, material prepared, and people on the shop-floor are practically working on the plan. The planner actually works on the *open* section, which is much longer then the frozen horizon.

The MPS process is usually updated weekly, which means that the week that just ended is removed from the beginning of the planning horizon and another week is added to its end, and requirements (demands and forecasts) related to the whole MPS are estimated again. The first week of the old open section is then set to *frozen*. This is called "rolling-horizon" procedure.

Therefore, the first step in an MPS process is, whenever appropriate, to roll the horizon. Then, the master scheduler should read (update) the MPS input information, such as gross requirements, on-hand inventory levels, expected material arrival, maximum number of overtime allowed (if any), costs and setups – in case these have been changed.

Roughly, the rest of the execution logic is this: starting from the first time bucket in the open planning horizon, an MPS system calculates net requirements based on gross requirements, initial inventories, lot sizes, minimum and maximum inventory levels. Then, based on capacity constraints, it calculates the master production row. If capacity is readily available, the MPS row will equal net requirements (NR), otherwise, MPS will be lower than NR. In this case, if backlogging is permitted, requirements not met can be transferred to later periods. This idea repeats to the following periods. The system can also build inventory in advance to meet future demand in periods of high demand. This requires the system to repeat the above steps several times until it reaches an acceptable plan. After a plan is proposed and the scheduler makes final adjustments, some periods should be frozen.

When different resources can be chosen and, at the same time, different setup times and production rates are involved, the system will need to be intelligent enough to make appropriate allocations: How much of ending inventory, service level, overtime, or setup time is acceptable? What would be the best alternative (plan)? That is, given penalties or cost values for performance indicators – like those just mentioned – how can the system define a good schedule? Imagine, for instance, that just for one period, ten different SKUs need to be scheduled to one of four possible - but different - production lines and that, depending on the line chosen, different changeover times can incur? What if these lines have different processing rates? What if some products cannot be scheduled simultaneously because the use same tools, pallets or fixtures? What if some products should be scheduled only after others? Consider in this scenario that even with these four lines, the shop floor cannot yet manufacture all the requirements – which products should the system schedule first? Expand these questions to a scenario with three hundred different SKUs, forty production lines, and a planning horizon with fifteen time buckets – what do you do? How many different solutions can exist? Based on importance weights, can a system find the best (optimal) solution – a solution that maximizes profits?

These are just some of the dilemmas involved in the master scheduling process that an MPS information system must consider – especially if some optimization is desired. One of the following sections mentions some techniques that can be used in an master production scheduling system – focusing and exemplifying on the use of two artificial intelligence techniques: genetic algorithms (GA) and simulated annealing (SA).

4. ILLUSTRATING THE MPS PROCESS COMPLEXITY

By following an illustrative heuristic for an MPS process (see Figure 7-4), the reader can begin to grasp how complex the master scheduling task is. Later, other examples will strengthen this point.

The following production scenario exemplifies the MPS process using a simplified heuristic. To meet all objectives previously described, the chosen scenario contains the most important characteristics involved in the search for a good master schedule.

Imagine a small factory composed of three production lines: L1, L2 and L3. This factory can make four different end items: A, B, C, and D. The factory uses a single MPS to plan short and mid-term production. For this, they use a mixed planning horizon, with three weeks and two months. Therefore, the planning horizon is composed of five non-homogenous time buckets: P1, P2, P3, P4, and P5.

A total of 40 hours per week and 170 hours per month are available as regular capacity. The factory operates in an eight-hour shift per day and has 45 employees, who are allocated to production lines depending on the product type and quantity assigned to each line. (Assigning the workers to the production lines is by itself a complex problem that will not be considered in this illustration.) Hence, production capacity is related only to production lines hours, more specifically, to the quantity of available and used hours per period.

The usage rate (parts/hour) is described in Table 7-2.

Therefore, products A, B, C can be made at any resource (at different speeds) but product D can only be made at L2 and L3, at 10 and 15 parts per hour, respectively.

As said above, regular capacity is eight hours per day, however, the MPS can use overtime. In this illustration, up to 1.6 extra hour can be used per day (8 hours a week), and, during the monthly periods, overtime can be up to 40 hours – that is, overtime is limited to a maximum of 20% of regular capacity.

Following the heuristic, Table 7-3 shows the ideal case, where, if capacity is widely available, the Total MPS equals Net Requirements (initial step of the heuristic).

Following the heuristic's logic described at Figure 7-4:

1. Time bucket: t=1, meaning Week 2;
2. Product selection. Product D has more restriction since it can be made only at L2 and L3; whereas the other products can be made at any resource. Hence p = D, and $p(r)$ = D(600).
3. Resource selection. Lines L2 and L3 do not have any "exact spare capacity left in *t*." On the contrary, both have all of their capacity (40h) available in *t*. L3 is chosen since it is the fastest one.
4. Pre-assigning D to L3 would consume 40 hours (600/15).
5. Since L3 has enough capacity to make D(600), then the pre-assignment is confirmed.

The process repeats, since product B is still left. However, since there is no resource with available capacity left, B(600) will not scheduled in t. Part of B's gross requirements can be postponed to the next time bucket, however, in this illustration, backlogging is not allowed. The process continues for the remaining periods. The final results are shown in Table 7-4.

In Week 3, B's net requirements are 1200, which can be made by L1 or L2. Because the rate is 20 units/hour for either line, and 40 hours is the maximum regular capacity available per line, one could schedule only one line, in which case, 800 (40x20) units could be made. All of the requirements would be met, but there would be ending inventory below safety stock; or another 20 h of the other production line could be used. This would result in requirements not met for the other two products.

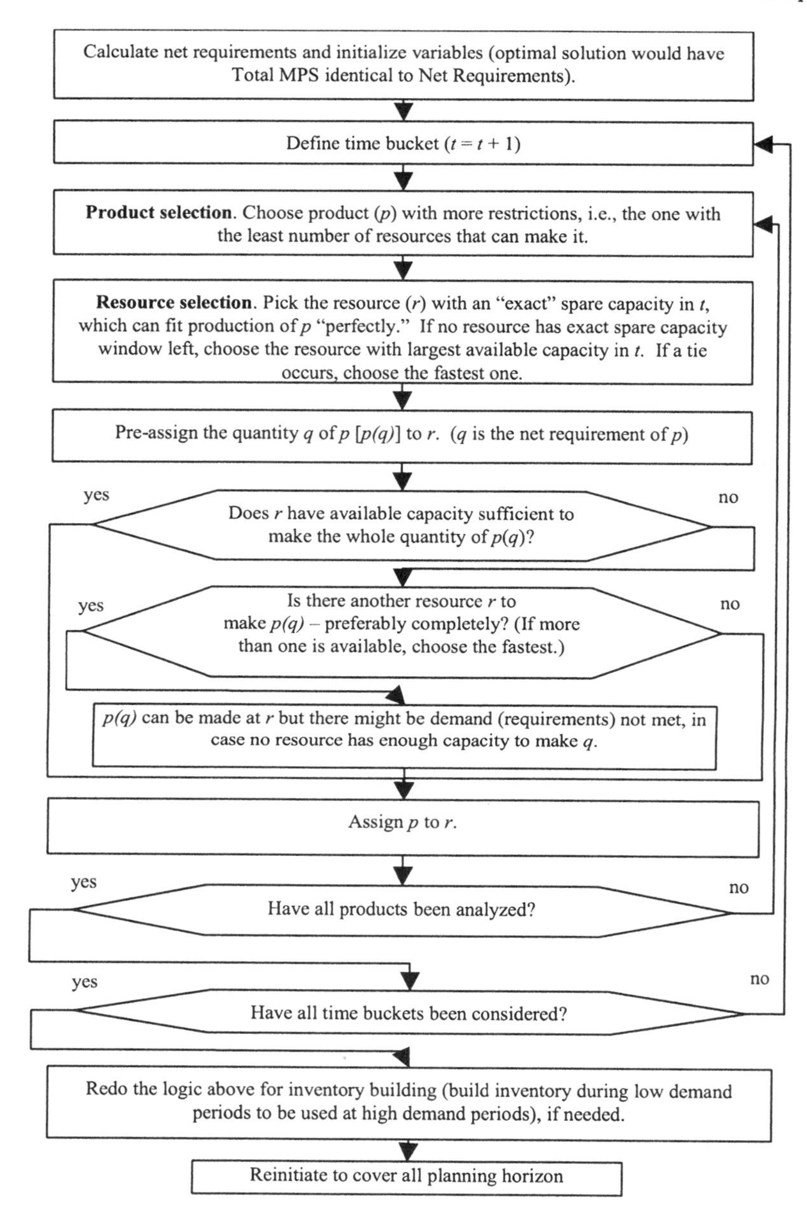

Figure 7-4. An MPS creation heuristic

Table 7-2. Usage rate matrix for the example scenario

	L1	L2	L3
A	15	20	25
B	20	20	25
C	15	30	20
D	-	10	15

Table 7-3. Item–table: Ideal case: Total MPS = Net Requirements (first 2 weeks only)

		Week 2				Week 3			
		A	B	C	D	A	B	C	D
On-hand		100	300	250	350				
Initial inventory		100	300	250	350	500	500	550	450
Gross requirements		400	400	500	500	600	650	650	600
Standard lot size		200	200	200	200	200	200	200	200
Safety inventory (SI)		400	500	400	300	400	500	400	300
Net requirements		**800**	**600**	**800**	**600**	**600**	**800**	**600**	**600**
MPS	L_1								
	L_2								
	L_3								
	Total	**800**	**600**	**800**	**600**	**600**	**800**	**600**	**600**
Requirements met		400	400	500	500	600	650	650	600
Requirements not met		0	0	0	0	0	0	0	0
Service level		**1**	**1**	**1**	**1**	**1**	**1**	**1**	**1**
Average service level in the period		**1**				**1**			
Ending inventory		500	500	550	450	500	650	500	450
Average inventory in the period		300	400	400	400	500	575	525	450
Total avg inventory in the period		**1500**				**2050**			
Below safety inventory		0	0	0	0	0	0	0	0
Below SI in the period		**0**				**0**			

The process repeats:

6. Product selection. Since all products left can be made at any resource, A is randomly chosen: A(800).
7. Resource selection. Any resource can make A. L3 is the fastest but has no spare capacity. L2 is chosen then.
8. Pre-assigning A to L2 would consume 40 hours (800/20).
9. Since L2 has enough capacity to make A(800), then the pre-assignment is confirmed.

The process repeats:

10. Product selection. Either B or C can be picked. Since the net requirements of C are the largest, that is chosen: C(800).
11. Resource selection. L1 is the only resource left with available capacity.
12. Pre-assigning C(800) to L1 would consume 53.33 hours (800/15). Maximum overtime is 8 hours making up to 48 hours of maximum available capacity. (Requirements not met will occur even if overtime is used).
13. Since there is no other resource left, C(800) is assigned to L1 – over time (8 h) will be considered but there will still be 80 units from net requirements that will not be met.

Table 7-4. Final results using the MPS creation heuristic (the first 2 weeks)

		Week 2					Week 3				
		A	B	C	D	cap. used	A	B	C	D	cap. used
On-hand		100	300	250	350						
Initial inventory		100	300	250	350		500	0	470	450	
Gross requirements		400	400	500	500		600	650	650	600	
Standard lot size		200	200	200	200		200	200	200	200	
Safety inventory (SI)		400	500	400	300		400	500	400	300	
Net requirements		**800**	**600**	**800**	**600**		**600**	**1200**	**600**	**600**	
MPS	L_1			720		48			600		40
	L_2	800				40	300	650			47,5
	L_3				600	40				600	40
	Total	**800**	**0**	**720**	**600**		**300**	**650**	**600**	**600**	
Requirements met		400	300	500	500		600	650	650	600	
Requirements not met		0	100	0	0		0	0	0	0	
Service level		**1**	**0,75**	**1**	**1**		**1**	**1**	**1**	**1**	
Average service level in the period		**0,94**					**1,00**				
Ending inventory		500	0	470	450		200	0	420	450	
Average inventory in the period		300	150	360	400		350	0	445	450	
Total avg inventory in the period		**1210**					**1245**				
Below safety inventory		0	500	0	0		200	500	0	0	
Below SI in the period		**500**					**700**				

From the results presented in Table 7-4, one can also see that overtime is used in all periods but the last, being mostly adopted in Week 4 and Month 2.

Although this heuristic does not include any optimization principle, the reader can rapidly begin to see the complexity in an MPS process, especially if the production scenario contains a large number of products, periods and resources. The real difficulty in this MPS creation heuristic is, therefore, located at three main points of the logic: at the product and resource selection and at the inventory building.

5. ARTIFICIAL INTELLIGENCE TECHNIQUES FOR MPS PROCESS IMPLEMENTATION

This work details the implementation of two artificial intelligence techniques used in the optimization of master production scheduling problems: genetic algorithms and simulated annealing. It briefly mentions other techniques like linear programming, hill-climbing and tabu search.

When solving scheduling problems, either master scheduling or shop floor scheduling, two approaches can be used: Optimal and approximate approaches (heuristics and meta-heuristics). Optimal approaches are used in small size problems while heuristics and meta-heuristics have generated good results for larger problems under reasonable computer time. Among the approximate methods, one can use local search algorithms, tabu search, hill-climbing search, genetic algorithms, and simulated annealing, among others (Brochonski, 1999). Optimal approaches would mainly consider linear and non-linear mathematical programming.

Some of the approximate techniques present very good results and end up being called optimization algorithms, although they do not guarantee optimality. (They can probably also be called pseudo-optimization strategies.)

In this work, genetic algorithms and simulated annealing are said to be used for "optimization" of MPS problems and are explained in detail later.

5.1 Linear programming

Linear programming is a mathematical tool usually used to find maximum profit or minimum cost in situations with several alternatives subject to restrictions or regulations. The mathematical representation of the problem includes the objective function, constraints and decision variables.

In practice, linear programming has been applied to several fields, such as in transport routings, production planning and scheduling, agriculture, mining, industrial layout problems. (Prado, 1999). One of the main disadvantages of this method is the computer time during the solution search due to the combinatorial explosion that might happen (Tsang, 1995).

5.2 Hill-climbing search

Hill climbing algorithms try to continuously improve a solution initially generated by a constructive heuristic. The main limitation of this method is the possibility of being trapped in a local optimum. For instance, consider Figure 7-5, the optimal solution – when minimization is the objective – is on point C, however, the hill-climbing search can get trapped on A or B and propose it as the optimal solution.

The algorithm can also be "spinning in circles" if the search process iterates in a flat part of the solution curve (or space). To solve the problem of local minimum and maximum, new approaches have appeared, as tabu search, GA and SA.

5.3 Tabu search

Tabu search is a control strategy for local search algorithms. A tabu list imposes restrictions to guide the search process, avoiding cycles and allowing the exploration of other regions of the solution space. In tabu search, the best neighbor solution is chose, among those that are not prohibited (those who are not in the tabu list). The list of prohibited moves is created by opposed moves most recently performed. The move stays in the tabu list for a limited number of steps. Then it is removed from this list, meaning it goes back to the "allowed moves" list. Therefore, in each iteration, some moves to a certain set of neighbors are not permitted (Brochonski, 1999).

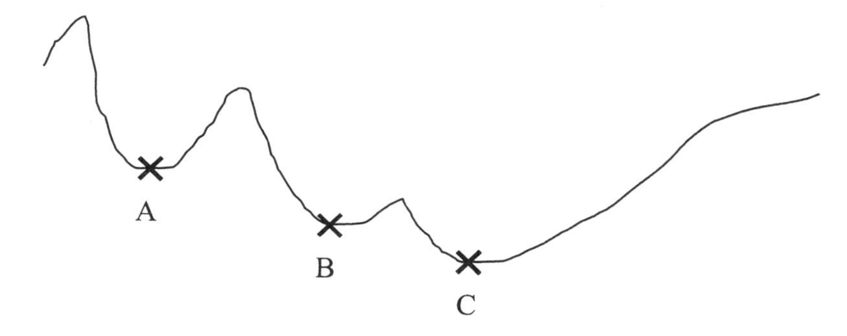

Figure 7-5. Illustrating the hill-climbing algorithm

5.4 Genetic algorithms

Genetic algorithms are an artificial intelligence search method based on natural evolution. In a GA, a population of possible solutions (individuals) evolves according to probabilistic operators conceived from metaphors to the biologic processes, namely, genetic crossover, mutations, and survival of the fittest. As the evolution process progresses, the fittest individuals survive, which represent better solutions to the problem, while the least fit individuals disappear (Tanomaro, 1995).

The decision variables to be optimized are coded in a vector (or object) called a chromosome. A chromosome is composed of a finite number of genes, which can be represented by a binary alphabet or an alphabet with greater cardinality. Selection, crossover, and mutation operators are applied to the individuals of a population. This mimics the natural process, guaranteeing the survival of the fittest individuals (solutions) in successive generations (Brochonski, 1999). In general, genetic algorithms have the following characteristics:

- Operate in a population (set) of points and not on an isolated point;
- Operate in a space of coded solutions and not directly in the search space;
- Only need information about the value of an objective function for each individual and do not require derivatives;
- Use probabilistic transitions and not deterministic rules (Tanomaro, 1995).

Generally, genetic algorithms have better chances of finding solutions closer to the optimal one compared to other search algorithms like hillclimbing since such methods operate on bigger search spaces. However, the cost for efficiency improvement is due to an increase in processing time (Tsang, 1995).

Execution of a genetic algorithm approach can be generically represented by the flowchart in Figure 7-6 (Wall, 1996). Starting from the use of any good heuristic, an initial population of individuals is created. A fitness function measures how good (fit) a solution (an individual) is. The process simulating the natural selection starts as each individual has its fitness factor calculated. At this phase, individuals with low fitness are removed and replaced by individuals with higher fitness, so that the population always remains the same size.

The next step consists of mating among individuals. It is important to highlight that individuals with higher fitness will have greater chances for being selected for the crossover; consequently, there will be a tendency to propagate their genes for the next generations.

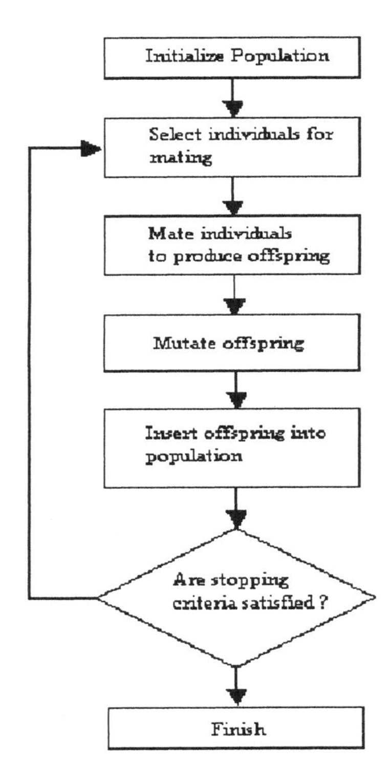

Figure 7-6. Generic genetic algorithm flowchart (Wall, 1996)

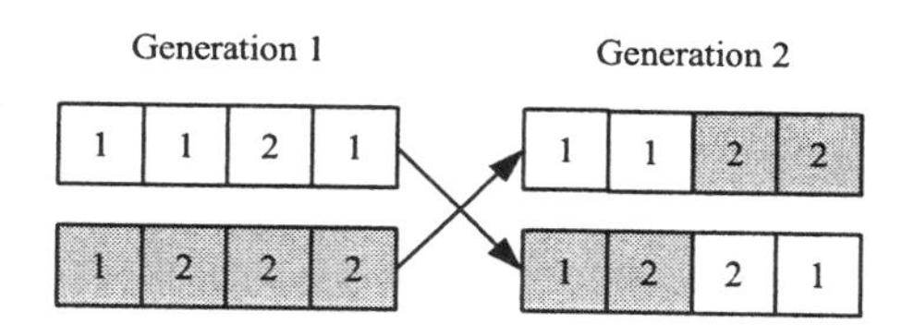

Figure 7-7. Crossover Operation

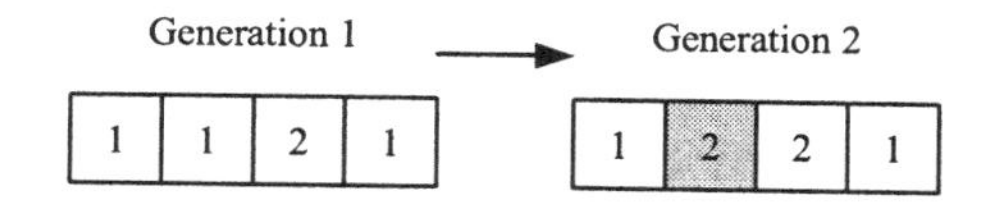

Figure 7-8. Mutation operation

After the crossover (represented at Figure 7-7), a mutation operation can occur to the chromosomes according to some probability.

Mutation consists of randomly altering a gene from any individual of a population to a value possible to be found. In the example shown at Figure 7-8, only two values are possible: 1 and 2.

Analogous to nature, genetic mutation has a very small probability of happening and will affect only a small number of individuals, altering a minimum quantity of genes. This way, the selection process forms a population where the fittest survives, mimicking the natural selection process. The crossover combines genetic material from two individuals in the search for individuals with better genetic characteristics than the parents. In case selection was not applied, GAs would have a similar behavior to the random search algorithms. The mutation, crossover and selection combinations provide a local search selection in the proximity of the fittest population individuals. Without the mutation, the search would be restricted to information from a single population, and genetic material eventually lost in a selection process would never be recovered (Tanomaro, 1995).

5.5 Simulated annealing

The physical annealing phenomena works through gradual temperature cooling of a high temperature metal, where each temperature level represents an energy level. Cooling finishes only when the material reaches the solidification point, which will correspond to the minimum energy state. If cooling occurs too rapidly, as the material reaches solidification, it will present imperfections, which compromise its resistance, also meaning that it did not reach the minimum energy state.

In simulated annealing, the energy state (level) is represented by an objective function to be minimized. Therefore, the minimum energy level represents the optimal solution and the temperature is a control parameter that helps the system to reach this minimum energy.

SA works similarly to a local search method or hillclimbing: it looks for neighboring solutions and accepts them if they are better than the current one. However, contrary to local search, which easily gets trapped in a local minimum, SA tends to escape from such minimums through the acceptance of worse solutions. The probability of accepting a worse solution depends on the temperature (the higher the temperature, the greater the probability) and on the variation of the objective function given by the solution being evaluated (the less the variation, the greater the probability).

Metropolis & Rosenbluth (1953) have introduced an algorithm that simulates the possibility of atoms movements based on energy gains at a given temperature.

Kirkpatrick *et al.* (1983) applied Metropolis' concepts in problem optimization, being considered as the precursor of a series of studies about simulated annealing. The problems solved by this work were the traveling salesman and the printed circuit board layout. (Bonomi & Lutton (1984) also applied this algorithm to the traveling salesman problem.)

McLaughlin (1989) compared simulated annealing with other meta-heuristics successfully in a cards game, and Connolly (1990) improved the algorithm by introducing innovations to the temperature change procedure and to the initial temperature acquisition in a quadratic distribution problem.

More recently, several other researchers have used simulated annealing in manufacturing problems. Radhakrishnan & Ventura (2000), for instance, have applied simulated annealing in production scheduling problems, with earliness and tardiness penalties and sequence-dependent setup times. Moccellin *et al.* (2001) have used a hybrid algorithm combining simulated annealing and tabu search to the permutation flow shops problem. Zolfaghari & Liang (2002) made a comparative study of simulated annealing, genetic algorithms and tabu search applied to products and machines grouping, with simulated annealing given the best results out of the three techniques considered.

The simulated annealing meta-heuristic tries to minimize an objective function that incorporates to the hill climbing approach concepts from physical metal annealing process. To get out of a local optimum, SA allows the acceptance of a worse solution according to a certain probability given by: $P(\Delta E) = e^{(-\Delta E/kT)}$, where

P(.) is the acceptance probability,
ΔE is the objective function increase,
T is the current temperature, and
k is a system constant.

In the real metal annealing process, temperature must decrease gradually to avoid defects (cracks) in the metal surface. In simulated annealing, such defects will correspond to reaching a poor solution.

6. FINAL THOUGHTS

Although widely used and known by industry and academia for many years, master production scheduling is still a field that needs considerable research. Its complexity comes from the fact that the MPS can drive an industry to success or failure, depending if it is good or poorly developed. The MPS is responsible for what the company will produce in the near and coming future; it defines how much inventory the corporation will carry, the service level it will be able to provide for its costumers, how much of its resources it will utilize, and the requirements to pass on to material and component suppliers. The master schedule defines the input data to day-by-day shop-floor scheduling and control activities. Based on all this, the MPS strongly impacts final product costs, a decisive measure for being competitive.

A good master production schedule provides the company with lower production costs, mainly by better use of resources and increasing savings in inventory levels, and consequently, greatly contributing to the increase and maintenance of the company's profit margins. But why is the MPS process so complex? Is it always so? As a matter of fact, the MPS process is not always complex – especially if one considers the scenarios found in the literature. When production capacity is considered widely available, then the MPS process can be solved quickly and easily, just using regular spreadsheet software. In this scenario, restrictions or requirements like safety stock, minimum production lot sizes, or standard lot sizes are easily met. In this fictitious environment, production quantities can be assigned to any resource that can manufacture the product type and changeover times are not of a concern since capacity is considered abundant. Consequently, demand is completely met, resulting in a maximum service level with minimum inventory costs.

This, however, is not the real industry scenario. On the contrary, the reality is the other way around, where production capacity is limited and should be used intelligently, setup times are to be minimized, since they consume valuable resource time and do not add value to what is being manufactured. The right number of employees to be assigned to each work center (production cell or line) should also be wisely determined. Inventory levels should be kept minimal (certainly the minimum level should be appropriately estimated). It is in this restrained scenario that production needs to be scheduled, that is, the definition of which products to manufacture, when, where, and how many units of each, need to be appropriately established. On the top of all that, service level should be maximized since meeting customer demand is just what will bring the real money to the company.

Replanning or rescheduling is also an important issue. Frequent changes to schedules disrupt production on the shop, disturb orders placed to suppliers, generate stops to current jobs being executed, and complicate the financial aspect of the corporation (and supplier's operations, consequently) by increasing costs, and modifying purchases already made, and generating unexpected ones that need to be accomplished.

Because of the difficulty in creating a good master schedule in industrial environments, researchers and developers are implementing new computer algorithms for the MPS process, either with heuristics or optimization techniques. Some of the techniques being used or, better said, that can be used are based on linear programming, hill-climbing and branch-and-bound methods, and meta-heuristics with artificial intelligence, such as tabu search, genetic algorithms, and simulated annealing. AI techniques do not guarantee optimality but are usually efficient in terms of computer time and produce good results (maybe even optimal ones in some cases). This work described the use of two meta-heuristics, namely, genetic algorithms and simulated annealing, in the "optimization" of master production scheduling problems. (In fact, the author could not find a single work on the literature considering these heuristics to the MPS problem.)

These techniques were implemented in C++ programming language. Several examples of productive scenarios were used for illustration and analysis. For these techniques, the main characteristics of a real MPS process and production scenario were considered. Other examples were developed using the optimization techniques. Starting from the objective function, five performance measures were considered: service level, inventory level, overtime, chance of occurring stockouts, an setup times. Results from some computer experiments were satisfactory, although no benchmarking was performed in this study. Computer time was also acceptable, ranging from seven to twenty minutes, depending on the AI technique used and the problem size.

As for future studies, there are still several questions to be answered, like:

- What is still missing to make the MPS process more easily solvable in today's marketplace?
- Considering the advance in computing speed, in which scenarios is the search for optimal MPS solution feasible?
- Based on the AI techniques implemented, which one can produce better results? In which scenarios?
- What is still missing to artificial intelligence approaches to be tested? Can new techniques like, such as ant colonies, also be applied?
- In fact, are genetic algorithms and simulated annealing AI techniques that can always be applied to MPS problems? When is one approach better than the other?
- When will branch-and-bound and beam-search methods provide better results than AI methods? As a matter of fact, how can these techniques also be applied to the MPS problem?
- Both search heuristics presented in this study considered algorithm for MPS creation (or adaptation) based on some criteria – their algorithm however, did not started from the net requirements. Future research should consider net requirement as important information to the MPS creation.

ACKNOWLEDGEMENTS

The author would like to thank *Conselho Nacional de Desenvolvimento Científico e Tecnológico* (CNPq), the *Coordenação de Aperfeiçoamento de Pessoal de Nível Superior* (CAPES) and the Pontifical Catholic Univiersity of Paraná (PUCPR) for funding this study. The author would also like to thank Márcio Morelli Soares and Paulo César Ribas for helping in the development of computer experiments, and Dr. Jeffrey W. Herrmann for his contribution to the author's Doctorate in production scheduling and for his invitation to be part of this book.

REFERENCES

Bonomi, E.; & Lutton, J. "The N-City Traveling Salesman Problem: Statistical Mechanics and the Metropolis Algorithm." SIAM Review, 26, 551-568, 1984.

Brochonski, P. C. "Um Sistema para Programação da Produção de Máquinas de Composição SMT", Curitiba. 1999.

Cavalcanti, E. M. B.; & Moraes, W. F. A. de. Programa-mestre de produção: concepção teórica X aplicação prática na indústria de cervejas e refrigerantes. Anais do ANPAD. Foz do Iguaçu – PR, 1998.

Connolly, D. T., *"An Improved Annealing Scheme for the Quadratic Assignment Problem."* European Journal of Operation Research, 46, 93-100, 1990.
Cox III, J. F.; & Blackstone Jr, J. H. "APICS Dictionary" – Ninth Edition. 1998.
Fernandes, C. A. O.; Carvalho, M. F. H.; & Ferreira, P. A. V., *"Planejamento Multiobjetivo da Produção da Manufatura através de Programação Alvo"* 13th Automatic Brazilian Congress, Florianópolis, Brazil, 2000.
Gaither, N.; & Frazier, G. Production and Operations Management. 8th Edition. South-Western Educational Publishing. 1999.
Garey, M.; & Johnson, D. "Computers, Complexity and Intractability. A Guide to Theory of NP-Completeness." Freeman, San Francisco, USA, 1979.
Higgins, P.; & Browne, J. "Master production scheduling: a concurrent planning approach." *Production Planning & Control.* Vol.3, No.1, 2-18. 1992.
Kirkpatrick, S., Gelatt, C. D. Jr.; & Vecchi, M. P. *"Optimization by Simulated Annealing."* Science, Vol.220, No.4598, 671-680, 1983.
McClelland, M. K., "Order Promising and the Master Production Schedule." *Decision Science*, 19, 4. Fall 1988.
McLaughlin, M. P. *"Simulated Annealing."* Dr. Dobb's Journal, 26-37, 1989.
Metropolis, N.; Rosenbluth, A.; Rosenbluth, M.; Teller, A.; & Teller, E. *"Equations of State Calculations by Fast Computing Machines."* J.Chemical Physics, vol. 21, 1087-1091, 1953.
Moccellin, J. V., dos Santos, M. O., & Nagano, N. S., *"Um Método Heurístico Busca Tabu – Simulated Annealing para Flowshops Permutacionais"* 23rd Annual Symposium of the Brazilian Operational Research Society, Campos do Jordão, Brazil, 2001.
Prado, D. Programação Linear – Série Pesquisa Operacional – 1 Volume. 2a. Edição, Editora DG. 1999.
Radhakrishnan, S. & Ventura, J. A. *"Simulated Annealing for Parallel Machine Scheduling with Earliness-Tardiness penalties and Sequence- Dependent Setup Times."* International Journal of Production Research", vol.38, nº 10, 2233-2252, 2000.
Slack, N.; Chambers, S.; & Johnston, R. Operations Management. Prentice Hall – 3rd Edition - 2001.
Tanomaro, J. "Fundamentos e Aplicações de Algoritmos Genéticos." Segundo Congresso Brasileiro de Redes Neurais.
Tsang, E. P. K. "Scheduling Techiques – A Comparative Study." BT Technol J Vol-13 N. 1 Jan 1995.
Vieira, G. E.; Soares, M. M.; & Gaspar Jr, O. "Otimização do planejamento mestre da produção através de algoritmos genético. In: XXII Encontro Nacional de Engenharia de Produção, 2002, Curitiba. Anais de Resumos ENEGEP 2002. São Paulo: ABEPRO, 2002.
Vollmann, T. E., Berry, W. L., & Whybark, D. C. Manufacturing Planning and Control Systems. Irwin – Third Edition. 1992.
Wall, M. B. "A Genetic Algorithm for Resource-Constrained Scheduling." Department of Mechanical Engineering - Massachusetts Institute of Technology. 1996.
Zolfaghari, S.; & Liang, M. "Comparative Study of Simulated Annealing, Genetic Algorithms and Tabu Search of Solving Binary and Comprehensive Machine-Grouping Problems." International Journal of Production Research, vol.40, nº 9, 2141-2158, 2002.

Chapter 8

COORDINATION ISSUES IN SUPPLY CHAIN PLANNING AND SCHEDULING

Stephan Kreipl, Jorg Thomas Dickersback, and Michael Pinedo
SAP Germany AG & Co. KG; SAP AG; New York University

Abstract: Network planning, production planning, and production scheduling are topics that have been discussed in the supply chain literature for many years. In this chapter we first provide an overview of all the different planning activities that can take place in supply chains while considering the existing functionalities that are available in commercial supply chain planning software. As a second step we consider the coordination and integration of these different activities in the implementation of a supply chain planning solution, which comprises network planning, production planning, and production scheduling. We conclude this chapter with a detailed discussion of an implementation of a supply chain planning solution at the tissue producer SCA Hygiene in Sweden.

Key words: Supply chain management, production planning, production scheduling

1. INTRODUCTION

Existing commercial supply chain planning software tools such as APO (Advanced Planner and Optimizer) from SAP, TradeMatrix Production Scheduler from i2 and the NetWORKS Scheduling system from Manugistics, consist typically of various different modules that are closely integrated with one another. In general, a supply chain planning and scheduling system consists of the following modules:

- Demand Management or Forecasting modules,
- Supply Network Planning modules (for distribution and medium term production planning),
- Production Planning and Scheduling modules,

- Outbound and Transportation Planning modules (short term distribution planning / deployment, vehicle scheduling),
- Order Fulfilment modules and
- Collaboration and Internet Planning modules.

A **Demand Management or Forecasting module** is supposed to estimate and determine future demand. The level of detail at which a forecast is made depends on the type of business: it may be on a product level, or it may be on an aggregate level such as a product group. Statistical forecasting methods like univariate forecasting methods or causal methods provide support to the planner in determining forecasts of the future. Demand Planning is mainly concerned with long term and medium term planning.

The main task of a **Supply Network Planning module** is to propagate the demand to the factories, make sourcing decisions, generate production proposals (while taking the capacities of the various factories into consideration), and generate distribution plans. Supply Network Planning mainly focuses on a medium term planning horizon and considers the entire supply chain. Due to the complexity of the task and due to the planning horizon, Supply Network Planning usually does its planning in time segments (buckets) using simplified master data (for example, ignoring sequence dependent set-up times). Supply Network Planning may be considered as the main coordination tool in a supply chain planning solution as it considers the entire supply chain and creates an overall production and distribution plan proposal. These planning activities are the first steps in the scheduling process of a supply chain.

After the Supply Network Planning module has produced an allocation of the overall demand to the various factories and has generated a rough cut production plan, the main task of a **Production Planning and Scheduling module** is to come up with a feasible short term production plan for each one of the factories. Whereas the Supply Network Planning Module considers the entire supply chain, the production planning and scheduling modules focus on the individual factories. Because the objective is to create a feasible and executable production plan, a Production Planning and Scheduling Module does its planning in continuous time on a detailed product level as well as on a resource level using detailed master data as inputs. While production planning focuses mainly on products and lot sizes, the subsequent production scheduling process focuses on resources and operations.

An **Outbound and Transportation Planning module** is responsible for the distribution/replenishment in the short term period. It updates the proposed distribution plan generated by the Supply Network Planning module using actual production and inventory figures. In case of shortages or excess production, decisions have to be made on how to split the available quantities between the different customers and locations. The result of the Outbound Planning is a short term distribution plan that is fixed for subsequent medium term planning runs. The transportation planning process generates shipping orders between the locations based on the established distribution plan and/or on different objectives and constraints such as the cost of different carriers, full truck loads or delivery windows at the different locations.

The main task of an **Order Fulfillment module** is to set a date and a quantity for the customer during the sales order entry. The confirmation of the sales order is based on the check for available quantities (usually referred to as an ATP check, where ATP stands for Available-to-Promise). An ATP check can, for example, verify the amount of available stock (i.e., stock that is free to use and not reserved for any other customer) and if the available stock is higher than the sales order from the customer, it can confirm the sales order to the customer. If there is no stock available, the ATP check could look at the planned production and verify when the desired product will become available. The final production date can then be confirmed to the customer.

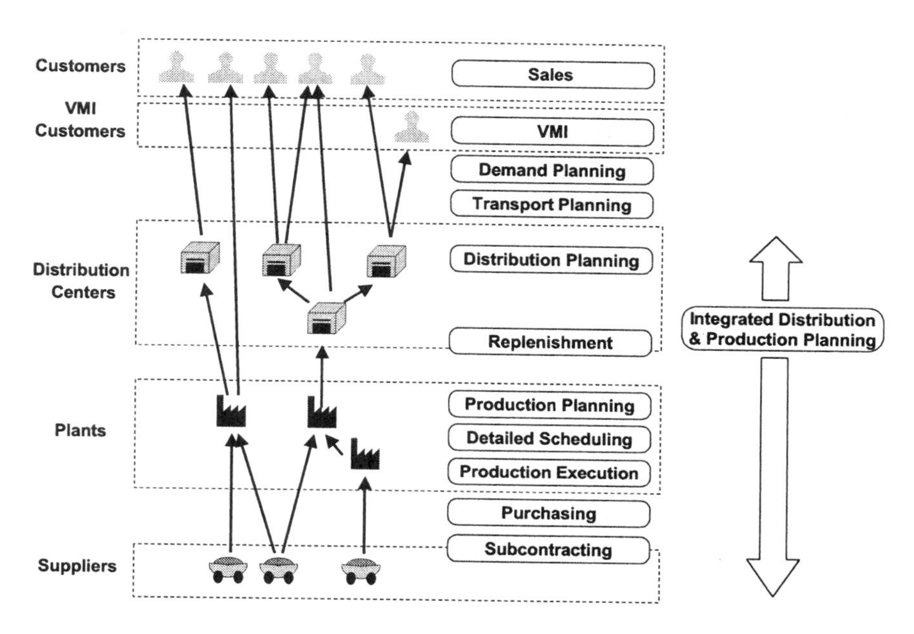

Figure 8-1. Supply chain structure

A **Collaboration and Internet Planning module** may be used in various different planning areas. They are used in VMI (Vendor Managed Inventory) for outbound sales and SMI (Supplier Managed Inventory) for inbound procurement scenarios as well as collaboration in the forecast calculation (Seifert, 2002; Chopra and Meindl, 2003).

The configuration of a supply chain may depend on the type of industry, the value of the products, and of course the type of company. Figure 8-1 shows a generic supply chain with two levels of production (i.e. the assembly groups are produced in a different plant than the finished products), alternative suppliers including subcontractors and two levels of distribution centers to be able to optimize the physical distribution and react quickly to changes in market demand.

Some customers – usually those with a very high and regular demand – are supplied in a collaborative manner (Vendor Managed Inventories or VMI, with the vendor being responsible for the inventory at the customers' site) based on his gross demand information. This kind of collaboration is getting more and more popular and may also apply in a company to the purchasing processes (Supplier Managed Inventories or SMI).

Typically, the consumer product industries, the chemical industries, and the pharma industries come closest to such a 'full blown' supply chain. Manufacturing industries on the other hand often have supply chains that are centered around main production sites with the distribution processes being of lesser importance. In what follows we assume a full blown supply chain and focus on the problem of coordinating two different planning processes: Supply network planning with the focus of covering the demands throughout the entire supply chain and while taking into account the conflicting targets of service levels and inventory costs on the one hand and of production planning and scheduling which focuses on an optimal performance (i.e., high resource utilization, high output and low set-up costs) on the other hand.

In Section 2 we describe the supply network planning process in detail, while in Section 3 we focus on the short term production planning and scheduling processes. In Section 4 we discuss the general issues and problems that are of importance in the coordination of network planning, production planning and production scheduling. Section 5 describes how different companies are coordinating and scheduling the supply chains over medium term and short term planning horizons. Section 6 describes a real-live supply chain planning implementation focusing on the coordination and scheduling of the medium and short term planning. Section 7 provides a summary of the paper.

2. THE MEDIUM TERM PLANNING PROCESS IN A SUPPLY CHAIN (SUPPLY NETWORK PLANNING)

2.1 Overview

The master input data for the supply network planning module comprises all the physical characteristics of the actual supply chain, i.e., the production plants, distribution centers, customer groups, and suppliers are known in advance. Furthermore, all transportation alternatives, like direct delivery or transport via a main distribution center, are specified. Bills of material and routings of products made in-house, information records concerning products externally procured, resource and capacity information are assumed to be available as well.

The main tasks of a supply network planning module include the following:

- The allocation of the demand at the different locations in the network to the various factories,
- The generation of a production plan proposal, and
- The generation of a distribution plan proposal.

The supply network planning must know the demand situation throughout the network (these demands include not only the sales orders and forecasts, but also the desired safety stocks at the different locations) and the current established supply situation (i.e., inventory figures throughout the network, the confirmed short term distribution and transport orders, and the fixed production plans in the different factories). The inventory information is necessary in order to compute the correct net demands at the factories; the difference between the net demand and the fixed production plan is the quantity that has to be produced in order to fulfill all the demands.

Using the above information, the main business objectives of the supply network planning module include the following (see Stadler and Kilger, 2002; Miller, 2002):

- Optimization of sourcing decisions,
- Reduction of production costs,
- Assignment of production to plants (where?, when?, how much?),
- Minimization of transportation costs,
- Minimization of storage and inventory costs,
- Minimization of material handling costs,
- Improvement of customer service levels, and
- Computation of EOQs (economic order quantities).

A supply network planning run generates purchase requisitions (to suppliers and vendors), planned distribution orders between the different locations, and planned production orders.

2.2 Dimensions of supply network planning

Because of the complexity of a global supply chain, the dimensions of a supply network planning process are typically very different from the dimensions of a production planning and scheduling process.

With regard to the time dimension, we use in a supply network planning process time buckets; the smallest time bucket considered is typically one day (24 hours). It is possible to use time buckets of different sizes for different periods; for example, we may plan the first 2 weeks in daily buckets, the subsequent 6 weeks in weekly buckets, and the remaining 4 months in monthly buckets. The closer we get to the current time, the more detailed we do the planning. Since supply network planning is usually done as part of the medium term planning process, it is in general not necessary to do such planning in minutes or seconds.

One important consequence of using time buckets in supply network planning is that no sequencing of operations is done. The outcome of a supply network planning run at the plant level includes planned production quantities for every time bucket, all with the same duration independent of the production quantity. We illustrate the scheduling behavior in a supply network planning process through an example:

Example 1: We have in the medium term planning process resource time buckets with a daily capacity of 1000 resource units. A given product requires 100 resource units of the available capacity in order to make one unit. If we have on day n a demand of 45 units, we create a minimum of five planned production orders: Assuming that capacity is free every day, we create for day n one planned order with an output of 10 production units using the total daily capacity of 1000 resource units. A total of 35 production units remain to be made and we create three more planned production orders on the days n-1, n-2 and n-3 – each with quantity 10. The remaining 5 units are produced on day n-4 using 500 resource units of the daily capacity. All five planned orders have the same duration of one day, but apply different capacity loads to the resources.

Another dimension in which supply network planning is different from production planning involves the master data. Supply network planning uses often a simplified form of production modeling. Instead of modeling for example all n operations to produce a certain product, only the bottleneck or the operations with the longest processing times are modeled. Even combining different smaller operations into one single dummy operation (which is longer) can be done to simplify model complexity. Another possible simplification for supply network planning is the possibility to ignore given lot sizes for production and distribution. These lot sizes are then considered in a later stage of the production planning process or in the short term distribution (replenishment) planning process.

Supply network planning can also use aggregated master data, like product groups or resource groups. Using aggregated master data reduces the complexity of the supply chain model and improves the performance of the supply network algorithms (Chopra and Meindl, 2003; Axsäter and Jönsson, 1984; Miller, 2002). We will discuss in Section 4 potential issues that arise when using aggregated master data in the medium term supply network planning process and detailed master data in the short term production planning and scheduling processes.

2.3 Unconstrained and constrained network planning

In commercial software tools for supply chain planning optimization, we typically find various different supply network algorithms. These algorithms can be grouped in two categories:

1. Algorithms that do not consider the actual capacity limitations of the resources can therefore create medium term production plans that have machine overloads (**unconstrained network planning**).
2. Algorithms that do take the capacity limitations of the resources into account can create finite and synchronized production and distribution plans (**constrained network planning**).

Examples of unconstrained network planning tools include heuristics that propagate the demands from the different locations based on fixed ratios down to the factories. These algorithms do not consider the actual capacity loads of the factories, but only consider ratios that determine how a demand is split between the different sourcing alternatives. The consequences of such an approach may be the following: Since the algorithm does not consider the capacity limitations, it may create an overload on the resources. The overload of the resources at the plant has to be resolved then in a later phase. This results in new scheduling dates and order quantities for the planned production orders, which may not be consistent with the original distribution plan. It is therefore necessary to re-plan the distribution using the updated production plan in order to obtain a feasible distribution plan. A separation of the production and the distribution planning is one characteristic of unconstrained network planning.

Examples of constrained network planning tools include priority rules and optimization algorithms. A priority rule uses the priorities of the demands in order to determine the sequence of the jobs. The priorities of the demands may be based on specified criteria (e.g. the importance of the customer or the profitability of the product). Since a priority rule may consider actual capacity loads, we have to specify also given procurement alternatives as well as their priorities. The priorities for the procurement alternatives may depend, for example, on production costs, inventory costs and/or transportation costs. A priority network algorithm would therefore consider in its first step the demand with the highest priority and try to fulfill it via the procurement alternative with the highest priority. If the procurement alternative with the highest priority is not available, then the next procurement alternative is considered, and so on. Once a demand is fulfilled, the demand with the next highest priority is considered. If one demand cannot be fulfilled, then the algorithm continues with the next demand and goes through its procurement alternatives. The algorithm stops after it has processed all demands.

In the case of optimization algorithms the supply network problem is modeled as a linear or a mixed integer program (see Miller, 2002; Kreipl and Pinedo, 2004). The algorithm searches through all feasible plans in an attempt to find the most cost effective one in terms of total costs. These costs may include, among others:

- Costs of production, procurement, storage, transportation and handling operations
- Costs of capacity expansions (production, transportation, etc)
- Costs of violations of safety stock levels (inventories falling below safety stock levels)
- Costs of not meeting committed shipping dates (late deliveries)
- Costs of stock-outs.

The capacities of the resources (production, storage, handling etc.) are modeled as constraints in a linear or mixed integer program. The optimizer uses linear or mixed integer programming methods to take into account all planning-problem-related costs and constraints simultaneously and calculates an optimal solution. As more constraints are activated, the optimization problem becomes more complex, which usually increases the time required to solve the problem. Decomposition methods are often used when the optimization problem is simply too large to handle within an acceptable running time. Another way to reduce the complexity of an optimization problem is to reduce the discretization. It is for example possible to activate certain lot sizes (like minimal lot sizes) only for the near future, while ignoring lot size settings beyond a certain period. Ignoring lot sizes in supply network planning can of course have consequences for production planning and scheduling, where lot sizes are of more importance (see Section 4).

The production plan proposal generated by the supply network planning module is, due to the simplified master data, not necessarily a feasible or executable production plan, even when using constrained network planning algorithms. A production planning and scheduling module takes the proposed solution from the supply network planning module and creates a feasible production plan for the short term.

3. SHORT TERM PRODUCTION PLANNING AND SCHEDULING IN SUPPLY CHAINS

3.1 Overview

The main objective of the short term production planning and scheduling module is to generate a feasible production plan. Depending on the complexity and the requirements of the business, a feasible plan may be a list of orders per shift – the sequence to be determined on a short notice and by the material flow from previous operations – or an exact sequence of the operations for each resource. In what follows we assume that for each resource an exact sequence of operations is required.

The steps required in the production planning and in the production scheduling are in general separate from one another. The production planning process generates planned orders (resp. production requisitions for external procurement) in order to cover the factory demands taking the lead time into account. The planned orders are scheduled in general assuming unlimited resource capacities (i.e. they do not consider the actual capacity of the resources; however, the orders do consume capacity of the resources). A production planning run can therefore result in a production plan that is not feasible. Production scheduling on the other hand takes the planned orders as an input and focuses on their rescheduling in order to generate a feasible plan. This two-step approach is in line with the classic MRP II approach, where material availability and capacity planning are considered in planning steps that follow one another (see Söhner and Schneeweiss, 1995; Zäpfel and Missbauer, 1993). Though modern APS (Advanced Planning & Scheduling) systems can consider material availability and capacity simultaneously, experience shows that the use of an APS system based on the two-step approach with an infinite production planning in the first step and a finite production scheduling in the second step is more than adequate. In Section 3.4 we describe the possible consequences of doing production planning and scheduling in one single planning step.

Using a production planning and scheduling system to create a sequence of operations for every resource becomes more important with a more complex production process. Issues that determine the complexity of the production planning and scheduling processes include the following:

- The number of bill of materials levels and the number of operations at each level,
- The number of finite resources per order (usually not all resources are planned taking into account their actual capacity),
- Complex and cost based lot sizing procedures,
- Sequence dependent set-up durations,
- Use of alternative resources and priorities for such alternatives,
- Coupling of alternative resources (e.g., if the first operation is produced on alternative 1, then the second operation has to be produced on alternative 1 as well),
- Use of multiple resources in production processes where several operations have to be processed in parallel. In such a case, an operation may take time and also occupy another resource (e.g., space). Examples are heating processes (ovens) for the curing of steel or tires.
- Use of secondary resources in scheduling for simultaneous planning of resources and labour or for the modeling of tools and fixtures with limited availabilities (e.g. dies for moulding),
- Complex material flow on the shop floor (job shop),
- Continuous material flow between orders (e.g. while the processing of the operation is still on-going, the first output quantities can already be used in succeeding operations),
- Order networks, which are used e.g. in the pharmaceutical industry because of the requirement for batch pureness. The difficulty in this case is that the link to subsequent orders has to be kept in scheduling.
- Time constraints (like minimum and/or maximum times) between orders and operations,
- Shelf life of products, and
- Production processes that require containers (tanks) to store (intermediate) fluids (which are common in the process industry).

Since most of the issues mentioned above (which tend to increase problem complexity) are usually not of any concern in a medium term planning process, the bucket plan generated in the medium term planning process may differ significantly from the detailed continuous time plan created in the short term planning process. The other implication is that the more complex the production process is, the more opportunities there are for optimization and the more important the production planning and scheduling is for the supply chain planning.

Comment: Planned production orders may often have different statuses – a planned order created in a production planning run may not yet be relevant for the production execution and may still be changed or deleted at some time during the planning process, whereas a production order that is ready for execution will not be subject to any changes in the production planning run. When an order becomes relevant for production, it changes its status. In SAP this is the conversion from a planned order to a production order. The production order is firmed up and will not be subject to any changes (with respect to its quantity). There are different requirements whether it should still be possible to change the schedule of the production order – if the paperwork for an order has already been printed, then it may not be desirable to change the schedule of the production order. Often it is possible to control whether automatic rescheduling is allowed via an additional status.

3.2 Production planning

The task of the short term production planning module is to create planned receipts – planned orders for in-house production and purchase requisitions for external procurement – and generate the dependent materials requirements through the given Bill-Of-Material (BOM) relationships between the finished product and its component parts. Another important element of the production planning run consists of the lead time computations using the available hours of the resources and the information concerning the routing and the durations of the operations. The lead time adds a timing element to the component quantity data provided in the bill of materials and the given lot size rules (see Zäpfel and Missbauer, 1993; Miller, 2002; Stadler and Kilger, 2002).

A production planning run in an APS system in general performs the following four steps:

1. Net demand calculation: In a first step existing receipt elements (stock, fixed production, fixed purchase orders, etc.) are compared to the existing demand elements (e.g. forecast, sales orders, distribution demand, dependent demand, safety stock requirements, etc.) and if the demand exceed the receipts, the difference equals the net demand.
2. Lot size calculation: The second step takes the net demand and computes the quantities of the planned orders based on the specified lot size rules for each product.
3. Bill-Of-Materials explosion: Assuming that the procurement type of the actual product is in-house production, a Bill-Of-Materials explosion takes place and the dependent material requirements are determined.
4. Scheduling: The last step of the production planning run schedules the planned order. The hours that the resources are available as well as the

duration of the operations are taken into account and usually a backward scheduling – starting with the delivery date – takes place without considering the existing capacity load of the resources (what we refer to in what follows as infinite scheduling). As an option it is also possible to consider the availability of the components in this step. In modern APS systems it is also possible to consider in this step the actual capacity load (what we refer to in what follows as finite scheduling), but as we will see later in Section 3.4, this can have several disadvantages.

Lot size determination plays an important role in production planning. Having exact lot sizes to cover the demand is fairly typical in a Make-To-Order environment but less typical in a Make-To-Stock environment. Common lot size procedures include:

- Fixed lot sizes
- Periodic lot sizes
- Minimum and maximum lot sizes
- Rounding lot sizes.

Figure 8-2 visualizes the differences between these lot size procedures.

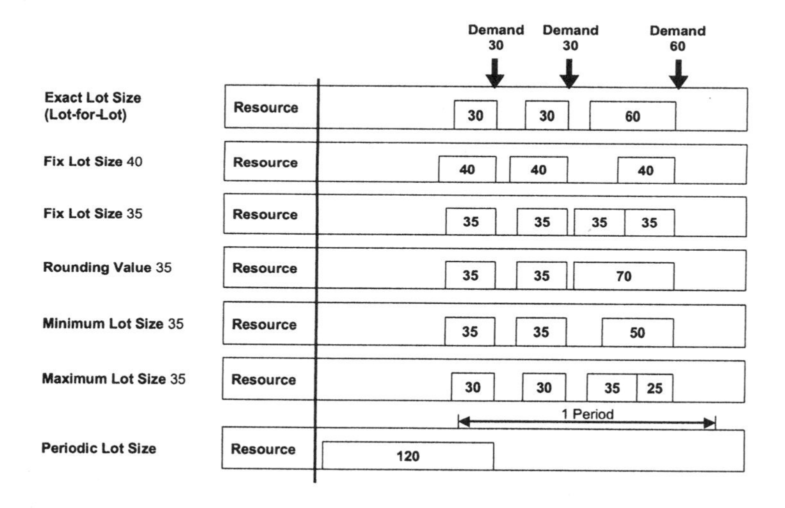

Figure 8-2. Common lot size procedures

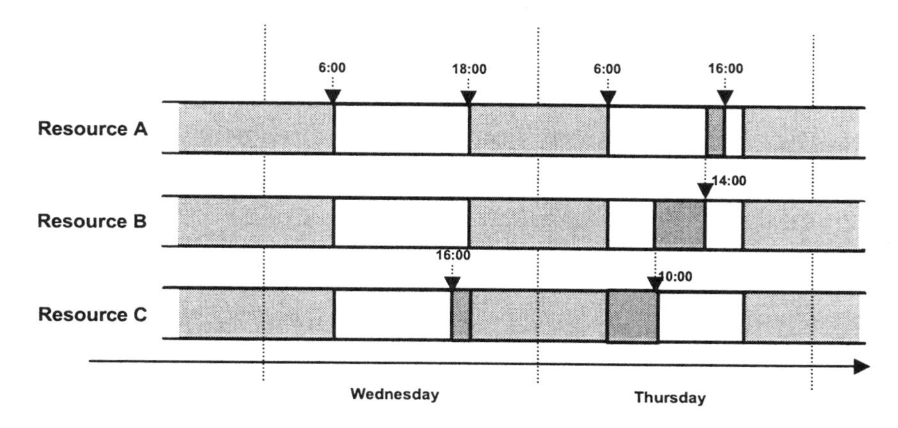

Figure 8-3. Backward scheduling in a production planning run

After the lot size calculation is done the Bill-Of-Material explosion takes place. A simple example describes this step: Assume that we produce a notebook and need the following components: 1 LCD screen, 1 hard drive, 1 DVD drive, 1 processor and an operating system. If, for example, we have to produce 120 units (this figure is determined based on the net demand and the lot size calculations), we get the following dependent material requirements: 120 LCD screens, 120 hard drives, 120 DVD drives, 120 processors and 120 operating systems.

One important criterion for production planning is to plan in descending order of the bill of material level to cover all requirements without planning products more than once during a production planning run. In most cases, an additional planning step before the production planning run is necessary in order to compute the bill of material levels for every product. Especially with the use of by-products it must be ensured that there are no cycles.

The last step of the production planning run schedules the planned order on the resources. Let us consider again the notebook example. Assume that we have three operations in the production of the notebook. The first operation combines the LCD screen, the hard drive, the processor and the DVD drive and takes 3 minutes per unit. The second step installs the operating system on the notebook and takes 2 minutes per unit. The last step packs the notebook into a carton for transport and takes 1 minute per unit. The first operation takes place on resource A, the second on resource B and the third on resource C. All resources are available every day from 6:00-18:00. The due date for the production of the 120 notebooks is in our example Thursday 16:00. If we assume the standard backward scheduling without considering the existing capacity load of the resources, we get the following scheduling results:

We start with the packaging operation. We have 120 units to produce and it takes one minute per unit. So the duration for this packing operation is 2 hours. This results in a start time for the third operation at 14:00 on Thursday, which equals the due date for the second operation. The second operation has a total duration of 4 hours, which results in a due date for the first operation at 10:00. The first operation extends into Wednesday 16:00, as the operation takes 6 hours.

The planned orders generated in the production planning run – even though they are scheduled assuming unlimited resource capacities – specify the capacity usages and also give some indication about the overall feasibility of the plan.

3.3 Production scheduling

The input for the production scheduling process is in general a production plan (generated assuming unlimited resource capacities) which was the result of the production planning process. The main task of production scheduling is to create a feasible production plan that considers all modeled constraints (like resource capacity, minimal and maximal time constraints between operations, secondary resources and multi-activity resources, continuous material flow etc.). No new planned orders are generated, deleted or changed during the scheduling process. In order to gain a competitive advantage a production schedule may be optimized by considering criteria such as sequence dependent set-up times, latenesses or cycle times.

In the production scheduling literature (see Blazewicz et al., 1993; Brucker, 2004; Pinedo, 2002) there is a classification of scheduling problems that is based, among other things, on the number of machines, the material flow and the objective function. Production scheduling problems are in the literature categorized as single machine problems, parallel machine problems, flow shop problems, job shop problems and open shop problems. Different objective functions and additional criteria like priorities, sequence-dependent set-up times or parallel resources lead to a huge number of scheduling problem classes. Each class of scheduling problems can be dealt with through simple priority rules, such as the weighted shortest processing time (WSPT) rule or through sophisticated algorithms. Examples of famous scheduling algorithms include Johnson's algorithm for minimizing the makespan in two machine flow shops (Johnson, 1954) and the Shifting Bottleneck Procedure (Adams et al., 1988) which was developed to minimize the makespan in larger job shop problems. For an overview of different priority rules, see for example (Panwalker and Iskander, 1977) and for more complex scheduling algorithms, see Blazewicz et al. (1993), Brucker (2004), and Pinedo (2002).

APS systems are usually designed to cover all classes of scheduling problems and are able to consider many real-world constraints like continuous material flow, shelf life etc. It is therefore not that common to find the well-known specific scheduling algorithms in an APS system, but more general approaches. Still, APS systems usually have interfaces that allow the user to link his own algorithms. In the following we describe the common options of an APS system to support the creation of a feasible production schedule:

- Interactive manual planning allows the planner to change the schedule of operations and orders, e.g. via drag and drop in a Gantt chart. The advantage here is that the impact of a planning step is immediately visible. Often the APS system allows defining whether adjacent operations are rescheduled automatically as well. Interactive manual planning in a Gantt chart plays an important role in production scheduling because the planner needs an option to change the plan manually in cases of exceptions and urgencies.
- Scheduling rules offer the possibility to design relatively complex rules for scheduling sets of operations. A simple rule can be based on one or more criteria like the order priority or set-up groups. The approach is resource by resource and operation by operation – each operation is scheduled only once. Using scheduling rules it is in general possible to model simple priority rules, like Shortest Processing Time first (SPT) or Longest Processing Time first (LPT). The advantages of this option are that the results are understandable and that customer-specific rules can be implemented easily (see Section 6). On the other hand, if the complexity increases, the scheduling rules may not be able to provide acceptable results any more.
- Most APS systems offer optimization tools for scheduling. Since most scheduling problems are NP-complete (Garey and Johnson, 1979), the optimization algorithms most of the time cannot provide real world examples with the actual optimum within an acceptable runtime. Common scheduling optimization algorithms in APS systems are genetic algorithms (Della Croce et al., 1995) and constraint based programming (Brucker, 2002). One big advantage of using optimization algorithms in scheduling is that they are able to provide good solutions for complex production environments with multiple interdependencies between the operations. One disadvantage of using optimization algorithms is that even if it is possible to influence the result via the weights of the objective functions and the activation and/or deactivation

of different constraints for the optimization algorithm, the results will be in most cases difficult to retrace.

The importance of an understanding how the resulting production schedule is calculated is often underestimated. The robustness of a plan and the ease of understanding a plan - key properties for its acceptance and usability in real life - have to be considered as well. One reason for this is that not all constraints can be modeled – usually they are not even all known. Therefore there will always be cases where the planner has to interfere and overrule the schedule suggested by the system. Another reason is that in many businesses the environment is not stable – i.e., from the demand side sudden demand changes and from the supply side production backlogs may require immediate adjustments to the plan. Therefore the creation of a feasible production plan that meets the constraints (technical and business) is not the only task for the APS system in the area of production planning and detailed scheduling. Other tasks include:

- Real-time integration of the shop floor execution, especially regarding backorders, in order to have the information about the actual capacity situation,
- Transparency of the planning situation, in order to assess the implication of manual interventions – e.g. for rush orders, and
- Support for exception handling (e.g. via alerts that notify the planner about imbalances and lateness).

A more detailed explanation of these tasks can be found in Dickersbach (2003).

3.4 Coordination of production planning and production scheduling

In general, planned orders are scheduled assuming infinite resource capacities in the production planning process (i.e., they do not consider the actual capacity loads, but they do consume capacity of the resources), because in most cases it is simply not possible to create an acceptable production schedule in the production planning process. This is due to the fact that the criteria for production scheduling – e.g., minimizing set-up durations, lead-time reductions, etc. – are not considered in the production planning run.

Modern APS planning systems allow simultaneous material availability planning and capacity planning. This allows the scheduling step in a production planning run to consider the actual capacity load (finite scheduling) and a feasible production schedule is the result. Based on our experiences we would recommend in most cases that using finite scheduling in production planning should be avoided because the result may be a scattered capacity loading of the resources. A scattered capacity loading may have as a consequence that many operations are scheduled far out into the future because they do not fit into the gaps between already scheduled operations (assuming that already scheduled operations are not rescheduled again in the production planning run). This could make it difficult for the schedule planner to understand the result, as he may not expect any operation to be scheduled so far out in the future. This scheduling behavior is especially surprising to the planner if the total free capacity on the resource is higher than the capacity requirement of the single operation, but the duration of the operation does not fit in any existing gap in the near future. Closely connected with this point are 'loser products' – those products which are planned last would get the dates far out in the future or may not even be planned at all due to a lack of capacity.

The normal behavior in a production planning run is that operations already scheduled are not touched again. However, in some APS systems it is possible to configure the production planning run in such a way that the creation of a planned order causes other planned orders to be rescheduled. The previously mentioned disadvantage of a scattered resource load can thus be reduced, because the actually scheduled operation could widen for example an existing resource gap between two already scheduled operations by rescheduling them in such a way that there is enough free capacity for the actual operation to be scheduled in between. From our experience we would not recommend this option because under normal circumstances it will inevitably cause severe performance problems as one operation may cause the rescheduling of all other operations that have been planned earlier. Furthermore, it is likely that the result will not be acceptable either since products that are planned closer to the end of the production planning run, are scheduled closer to their demand dates, while earlier planned products may have to be rescheduled completely.

A normal planning cycle assumes that on a regular basis first the production planning over a given horizon is done and then the detailed scheduling for either the same period or a shorter planning period is done. As the result of the scheduling, planned orders may cover their demands late. This does not have any impact on the following production planning runs because additional receipts should only be created in case of shortages in the planning period and not due to lateness. It is also possible to define the production planning run in such a way that the results of the scheduling are deleted unless the operations are firmed regarding their schedule.

4. GENERAL COORDINATION ISSUES BETWEEN MEDIUM TERM AND SHORT TERM PLANNING IN SUPPLY CHAINS

4.1 Overview

Two issues are of concern in the coordination of medium term and short term planning:

- The transition from a bucket-oriented planning system to a continuous time planning system,
- The transition from a supply chain network perspective to a plant specific perspective.

A transition from bucket-oriented planning to continuous time planning implies changes in the master data because bucket-oriented planning and continuous time planning require different types of master data. In a continuous time scheduling sequence certain dependencies (e.g., in set-ups) as well as constraints that are pertinent to the production process (e.g., shelf life) may have to be considered. In medium term planning an aggregation of products and/or resources may be considered as an option in certain cases (see Stadler and Kilger, 2002; Chopra and Meindl, 2003; Miller, 2002). In our experience this is often the case when no APS planning system is used – production planning and distribution planning (not demand planning) are usually performed on a product level when an APS planning system is being used (see Sections 4.2. and 4.3).

The change of focus when going from a supply chain network level to a plant level implies that actual demands as well as changes in demand within the supply chain network (at other locations than the plant) are not taken into consideration any longer. The resulting distribution plan and the sourcing decisions (for in-house production) are fixed inputs for production planning (see Section 4.4).

Other coordination areas between medium- and short term planning are related to the planning horizons and the planning steps itself:

- The integration between medium term and short term planning has to occur at a certain point in time. There are two options: Bordering horizons or overlapping horizons (see Section 4.5).
- Finally the planning itself has to be coordinated – what should be considered as a data transfer from medium term planning to short term planning and what may be overwritten. The different options are explained later on in Section 4.6.

In general, the more complex the production process is, the higher the probability of having large differences between medium term and short term planning. The challenge in this case is to find out (there is no other way to do this than empirically) how big these differences typically are and to adjust the medium term master data and the configuration of the coordination (e.g. via the medium- and the short term horizon) to ensure sufficiently precise information from the medium term plan.

In what follows we assume that supply network planning (including the rough-cut production plan) is part of the medium term planning process, while production planning (with detailed master data) and production scheduling are parts of the short term planning process.

4.2 Master data

In order to support the different requirements for bucket-oriented medium term planning and time-continuous short term planning different sets of master data are required. The implications of this are the following:

- There is an additional effort in creating double master data and keeping it synchronized for medium term and short term planning. The amount of this effort depends on the degree of automation that can be used for the creation and maintenance of master data for medium term planning. This effort should not be underestimated.
- As the scope of supply network planning considers the total supply chain, including production and distribution planning with a medium term time frame, we may decide not to model every production step in the supply network planning process. Imagine for example a complex production process like the printed circuit board industry which has up to 60-80 processing steps. It does not make much sense to do the medium term planning with all these detailed processing steps. As a consequence manual decisions are necessary in order to model the medium term master data.
- The modeling issue gets more difficult to solve if many production steps are processed on bottleneck resources. If we decide not to model all production steps in supply network planning, we have to make sure that the capacity requirements of the non-modeled production steps are considered on the critical resources (for example by increasing the resource consumption of other production steps). Another approach would be to extend the short term period in order to see the correct resource loads for a longer period and accept the difference for medium term planning.

4.3 Aggregated master data for medium term planning

According to our experience the aggregation in medium term planning is usually limited to the time dimension (time buckets instead of continuous time) and a simplification of the master data (see the previous section). The master data is usually transferred to the APS system from the Enterprise Resource Planning (ERP) system at the lowest level of detail (i.e. products, resources, bill of materials and routings). It is therefore in general relatively simple to create also precise short term planning master data in the APS system, as we can copy the master data from the ERP system 1:1 to the APS system. Large volumes of master data (say, up to 100.000 products for certain industries) are therefore not an issue. An aggregation of the master data to product groups and resource groups for the medium term planning on the other hand requires manual decisions, like the following:

- Which products should be combined into aggregated products?
- Which production times and operations steps should be included in the routing of the aggregated product when the times and operations differ slightly?
- Which components and which quantities have to be included in the bill of material of the aggregated products, when the input quantities of the planning relevant components differ slightly?
- Which resources should be put together into aggregated resources?

No guidelines, that hold universally, can be established. It is therefore a time consuming process to decide about, create and maintain aggregated master data.

A second implication from aggregated master data arises when the medium term planning results have to be disaggregated in order to be used in production planning (Chopra and Meindl, 2003; Okuda, 2001; Miller, 2002). Here we get among others the following planning issues or additional planning steps:

- Assume the medium term planning process generates planned production orders and distribution orders for aggregated products. In order to use the planning results for the short term planning, we have to disaggregate the planning results and obtain results for detailed products. There is obviously no universal disaggregation strategy that in all instances leads to the desired results. A split based on the total demand quantities for the detailed products is a common approach.
- The disaggregation process becomes more complex when the planned production orders are also using aggregated resources. A second split concerning the resources has to be defined. But disaggregating both products and resources could lead to a situation, in which all products

after the disaggregation are produced on every one of the resources in small quantities; such a solution may not be appropriate.

- The medium term planning process needs to know the planning situation from the short term planning. If we use aggregated products and resources in the medium term planning process and detailed products and resources in the short term planning process, the following tasks have to be executed before the medium term planning process can start: The medium term planning process needs to know the capacity utilization of the resources and the demand and receipt quantities for the products. This information can be calculated by adding up the demand and receipt quantities for all products, which are combined into the aggregated product and similarly by adding up the capacity loads from all resources, which are combined into the aggregated resource.

4.4 Coordination of global supply chain planning and planning of individual plants

The two most common methods for integrating supply network planning with production planning are (see Figure 8-4):

- Limit the supply network planning process (distribution and production) to a medium term planning period (Method 1)
- Use supply network planning for distribution planning across short and medium term and restrict the rough cut capacity planning to the medium term (Method 2).

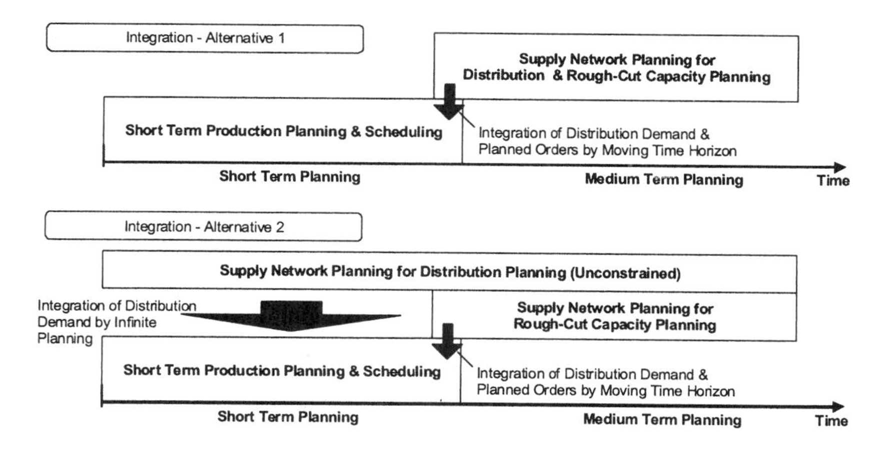

Figure 8-4. Alternative medium and short term integration approaches

The disadvantage of the first alternative is that the short term planner will not be able to respond to demand changes within the short term horizon since no data concerning demand changes are transferred to factories. In order to propagate demand changes across the supply chain to the factories, the second approach requires the use of an unconstrained network planning algorithm (because a constrained network algorithm would regard the current production plan as an additional constraint) with all the downsides regarding sourcing alternatives and feasible distribution plan.

By extending or compressing the overlap between short term and medium term planning it is possible to modify these properties to a certain extent. For example, if the overlapping horizon in the first alternative is extended and the current short term production plan is deleted outside a specified firmed planning horizon before the supply network planning run takes place, constraint network planning algorithms can also be used (Kreipl and Pinedo, 2004). In this case scheduling in the overlapping horizon has to be repeated after each supply network planning run.

4.5 Horizons for medium term and short term planning

The separation between medium and short term planning period is controlled by the medium term horizon (supply network planning plans outside the medium term horizon) and the short term horizon (production planning and scheduling plans within the short term horizon). If the medium term and short term horizons are different, it is possible to create an overlap between the medium- and short term planning horizons in order to reduce the problems at the boundaries.

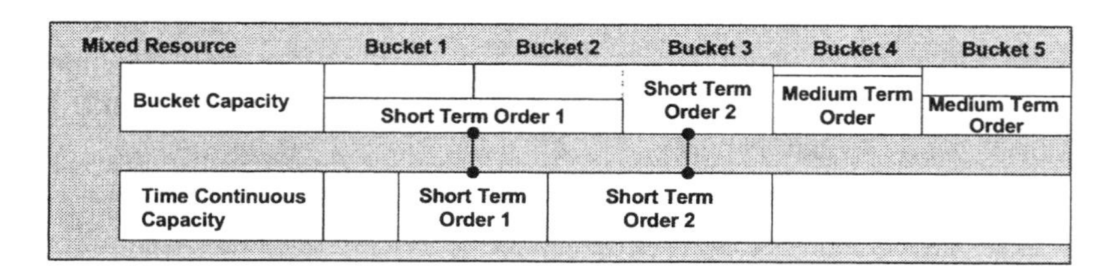

Figure 8-5. Mixed resource concept

The borderline between medium term and short term planning is apt to cause some problems, since planned medium term and short term production orders may be partially within the short term planning horizon and partially within the medium term planning horizon and will therefore consume capacity in both planning horizons. An approach to meet these concerns is the use of overlapping horizons implying that some orders may have to consume resource capacities both in the short term and in the medium term.

If we have short term and medium term planning periods that overlap, we need one common resource for both short term and medium term planning in order to avoid that one operation in the overlapping period has two consumptions of the same resource – one bucket capacity consumption and one continuous time capacity consumption. The concept of mixed resources implies that they must have, at the same time, a bucket capacity for the medium term planning and a continuous time capacity for the short term planning. In order to avoid a double consumption of the capacity, the short term production orders also have a bucket capacity consumption (which needs to be maintained in the master data), while the medium term planned production orders only consume bucket capacity.

Even with the use of a mixed resource there will be differences in capacity consumption between medium term and short term planning (the most obvious reason being sequence dependent set-up times). The bigger the difference in capacity consumption, the shorter the overlapping period should be (because there is no consistent view of the overlapping period).

One of the questions regarding the integration of medium term and short term planning concerns the time horizons over which supply network planning and production planning and scheduling are used. Since procurement can be triggered from supply network planning as well, the lead time of procurement should not be used as an indication for the length of the horizons. Better indicators for the appropriate horizon for a detailed scheduled production plan are the firmed horizons for production, the set-up times and the production cycles. The choice of the coordination between supply chain and plant level (see Section 4.4) and the choice of the coordination of the production planning steps (see Section 4.6) also have an effect on how far the medium term and short term horizons should overlap.

4.6 Coordination of the production planning steps

A feasible medium term plan may not lead to a feasible short term plan because some constraints were ignored in the medium term planning process. Two examples are described in what follows:

1. Assume we use an optimization approach in the constraint network planning process. Due to a complex supply chain and runtime limitations,

we have to remove some or all discrete constraints from our optimization problem. In this case the resulting capacity load is too small, because after converting the medium term planning orders (using the simplified master data) into production planning orders (using the detailed master data and the lot sizes), the planned order quantities increase (due to the lot size rules) and therefore also the resource load. What seemed to be a feasible production plan in the supply network planning process, suddenly becomes a production plan with capacity overloads.

2. A similar behavior can be observed with sequence dependent set-up times, as they are not modeled in supply network planning. A simple workaround for this problem is to reduce the available capacity in the medium term planning or to include a fixed set-up time, but in both cases we might get significant differences between the resource load in the supply network planning process and the production planning and scheduling processes.

The coordination of the production planning steps describes how the outcome of the medium term planning process is used in the short term planning and which application (medium term planning module resp. short term planning module) performs which steps in which horizon (medium- resp. short term). We present here three options to coordinate the medium term production planning result to the short term planning:

Option 1 (Order Conversion): The first option is to use a planned order conversion, i.e. the quantities planned by the medium term planning are converted into planned orders for short term planning regardless of the demand situation at the plants. This implies that no net requirements calculation is performed in short term planning. The advantage of this alternative is that decisions made in the medium term planning, which are based on a better overview regarding time and network are adhered to. The disadvantage is that the short term production planning will not be able to react to short time changes since no production planning run is done any more.

Option 2 (Short Term Production Planning): If the medium term planning is used rather for distribution planning, evaluation and feedback about the feasibility of a plan, there is the alternative to perform a short term production planning run within the short term horizon which will delete all medium term planned production orders and create short term planned production orders. In this case the medium term planning result has as only input to the short term production planning the resulting distribution demand (see also the case study in Section 6).

Option 3 (Different Planning Versions): The third option is a complete separation of the medium and short term planning. In this case the medium term planning is performed in a non-operative planning version by the supply chain planner. The result is a feasible plan for distribution and production. The feasible production quantities are transferred to the operative planning version as planned independent requirements for short term planning. If a medium term planner is responsible for creating a feasible master production schedule for the factories and there are local short term planners to realize the master production schedule, this option has an advantage in supporting a clear separation of the responsibilities: In the case of a shortage, it is transparent whether this is due to an infeasible demand (which is in the responsibility of the medium term or supply chain planner) or due to a low performance in the production (which is in the responsibility of the local or short term planner). Differing from the first option, the short term planner is still able to react to demand changes in the supply chain.

There is not one best option for coordinating medium term and short term production planning. All options have their advantages and disadvantages, and it depends on the type of business which one is the most appropriate.

The coordination between medium term and short term planning becomes more complicated if there are factories within the supply chain that are supplied by other factories – especially if alternative factories exist for sourcing. Medium term supply network planning is capable to decide upon the sourcing based on a global optimum. However, if option 2 or option 3 is used for the coordination, the production planning in the short term horizon might discard the sourcing decision by medium term planning.

As an example how the coordination between medium term and short term production planning impacts the coverage of the business requirements, we pick a business with seasonal demand (e.g. the beverage industry) which builds up inventory during the quiet seasons. Per definition the medium term planning considers a longer planning horizon than the short term planning. The medium term planning creates planned production orders for the plants based on the demand for the complete horizon. As the result of a constraint network planning approach we assume in this example that due to capacity constraints, we have to build up inventory in earlier periods to cover the demands for a later period. In the following figure we have a demand, which is covered by several planned production orders from a medium term planning run. The reason for having several planned production orders is in this case not due to lot size rules, but due to the fact that we use time buckets in medium term planning and that the planned orders have a fixed duration independent of their quantity. If the capacity of a bucket resource is full, a new order is created in the previous time bucket (see example 1 in section 2.2).

The consequence of the situation described in Figure 8-6 is the following: If we convert the medium planned production orders up to day N-2 (as described for the first option), we get short term planned production orders inside the short term production horizon, but no demand. This has no impact on the short term production planning (as it is not done), but in production scheduling it is not possible any longer to consider the due dates for example as optimization criteria.

Using the second option and executing a short term production planning run, the system will delete the medium term orders, but not create any new short term planned orders as no demands exist inside the short term planning horizon.

With the third option the medium term orders are transformed as independent requirements for the short term planning. The short term production planning run creates planned orders for the date and quantity that was determined in medium term planning.

Therefore the first and the third option would be appropriate in this example or the short term production planning run has to be enhanced in the case of the second option, e.g. it has to consider demands outside the short term horizon as well.

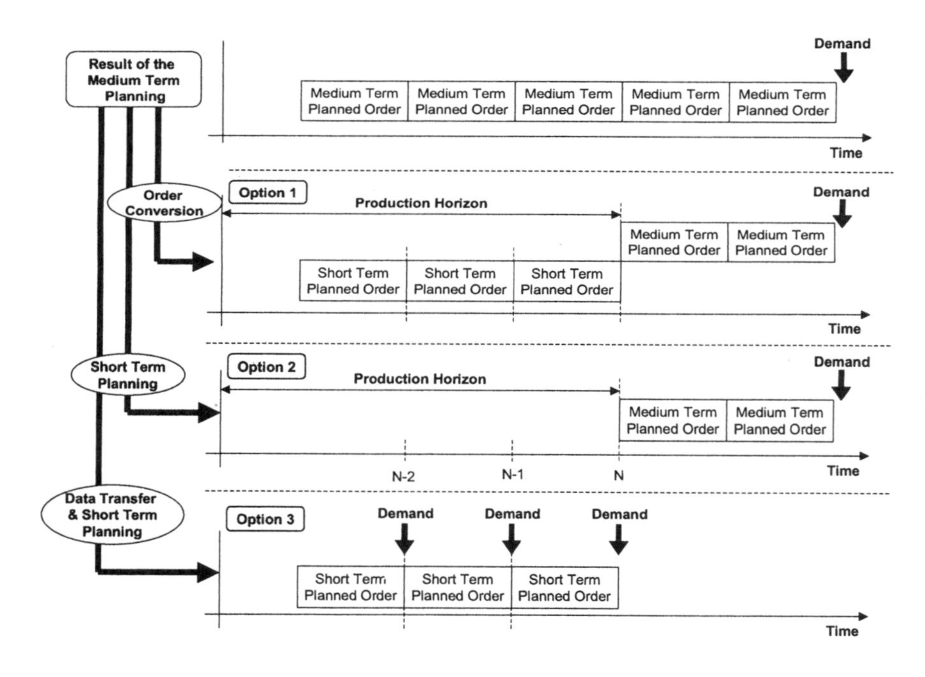

Figure 8-6. Impact of different coordination options on build-up of inventories

4.7 Criteria for good coordination between medium term and short term planning

In the following we describe some criteria that can be considered as guidelines for a good integration:

- Firmed horizon to provide a reliable basis for production: The execution of the production activities requires the staging of the components and the printing of the order papers. Therefore the flexibility in the very short term is limited.
- Sufficient degrees of freedom for sequence optimization (especially when many production constraints are added in short term planning) – e.g. if medium term planning provides a daily production proposal and the duration of an operation is one day, there is only limited sequence optimization possibilities.
- Flexibility to react in production to demand changes in the network.
- Consistent capacity view: The capacity load in medium term planning should be close to the corresponding short term capacity load. If this is not the case, the medium term master data should be adjusted.
- Feasible distribution plan based on a feasible production plan.
- The set-up of the medium and short term planning should support the responsibilities within the organization (the focus might be either on separation or on integration of the plan and the data basis).

Some of these criteria are contradictory and as companies have different requirements regarding the coordination – depending on their industry, supply chain, production complexity and organization, there is no optimal generic way to coordinate medium- and short term planning.

5. MEDIUM TERM AND SHORT TERM PLANNING IN DIFFERENT COMPANIES

Depending on the industry of the company the importance of supply network planning versus production planning and scheduling is regarded differently. Typically companies with complex production processes and many bill of material levels – e.g. in the engineering industry – place more emphasis on production planning and scheduling while companies which produce commodities – e.g. in the consumer goods industries – focus more on the medium term and supply network planning. The reason for this is that for commodities the planning complexity and the potential for economization is usually higher in the supply network planning (inventories, transports, service levels) than in the production planning and scheduling.

The perception of medium- and short-term is very relative in both cases, though companies with the focus on production planning and scheduling tend to have a longer short-term horizon. Nevertheless the short term production horizon does range between three days and three months.

The organization is another aspect which has a strong impact on the way medium and short term planning is performed in the company.

Companies with a strong emphasis on supply network planning often have a cross-location supply chain organization that focuses on the supply chain costs and targets. This organization takes the decision for sourcing alternatives and provides the production planning organization of the plants with monthly or weekly production requirements. These organizations might be more or less close together – sometimes they are even in different regions. The degree of cooperation between the planning organizations determines whether a clear separation of responsibilities (up to a separation of the data basis) or the transparency of the medium- and short-term plans across the organizations is more in focus. Companies within the consumer goods, chemicals and pharmaceutics industries have often such an organization. There are companies within chemicals which use medium term planning to plan the whole supply chain including the production up to a few days' horizon, and companies within consumer goods which have the strategy to align their production schedule on short notice to the demand changes in the supply chain and thus accepting sub-optimal production costs due to frequent set-up changes.

Another organizational model is that supply chain and production planning lies within one organization. This is mostly the case if supply chain planning evolved from the production planning organization. For these types of organizations the transparency up to a common data basis is most important. Depending on the business requirements to be able to react quickly to deviances in the demand or in the supply, option 2 (see Section 4.6 about coordination of the production planning steps) is favorable.

There are cases of heterogeneous companies where no common planning process is established across different business units. For these companies either option 1 or option 2 (see Section 4.6) is used, because these options allow more flexibility regarding different planning cycles and horizons, and the separation between medium- and short-term planning does not need to be shared by all parties.

Companies with a strong engineering background use medium term planning for production mainly for reporting purposes (if at all). Therefore they tend to use the second option for the coordination of medium and short term planning.

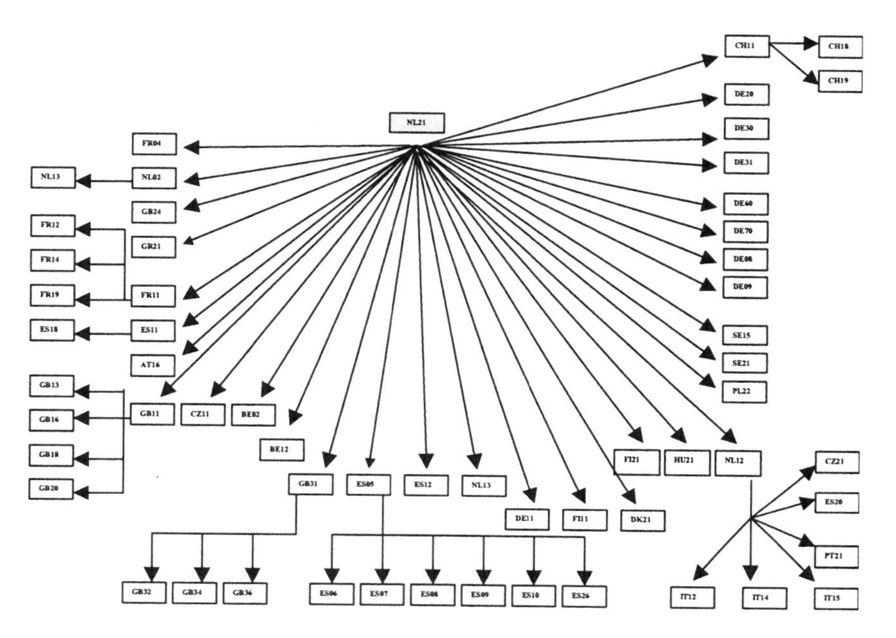

Figure 8-7. Simplified supply chain structure from SCA Hygiene Products for one factory

6. SCA HYGIENE: AN EXAMPLE OF COORDINATING NETWORK PLANNING, PRODUCTION PLANNING AND PRODUCTION SCHEDULING IN SUPPLY CHAINS

This section describes an implementation of SAP's APO supply chain planning software at the tissue producer SCA Hygiene Products in Sweden. SCA Hygiene Products is one of the leading tissue producing companies in the world. Its network consists of 55 production plants in 25 countries and more than 100 distribution centers (DCs) all over the world. Its net sales in 2004 were 45 billion SEK (approx. 6.7 billion US $). Total net sales of SCA including the packaging and forest divisions were 90 billion SEK in 2004 (approx. 13.4 billion US $). [SCA Annual Report 2004]

SCA Hygiene Products started in 2000 a supply chain project with the objective to have an integrated supply chain planning approach that increases the visibility of the entire supply chain. The scope of the project provides an example of how an actual supply chain planning implementation considers the coordination, planning and scheduling issues over all its stages. SCA decided to use among others the following planning modules: Demand planning, supply network planning, production planning and scheduling, distribution planning, and the transportation planning algorithms from the supply network planning module. The system has been operational in several pilot countries since 2002 and there is an on-going world wide roll-out taking place since that time. The full world wide implementation is planned to be completed by 2007.

The supply chain at SCA is structured as follows: Every product is produced in one factory. This simplifies the task for the supply network planning, since no real sourcing decisions have to be made. DCs are in general supplied by several factories, but in this case different products are transported between each factory and the DC. In general we have several stages between the factory and the DCs, as shown in the supply chain example for the one factory below. The total supply chain consists of 55 similar supply chains, which are all connected. The transportation durations between the locations depend on the origination sites and the destination sites.

The production process depends of course on what types of products are produced in each factory. SCA Hygiene products include baby diapers, feminine hygiene products and tissues. In what follows we describe a simplified production process in a tissue factory as it is modeled in the APS system. In general we have three stages:

- The first stage consists of the rewinding process. In the rewinding process the final product is produced.
- The second stage consists of a packing machine, which packs the goods into consumer packs.
- The third stage consists of a second packing machine, which packs the consumer packs into transport packs.

The rewinding resource can feed several packing machines at the second stage in parallel.

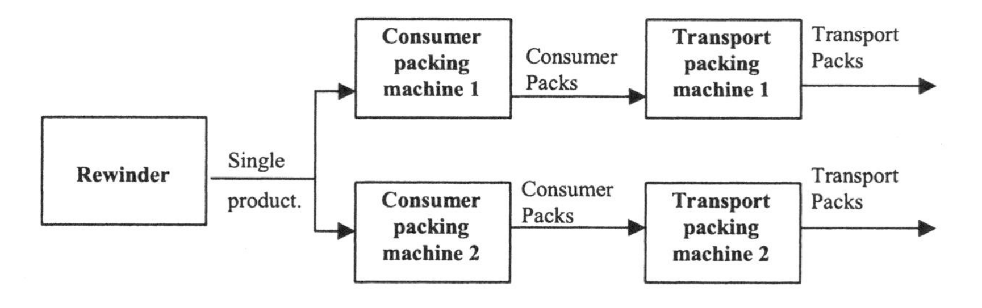

Figure 8-8. Simplified production process

The average number of production orders for a typical tissue factory ranges from 25 to over a 100 a day, which results in the same amount of operations on the rewinder.

The task of the supply network planning module is to transfer every day the demands (sales orders and forecasts) for all products from all DCs to all factories for the next 16 weeks. Due to the fact that no sourcing decisions are necessary (as every product is produced in exactly one factory), a simple supply network planning heuristic is used. As there is also a clear organizational separation between the production planning teams and the network planning team, no planned production orders are created in the network planning run. The only result from the network planning run is the creation of the demand for the various factories. The assumption hereby is that most of the factory demand can be produced on time. The advantage of this approach is that no medium term master data problems occur, because there is no need for any medium term bill of materials and routings. The following screenshot shows the planning tool for the medium term planner.

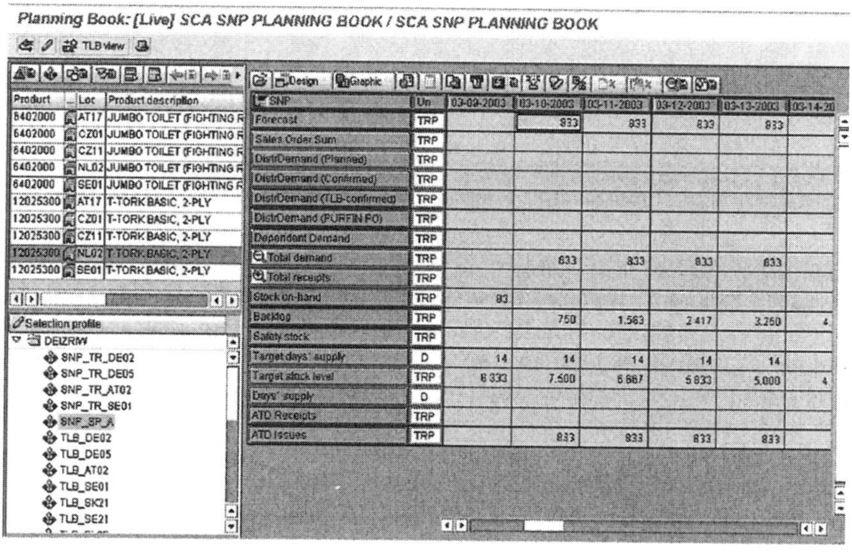

Figure 8-9. Medium term planning user interface

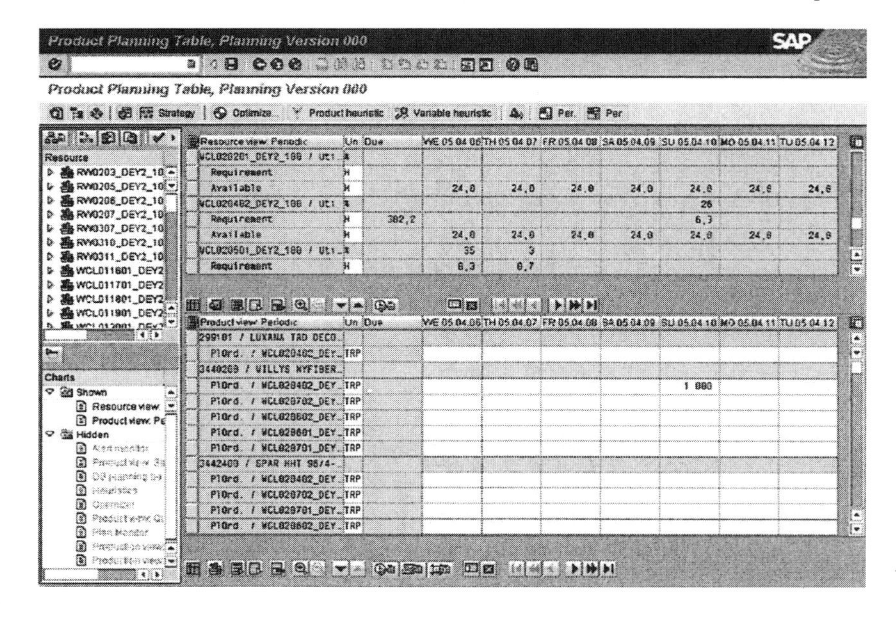

Figure 8-10. Production planning user interface

In the upper left window the user can select one or several products. The lower left window contains predefined selections, with every selection containing a set of products. The right window contains different key figures (like forecast, sales orders, distribution demand, stock-on-hand, backlog, etc.), which provides the user with the necessary planning information. The columns display the different time periods (in the example the time period is in days). In the example above we have a forecast demand of 833 for March 10, 2003. As it is the only demand, the total demand quantity on that day equals 833. We get a backlog of 750 on the same day, as we have only 83 units on stock to cover the forecast demand of 833. The value 14 for target days of supply is used to calculate the target stock level. In this case the demand of the next 14 days is written in the key figure target stock level.

In a second step production planning creates planned orders for all production levels based on the distribution demands (factory demand) for the finished goods. The creation of the production orders is done without considering the actual capacity load from the resources, i.e. a capacity overload is possible. No planned orders are created or deleted inside a fixed planning horizon, which in average is about 2-3 days. The period of the production planning run is identical to the medium term planning period (16 weeks). Due to long production times (sometimes several days), the use of continuous production planning algorithms was necessary in order to model the material flow. These heuristics allow managers to plan with the quantities which are produced and available during the time period of the production. The production planner uses a so-called production planning table for his tasks. Similar to the network planning user interface, we have still a periodic display (even if the orders are scheduled continuously) and a focus on products.

The production planning table can be customized to the needs of every planner. In the screenshot above we see in the upper left window the different resources, while the lower left window allows the selection of different charts. In the example above a periodic resource view (top) and a periodic product overview chart (bottom) have been selected. The resource view gives an overview on the available and required resource load. The product view loads automatically all products, which are produced on the selected resource and gives in the example above a weekly overview about the production quantities. In the example above we have available resource capacity on resource WCL020402_DEY2_100 of 24h on the 10th. April 2005. The actual production plan consumes 6.3 hours, which leads to a utilization rate of 26%. The 6.3 hour resource utilization is due to the planned order for product 3440200 on the same day (see lower part of the screen).

In the subsequent short term production scheduling process, the planner tries to create the schedule for the next two weeks. This step is supported via simple scheduling rules, which sequence the operations on the resources based on certain criteria. Two criteria are shown here as examples:

- In the material master a numerical key is maintained. The scheduler sorts the operations based on this key in ascending order. The keys are assigned in such a way that they minimize the set-up times.
- For every product so-called “cover weeks” are calculated. These cover weeks specify for every product, for how many weeks the existing stock covers the actual factory demand. The scheduler sorts then the operations based on the cover weeks in ascending order.

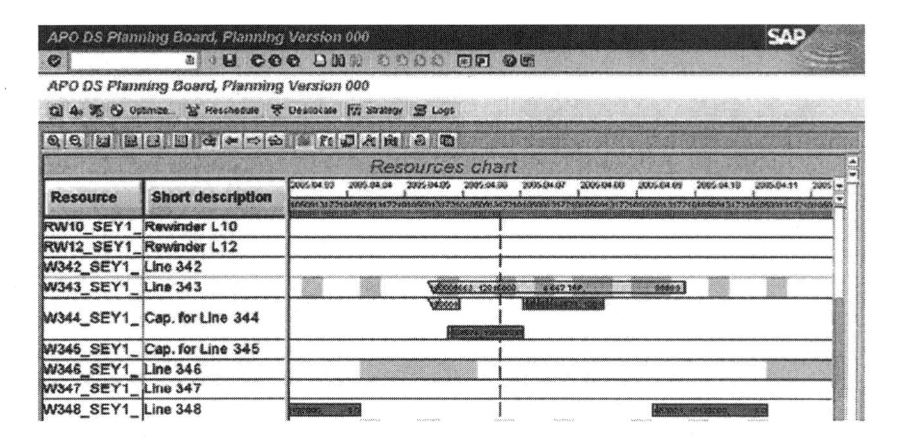

Figure 8-11. Scheduling user interface

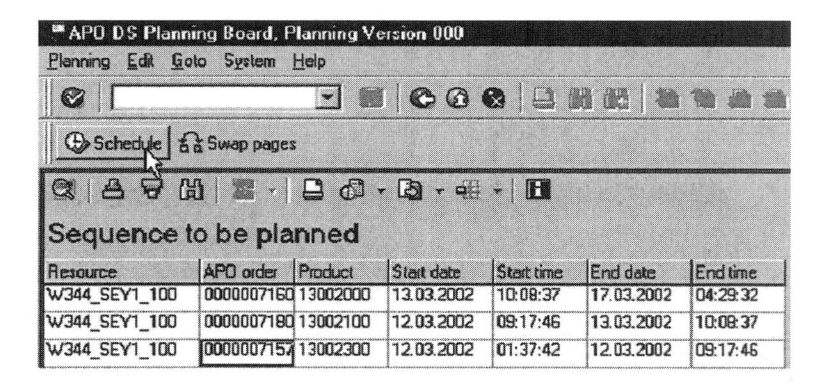

Figure 8-12. Manual sequencing of the operations

A simple manual sequencing in the Gantt-Chart (see Figure 8-11) – either using drag & drop or using a list (see Figure 8-12), is also part of the daily tasks of the production scheduling planner. In the Gantt chart we can see different operations among others on resource W344_SEY1. In Figure 8-12 we see a list display.

An important planning task at SCA Hygiene Products is the component planning. For several suppliers a just-in-time delivery was agreed upon. Collaboration scenarios provide the suppliers with the necessary information. This includes that the suppliers need to see the component demand on a continuous basis, i.e. if an order takes for example 4 days, it should display for every day the corresponding component demand (and not the complete demand on the first day). This was solved again with continuous production planning heuristics.

Due to the fact that SCA used an unconstrained network planning approach, it is possible that the calculated quantities and dates of the distribution orders from supply network planning can not be realized due to capacity overloads. In order to calculate a confirmed short term distribution plan, a so-called deployment step takes place for the next 3-4 days. Deployment runs every night after the supply network planning run. During deployment, the existing stock figures and incoming confirmed distribution orders are considered for every location, and a feasible confirmed distribution plan is calculated for the next days. Outside of the deployment horizon, all distribution orders are based on the ideal situation, that all demands can be produced on time.

In a final step, a simple transport planning step is executed every day directly after deployment and combines for all transportation lanes the existing confirmed distribution orders into full truck loads.

When the supply chain planners start their daily planning tasks, they check first the proposed transport orders for the day, make adjustments if necessary and send the orders to the connected R/3 ERP system for execution. Then they look into the alert situation for the short term distribution period and check if there are possible shortfalls. Outside the deployment horizon no shortfall alerts occur at the DCs, as an unconstrained supply network planning algorithm is used. At the factory level it is possible to compare the resulting factory demand from supply network planning with the planned production for the next weeks. If the factory demand is higher than the planned production, shortfalls in the future will occur if no further action is taken.

7. SUMMARY

We conclude this chapter with a summary of the main points:

There is no optimal generic way to integrate medium and short term planning, as each possible approach has advantages and disadvantages.

The common recommendation to plan with aggregated master data in medium term planning and with detailed master data in short term planning can cause several problems and should be evaluated carefully before following that approach.

One frequently mentioned advantage of APS systems relative to MRP II systems is that an APS system considers simultaneously material availabilities and capacities. This is mainly valid for medium term planning. In short term planning, simultaneous material availability and capacity planning (i.e. a finite scheduling during the production planning run) can lead to sub-optimal schedules, which may cause additional planning efforts due to shattered capacity loads, loser products and unexpected scheduling results.

REFERENCES

Adams, J., Balas, E., and Zawack, D., 1988, "The Shifting Bottleneck Procedure for Job Shop Scheduling," *Management Science*, **34**:391-401.

Axsäter, S., and Jönsson, H., 1984, "Aggregation and disaggregation in hierarchical production planning," *European Journal of Operations Research*, **17**:338-350.

Blazewicz, J., Ecker, K., Schmidt, G., and Weglarz, J., 1993, *Scheduling in Computer and Manufacturing Systems*, Springer Verlag, Berlin.

Brucker, P., 2002, "Scheduling and Constraint Propagation," *Discrete Applied Mathematics*, **123**(1-3):227-256.

Brucker, P., 2004, *Scheduling Algorithms*, Fourth Edition, Springer Verlag, Berlin.

Chopra, S., and Meindl, P., 2003, *Supply Chain Management: Strategy, Planning and Operation*, Second Edition, Prentice-Hall, Upper Saddle River, New Jersey.

Della Croce, F., Tadei, R., Volta, G., 1995, "A Genetic Algorithm For The Job Shop Problem," *Computers and Operational Research*, **22**(1):15-24.

Dickersbach, J., 2003, *Supply Chain Management with APO*, Springer Verlag, Berlin.

Garey, M.R., and Johnson, D.S., 1979, *Computers and Intractibility – A Guide to the Theory of NP-Completeness*, W.H. Freeman and Company, San Francisco.

Johnson, S.M., 1954, "Optimal Two and Three-Stage Production Schedules with Setup Times Included," *Naval Research Logistics Quarterly*, **1**:61-67.

Kreipl, S., and Pinedo, M., 2004, "Planning and Scheduling Issues in Supply Chains - An Overview of Issues in Practice," *Production and Operations Management*, **13**:77-92.

Miller, T.C., 2002, *Hierarchical Operations and Supply Chain Planning*, Springer Verlag, London.

Okuda, K., 2001, "Hierarchical structure in manufacturing systems – a literature survey," *International Journal of Manufacturing Technology and Management*, **3**(3):210-224.

Panwalker, S.S., and Iskander, W., 1977, "A Survey of Scheduling Rules," *Operations Research*, **25**:45-61.

Pinedo, M., 2002, *Scheduling – Theory, Algorithms, and Systems*, Second Edition, Prentice-Hall, Upper Saddle River, New Jersey.

Pinedo, M., 2005, *Planning and Scheduling in Manufacturing and Service Industries*, Springer Verlag, New York.

Seifert, D., 2002, *Collaborative Planning, Forecasting and Replenishment - How to create a Supply Chain Advantage*, Galileo Press, Bonn.

Söhner, V., and Schneeweiss, C., 1995, "Hierarchically integrated lot size optimization," *European Journal of Operations Research*, **86**:73-90.

Stadler, H., and Kilger, C., 2002, *Supply Chain Management and Advanced Planning – Concepts, Models, Software, and Case Studies*, Second Edition, Springer, Berlin.

Zäpfel, G., and Missbauer, H., 1993, "New concepts for production planning and control," European Journal of Operations Research, **67**:297-320.

Chapter 9

SEMICONDUCTOR MANUFACTURING SCHEDULING AND DISPATCHING

State of the Art and Survey of Needs

Michele E. Pfund, Scott J. Mason, John W. Fowler
Arizona State University, University of Arkansas, Arizona State University

Abstract: This chapter discusses scheduling and dispatching in one of the most complex manufacturing environments – wafer fabrication facilities. These facilities represent the most costly and time-consuming portion of the semiconductor manufacturing process. After a brief introduction to wafer fabrication operations, the results of a survey of semiconductor manufacturers that focused on the current state of the practice and future needs are presented. Then the chapter presents a review of some recent dispatching approaches and finally an overview of a recent deterministic scheduling approach is provided.

Key words: Scheduling, dispatching, semiconductor manufacturing

1. INTRODUCTION

In recent years, both the number of applications and market demand for integrated circuits have increased dramatically. Microprocessors, memory chips, microcontrollers, and other semiconductor devices have become part of everyone's lives: from fuel injection systems in modern automobiles, to personal computers, to cellular phones, to the projection system inside of television sets. This increased demand has in turn caused microelectronics factories (wafer fabrication facilities or "wafer fabs") to increase their efforts to provide high-quality, on-time deliveries of affordable products to their customers. Today's wafer fabs have been forced to become increasingly conscious of their due date performance, as dissatisfied customers now have a number of other manufacturers to turn to, should they need to find another supplier.

A number of different types of factories exist today, some of which are commodity based in that they produce a standard suite of products for general marketplace consumption. However, other factories produce application-specific integrated circuits (ASICs) for a wide array of customers. While a commodity wafer fab typically produces large quantities of a few different product types ("high-volume manufacturing"), ASIC fabs usually are tasked to produce lesser volumes of each customer's different product portfolio ("ASIC manufacturing"). Regardless of the type of wafer fab, microelectronics manufacturers strive to schedule the various orders (jobs) in their factory in such a way as to maximize on-time delivery to their customers. Companies that meet or exceed their customers' due date expectations generally have a better chance of retaining customers and receiving subsequent orders due to their previous performance. Obviously, some customer orders are more important than others.

According to the International Technology Roadmap for Semiconductors (ITRS, 2003), the cost of equipment is over 75% of factory capital costs. The ITRS indicates that, in order to utilize this equipment effectively, significant improvements in factory planning and scheduling are required. In addition to the cost pressures, today's highly competitive semiconductor markets place a greater emphasis on responsiveness to customers. In the past, competition has been primarily in the product design arena, but in the last several years high on-time delivery performance has become equally important for competitive success. Good delivery performance consists of order lead times that are both short and reliable. This can be achieved through either good production scheduling or using inventory to buffer customers against lengthy manufacturing delays. For a variety of reasons including holding costs and potential obsolescence, the latter option is becoming less attractive to semiconductor manufacturers. Thus, a recent thrust of manufacturing management has been on using effective scheduling techniques as a vehicle to achieve a competitive advantage.

Scheduling semiconductor manufacturing facilities is a very difficult problem and is among the most complex scheduling problems encountered today. Uzsoy, Lee, and Martin-Vega (1992) provide an excellent description of the semiconductor manufacturing process, placing scheduling in the context of production planning and fab performance evaluation. Wafer fab scheduling is a challenge that has yet to be tackled in a completely satisfactory manner. There are six main features that complicate scheduling these systems: large number of processing steps, re-entrant flows, batch tools, random equipment failures, sequence-dependent tool setups, and the fact that some processing steps require auxiliary resources (e.g. reticles).

In a typical wafer fab, there often are dozens of process flows. Each process flow contains 300-500 processing steps and more than one hundred machines. These machines are expensive, ranging in price from a couple of hundred thousand dollars to over fourteen million dollars per tool. The economic necessity to reduce capital spending dictates that such expensive machines be shared by all lots requiring the particular processing operation provided by the machine, even though they may be at different stages of their manufacturing cycle. This results in a manufacturing environment that is different in several ways from both traditional flow shops as well as job shops. The main consequence of the re-entrant flow nature is that wafers at different stages in their manufacturing cycle have to compete with each other for the same machines. The manner in which this competition is resolved has a clear impact on plant performance measures.

Furthermore, the nature and duration of the various operations in a semiconductor flow differ significantly. Some operations require 15 minutes or less to process a lot, while others may require over twelve hours. Many of these long operations involve batch processes. In reality, it is not uncommon for one third of the fab operations to be batch operations. Batch machines tend to off-load multiple lots (1 to 6) onto tools that are capable of processing only one lot at a time. This leads to the formation of long queues in front of these serial tools and ultimately a non-linear flow of products in the factory. The probabilistic occurrence of long tool failures results in large variability in the time a job spends in process. High variability in cycle times prevents accurate prediction of production cycle times, resulting in longer lead-time commitments. There are some machines, such as implanters, that require significant sequence-dependent setups. If not scheduled well, these tools can become bottlenecks. Finally, some processing steps require an auxiliary resource, such as a reticle in photolithography, in order to process the job. Some of these auxiliary resources are quite expensive, so only a very limited number of them are purchased. Therefore, the challenge is to ensure that the machine and the auxiliary resource are available at the same time. All of these factors combine to make scheduling wafer fabs quite challenging.

There is a great need to schedule semiconductor wafer fabrication facilities well. The next section discusses efforts to better understand the scheduling needs of semiconductor manufacturers.

2. SEMICONDUCTOR FAB SCHEDULING: SURVEY OF PRACTICES AND FUTURE NEEDS

In 2000, the Semiconductor Research Corporation (SRC) and International SEMATECH combined forces to jointly fund a three-year research effort called the Factory Operations Research Center (FORCe). The authors of this chapter were investigators on a project team funded in this program that focused on developing a scheduling and rescheduling methodology that could be used in a semiconductor wafer fabrication facility. The goal of this project was to investigate an approach capable of handling various real world fab situations such as equipment breakdowns, raw material unavailability, and processing time variability while providing quality solutions with reasonable solution times. More information on this project can be found in Fowler *et al.* (2002).

In order to understand the tools that are currently being utilized in the semiconductor wafer fabrication facility, a survey instrument was created and sent to each of the FORCe member companies. The survey was designed to ask specific questions regarding the types of scheduling methodologies currently implemented, the limitations of these methodologies, and the needs for future generation scheduling systems. In total, 16 respondents from 14 companies participated in this survey, representing fabs from Europe, Asia, and North America.

2.1 Survey questions and results

In order to ensure a common understanding of the terms used in the survey, the following definitions related to this survey were provided:

Planning: The development of detailed capacity and material plans that assess the fab's capability to meet market demands. Decisions include determining product mix, new equipment purchases, staffing levels, etc.

Order Release: The determination of when to release lots to the manufacturing floor.

Scheduling: The creation of a detailed plan that determines the order of how lots will be processed as they move through the fab. It is typically performed once per shift or once per day to determine the schedule for the whole time interval. This step is optional.

Rescheduling: The re-evaluation of a scheduling rule decision within the original scheduling time period. This is typically done either at fixed intervals or when a schedule deviates from its original plan.

Dispatching: The immediate assignment of a specific resource to one of several possible lots. It answers the question: which lot should be processed on this machine now? If scheduling has been performed, the goal of dispatching is to choose the lots that best meet the schedule. If scheduling has not been performed, dispatching rules (such as FIFO, critical ratio) are chosen that have been shown to work well for a given factory measure(s).

The specific questions asked in the survey and the responses or a summary of the responses are given below.

1. **Do you use scheduling / dispatching software within your fab? If so, what software do you use?**

95% of the respondents currently utilize a scheduling / dispatching software in the fab, and the 5% not utilizing scheduling / dispatching software are currently planning new installations. Of the respondents, 56% replied that they were using some form of Brooks software to accomplish dispatching. The remaining responses were almost uniformly distributed among Adexa, Workstream, and homegrown software solutions. However, it was apparent that the off the shelf versions of these software solutions were not completely sufficient for many manufacturer's needs as roughly 38% had installed custom add-on solutions to their standard off-the-shelf software packages.

2. **Do you know what type of scheduling / dispatching rules are used in your fab (i.e. FIFO, critical ratio, etc.)?**

Several survey responses indicated a single rule; these included FIFO, critical ratio, WIP balancing, and least slack. Several other responses indicated that multiple rules are used with different rules at different workstations. In some cases, there were indications that setup avoidance, starvation avoidance, priority for hot lots, and batching rules are used in conjunction with the dispatching rules.

3. **How often is a new schedule generated?**

The distribution of these responses is trimodal as shown in Figure 9-1: with one group that is real time (or almost real time), a group that schedules daily, and a group that schedules (plans) on a weekly basis.

4. **If you could schedule / reschedule very quickly, how often would you do this?**

The responses from this question were somewhat split, with slightly over 35% desiring rescheduling with every job movement. However, 65% of the respondents desired longer rescheduling intervals within a shift. For those desiring rescheduling to be performed at intervals within a shift (other than every job movement) a majority of the respondents desired a rescheduling interval of every 8 hours.

5. **Do you foresee any problems with rescheduling too often?**

Question #5 provided significant insight for challenges that may be faced by scheduling too often. While one respondent did not foresee any problems, others felt that rescheduling too often would create challenges for operators (setup, stability, confusion), would require significant hardware, would be computationally expensive, and would potentially lead to system stability issues.

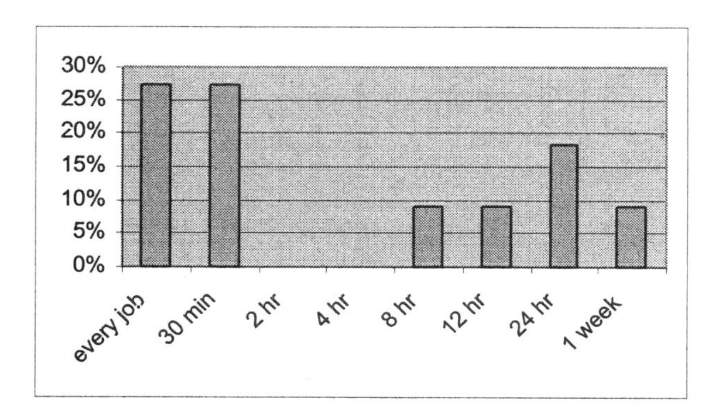

Figure 9-1. Distribution of schedule generation times

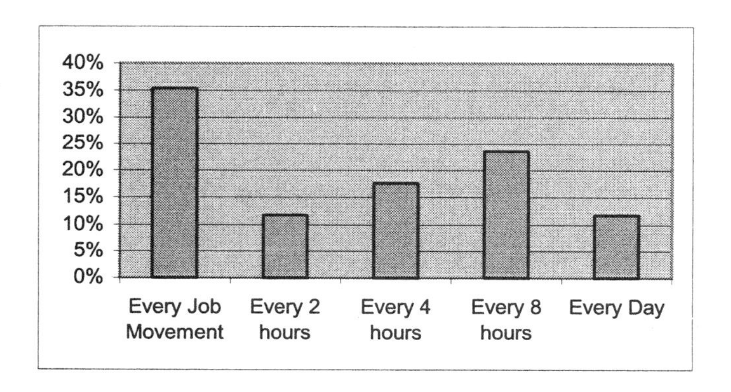

Figure 9-2. Desired rescheduling frequency

6. Listed below are some common fab performance measures. Please rank each of the following in terms of relative importance in your fab.

Of these performance measures, it appears that factory throughput, on-time delivery, and cycle time are consistently highly important fab performance measures. The lowest importance fab performance measures appear to be labor utilization and wafer starts.

We also received write-in performance measures that were listed as either the highest importance or second highest importance. Each of these had a single response (with exception of lot turn which had two). The list includes: Cost per wafer, Gross Margin, Coefficient of Variability, X-Factor, Line Yield, Static Lots, Hold Rate, Lot Turns × Die Yield.

Table 9-1. Ranking of fab performance measures

Performance Measure	Low		Moderate		High
Cycle Time	0%	7%	7%	20%	67%
Equipment throughput	0%	6%	13%	44%	38%
Factory throughput	0%	6%	6%	6%	81%
Inventory levels	0%	0%	31%	38%	31%
Labor utilization	31%	6%	31%	19%	13%
On-time delivery	6%	6%	6%	19%	63%
OEE	6%	19%	38%	13%	25%
Wafer starts	13%	7%	27%	13%	40%

7. Of these performance measures, which are the most three important performance measures in your fab?

Cycle time (15), factory throughout (12), and on-time delivery (12) were clearly the top three performance metrics. Three fabs use inventory turns as one of there top three measures and no other measures were mentioned more than once.

8. How would you rate your scheduling / dispatching system in terms of your fab's performance measures?

A majority (56%) of the respondents rated their scheduling / dispatching system as satisfactory. 19% rated their scheduling systems as poor or unsatisfactory, whereas 26% rated their scheduling system good or outstanding.

9. Please describe your reasoning for the rating in question 8.

The respondents that rated that their system as poor or unsatisfactory indicated that the system does not support operators sufficiently and that the bottlenecks in the line still starve occasionally. Most of the respondents that said the system is satisfactory indicated that the system has shown to improve factory operations. However, most of these respondents also indicated that there is room for improvement in their system. Finally, the respondents that rated their system as good or outstanding indicated that factory performance has significantly improved since the system was installed.

10. Is scheduling / dispatching perceived as beneficial to the factory? Why or why not?

All respondents indicated that scheduling / dispatching is beneficial to the fab either because it improves factory performance measures or because it controls how product moves through the floor. Performance improvements mentioned included: increased output, improved line/die yield, reduced cycle time mean, reduced cycle time variability, improved on-time delivery, maximizing utilization of the constraint tools, and improved line balancing.

11. What is the biggest challenge that your factory faces with scheduling/ dispatching?

While a majority of challenges were found to be of a technical nature (capturing real time data, choosing rules, changes in product mix, hot lots, etc.) a significant number of respondents identified management related issues as well (philosophy, system, operator compliance). The lack of capability to perform offline simulation of rules to understand the factory performance prior to actual implementation of the rules / policies in the factory was also mentioned as a major challenge.

12. What features of your scheduling / dispatching system do you use the most?

Respondents indicated that the most utilized features of the system are lot priority adjustment, reporting of fab conditions, and rule customization.

13. What features would you add to your software?

When asked what features they would like to add to the software, many respondents indicated that they would like better rules, linkage to their fab's simulation model, changes to the graphical interface of the software, and better reports.

14. Listed below are some events that could affect the quality of a schedule, please indicate the frequency of each event.

Table 9-2 describes the frequency of events.

15. For each previously identified event, please indicate how you think each event would impact the performance of a previously determined schedule.

Table 9-3 describes the perceived impact of events.

16. From the events listed in questions 14 and 15, which are the top three events in terms of frequency?

Of the events identified, respondents identified new job arrival, job goes on hold, and bottleneck machine breaks down as the top three events. Other events identified were personnel delay and planned schedule not being followed by personnel.

17. From the events listed in questions 14 and 15, which are the top three events in terms of impact?

Table 9-2. Frequency of events that impact schedule

Event	Every 15 min	Every hour	Every 4 hours	Every shift	Every day	Less often
Cycle time exceed estimates	14%	7%	0%	29%	21%	29%
Job cancelled (dropped / destroyed)	13%	0%	7%	13%	13%	53%
Job changes priority (i.e. becomes hot!)	13%	0%	13%	20%	27%	27%
Job goes on hold	27%	27%	7%	20%	20%	0%
Bottleneck machine breaks down	13%	13%	7%	13%	20%	33%
Non-bottleneck machine breaks down	13%	7%	13%	13%	20%	33%
New job arrival	53%	20%	7%	7%	0%	13%
Num. jobs in queue exceeds threshold	13%	7%	33%	40%	0%	7%
Over/under estimated processing time	7%	13%	13%	20%	7%	40%
Personnel delay	13%	13%	20%	27%	20%	7%
Schedule not followed by personnel	13%	7%	20%	27%	7%	27%
Preventative maintenance	7%	0%	20%	33%	40%	0%
Quality-related problems	7%	7%	7%	7%	40%	33%
Raw material delay	7%	7%	7%	7%	13%	60%
Transportation system delay	14%	21%	14%	21%	0%	29%

Of the top five events identified in question #16, two of them (bottleneck machine breaks down and job goes on hold) were both identified as being one of the top three events in terms of impact. Other factors identified were quality related problems, job changes priority, and job cancelled.

18. What type of line is your factory (production, R&D, pilot line)?

Slightly over half of the respondents identified their line as a production line. 20% identified their line as production and R&D, 14% identified their line as production / R&D/ pilot line, and one respondent was from an R&D line.

Table 9-3. Impact of events upon schedule

Event	No impact		Moderate impact		High impact
Cycle time exceed estimates	6%	13%	19%	38%	25%
Job cancelled (dropped / destroyed)	0%	19%	31%	19%	31%
Job changes priority (i.e. becomes hot!)	0%	0%	13%	44%	44%
Job goes on hold	0%	13%	25%	25%	38%
Bottleneck machine breaks down	0%	0%	6%	0%	94%
Non-bottleneck machine breaks down	0%	25%	50%	19%	6%
New job arrival	6%	19%	56%	6%	13%
Num. jobs in queue exceeds threshold	0%	25%	38%	13%	25%
Over/under estimated processing time	0%	13%	44%	31%	13%
Personnel delay	6%	13%	44%	31%	6%
Schedule not followed by personnel	0%	0%	19%	38%	44%
Preventative maintenance	13%	19%	38%	25%	6%
Quality-related problems	19%	6%	13%	13%	50%
Raw material delay	25%	13%	13%	25%	25%
Transportation system delay	7%	13%	40%	20%	20%

19. How long has your fab been operating?

Most responses (13) were from fabs that had been operating 5 years or longer with 3 operating between 2-5 years.

20. What products are manufactured in your fab (check all that apply)?

The data indicate that roughly 70% were ASIC's and Logic Manufacturers, and roughly 40% manufactured memory products. Obviously, both memory and logic are run together in some fabs.

21. Would you consider your fab to be....?

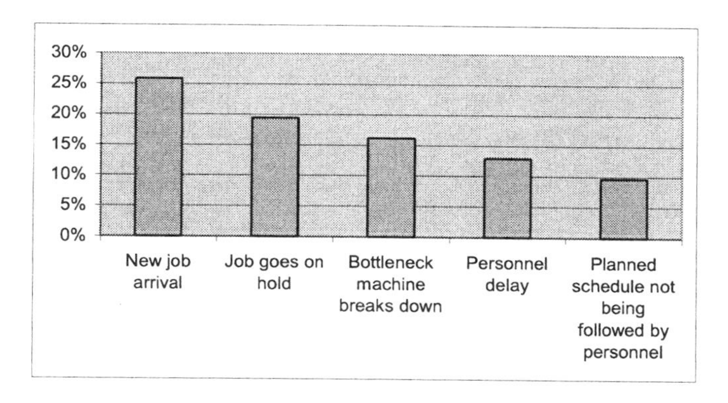

Figure 9-3. Top 3 events: frequency

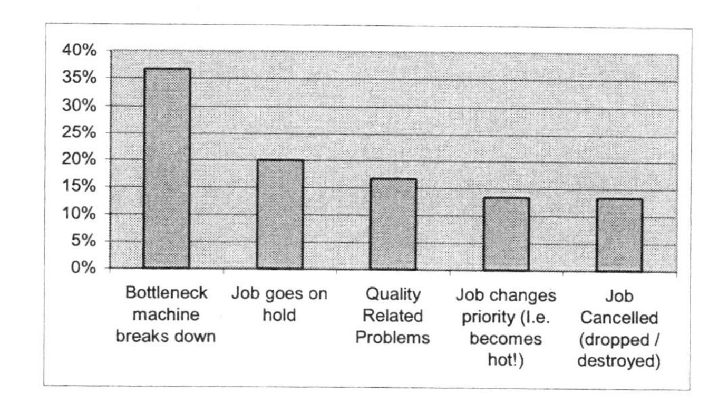

Figure 9-4. Top 3 events: impact

From the demographics, a majority of the respondents were from high volume plants with high product mix (10), followed by low volume with high mix (4), and high volume with low mix (2).

22. Which MES system is your company using? (This information is being collected to help us assess what types of data are available for scheduling / dispatching)

A variety of MES systems are being utilized. Workstream is the dominant MES system in our survey responses with 6, Promis is second with 4, and SiView, FactoryWorks, and an in-house system each with 2. We note that SiView was identified as being implemented in newer fabs.

23. How long has your company been using software supported scheduling / dispatching rules in the fab?

A majority of the respondents had mature scheduling systems in place for 5-10 years. In fact, over 90% of respondents had scheduling systems in place for 5 years or more.

24. What position do you hold in your company? How long have you held this position?

Survey participants were from areas relevant to planning, scheduling, and dispatching. Most respondents had been involved with their current position for a period of 1-3 years. Titles included: Fab Modeler, Industrial Engineer, Line Control Manager, Manager of Automation, Manufacturing Control Manager, Factory Integration and Modeling group, Supply Chain Manager, and Simulation Manager.

2.2 Survey conclusions

From the survey responses, we found that many dispatching systems are in place and are mature installations (install time greater than 5 years). These systems have been considered to be "satisfactory" in that benefits are being received, but the majority believes that more benefits are possible. Specifically, respondents indicated that better scheduling / dispatching rules, test environments, and reporting tools are needed.

Compared to the SEMATECH Measurement and Improvement of Manufacturing Capacity (MIMAC) Survey and Interview Results (Fowler and Robinson, 1995), cycle time and on-time delivery have gained significant importance in the fab. Respondents indicated that these performance metrics are most impacted by a bottleneck machine breakdown and jobs going on hold, which were also the two most frequently occurring events as indicated by respondents. Thus, scheduling/rescheduling methodologies which incorporate these events are needed.

With regard to the frequency of rescheduling, respondents had mixed opinions. While most respondents favored rescheduling either at every shift or within a shift, many cited management challenges (operator stability/morale, staging for setup) as well as technical challenges such as hardware support. However, over 35% of respondents would like to reschedule after every job movement.

A majority of the respondents were in fabs that were producing ASIC and Logic products in a production line with high volume / high product mix. Respondents from production / R&D and production / R&D pilot lines also participated in the survey as well as those from low volume / high product mix and high volume / low product mix.

3. STATE-OF-THE-ART SEMICONDUCTOR MANUFACTURING DISPATCHING

A large number of research papers have been published which describe theoretical, potentially applicable dispatching strategies for wafer fabrication facilities. Papers have focused on a wide variety of levels in the manufacturing environment hierarchy, from individual tools (e.g., a photolithography stepper), to toolgroups containing multiple, identical machines operating in parallel (e.g., a collection of medium current implanters), to entire manufacturing facilities (e.g., the front end wafer fab).

In a typical dispatching research project, researchers first develop a new dispatching methodology or rule in order to minimize or maximize some desired performance objective, such as average lot cycle time (i.e., the total time required to complete a lot's required processing from start to finish), tool or fab throughput (i.e., the rate in which lots are processed through a tool/toolgroup of interest or exit the fab), or work in process (WIP) (i.e., all unfinished lots current in the fab). Then, they test their rule's ability to achieve the stated performance objective by comparing the new dispatching rule to existing rules (e.g., first in, first out (FIFO), earliest due date (EDD), and so on). These tests are typically conducted using randomly generated data sets that are meant to imitate real world manufacturing conditions.

While a number of semiconductor datasets exist in the public domain [MIMAC Testbed Datasets, http://www.eas.asu.edu/~masmlab], these datasets have undergone some amount of desensitizing so that company-proprietary information is not disclosed to the general public. For example, while values for a tool's mean time to failure and its mean time to repair are widely available, additional information pertaining to the underlying probability distribution associated with each tool parameter is not furnished. Further, information pertaining to the inherent variability in these parameters' underlying distributions is not disclosed. Therefore, although an author's experimental results may suggest that a proposed dispatching rule is a superior approach, its true superiority strongly depends on two things: 1) the viability and practicality of the data sets that were used during experimental testing, and 2) the quality of the competition that the proposed dispatching approach was tested against.

In the best scenario, a proposed dispatching method's superiority is established through experimental testing using actual semiconductor manufacturing data. In this scenario, actual fab data often is extracted from the manufacturing execution system (MES) for use in developing and testing dispatching approaches. In fact, some semiconductor manufacturers use dispatching systems that communicate directly with their Manufacturing Execution System (MES) in near real time. Examples of these products are one from FabTime [http://www.fabtime.com] and Brooks Automation's Real Time Dispatch product [http://www.brooks.com]. Once the dispatching rule's efficacy is confirmed, the final step in the process is for the dispatching rule to be implemented in an actual wafer fab. Only through actual implementation and potential further refinements can a proposed dispatching rule actually impact a semiconductor manufacturer's desired performance objective(s).

Mitel Semiconductor and Advanced Micro Devices are two representative semiconductor manufacturers that chose to install real-time dispatching systems that integrate with their fab's MES. Pickerill (2000) describes the deployment of an online dispatching system at Mitel's Plymouth, United Kingdom facility which was implemented with the hope of improving visibility into line balance and schedule adherence, and ultimately, improving cycle time and on-time delivery of customer orders. In a two-phase approach, Mitel first implemented a just-in-time/kanban system that incorporated both the fab's MES and its dispatching system. Upon successful completion of this first phase, Mitel then was able to develop and implement a WIP control system using MES and dispatching rules which contained various priority or "service" levels for lots in order to help meet the company's business objectives and ensured bottleneck equipment was not starved of WIP for processing. Similarly, Appleton-Day and Shao (1997) describe the implementation of a real-time dispatching system at AMD's Fab 25 facility in Austin, Texas, providing practical examples of how the system provided tangible, measurable improvements in fab performance.

In the sections that follow, we first examine two basic priority-based dispatching rules whose use is quite common at a number of semiconductor manufacturing facilities. Next, some references to toolgroup-specific dispatching studies are provided. These types of dispatching analyses are typically performed on fab bottleneck tools, such as photolithography steppers. Finally, a detailed review of three companies' dispatching rule development and implementation efforts is presented. We highlight the fab performance objectives of interest, the dispatching strategies that were employed, and then discuss the success stories associated with each company's dispatching implementation.

3.1 Simple priority-based dispatching rules

Most semiconductor manufacturers use some sort of priority-based dispatching rule for the majority of their toolset. Examples of priority-based rules include Priority FIFO and Priority Critical Ratio. These simple rules can be employed both at an individual tool level, as well as on a global level across the entire wafer fab. Priority FIFO dictates that lots are to be processed in terms of decreasing priority or importance (e.g., hot lots should precede engineering lots, which in turn should precede production lots). However, when multiple lots of the same priority are in queue, ties are broken by selecting the lot that has been in queue at the tool of interest the longest. Ties in lot priority occur frequently in practice, as it is not uncommon for 75-85% of the lots in queue at a tool to be production lots.

Priority Critical Ratio is a widely used dispatching approach which first dispatches lots in terms of their priorities (like all priority-based dispatching approaches). However, in the event of priority ties, this rule compares lots in terms of their critical ratios, a measure of lot slackness with respect to its due date. Consider lot j and its associated due date d_j. As lot j is currently waiting to be processed at some step/operation of its associated process flow, the total amount of remaining processing time which lot j must undergo prior to finishing its process flow can be easily computed using either theoretical process time information, planned cycle time information, and/or actual cycle time information. If we denote the total remaining processing time of lot j as P_j, then the critical ratio of lot j at time t for the case when $d_j > t$ can be computed as

$$CR(j,t) = \frac{\textit{Lot Due Date - Current Date}}{\textit{Remaining Processing Time}} = \frac{d_j - t}{P_j} \tag{1}$$

A negative value of $CR(j,t)$ indicates a lot is already late, as its due date has passed. Further, a $CR(j,t)$ value between zero and one suggests that lot j will most likely be late, depending on which values of lot processing time were used in the P_j calculations (i.e., theoretical or actual). Finally, an on-time (ahead of schedule) lot will have $CR(j,t) = 1$ or $CR(j,t) > 1$. The lot with the minimum $CR(j,t)$ value is selected for processing at every time t when a machine becomes available to process a subsequent lot.

In addition to the two aforementioned priority-based rules, it is clear that priority-based versions of other common dispatching rules, such as EDD, shortest processing time (SPT), and so on, could easily be deployed in a semiconductor manufacturing environment. In fact, these simple rules (and their associated priority-based variants) are included in the standard suite of dispatching rule choices for discrete event simulation users. However, Priority FIFO is commonly used in manufacturing environments whose desired performance objective is cycle time minimization, whereas Priority Critical Ratio is more common in factories that are focused on on-time delivery of customer orders. While these two simple rules have proven to be reasonably effective in practice, they may not be the most effective method for dispatching lots throughout an entire wafer fab for two reasons.

First, simple, priority-based rules often focus on a small subset of the desired performance objectives of semiconductor manufacturing management. Second, they do not accommodate the various tool-specific processing characteristics associated with semiconductor manufacturing:

- sequence-dependent setup times (e.g., due to species changes in ion implantation)
- batch processing (e.g., diffusion operations)
- auxiliary resource requirements (e.g., reticles required for lot processing on photolithography steppers).

3.2 More advanced dispatching strategies

As is the case in most manufacturing environments, no single "simple" dispatching rule exists that is capable of effectively handling all of the processing characteristics inherent in semiconductor manufacturing. While a simple rule such as Priority Critical Ratio may indeed be applicable for 70-80% of a wafer fab's toolset, the aforementioned processing complexities of batch-processing and sequence-dependent setup are not appropriately accounted for under this rule. This complication is further exacerbated by the presence of multiple (often competing) fab performance objectives.

A number of papers in the open literature discuss complex dispatching rules. However, only a select few of these papers actually detail results from implementing the proposed, complex rules within a semiconductor wafer fab. In some cases, a set of dispatching rules are individually applied in one or more specific tool and processing environments, while in other cases, a blended or combination dispatching rule approach is employed throughout the entire wafer fab. Regardless of the method used, complex dispatching rules all share one common trait: they are developed in order to drive fab performance towards management's stated performance objectives.

3.2.1 Toolgroup-specific Dispatching Rules

Yang *et al.* (1999) discuss dispatching strategies that were developed for the bottleneck tool in the thin films area of Taiwan Semiconductor Manufacturing Company's (TSMC's) Fab 3 in Hsin-chu, Taiwan. The methodology underlying TSMC's dynamic dispatching model for bottleneck resource allocation works to reduce lost machine productive time on tools that are downstream from the bottleneck and to increase total moves for these downstream machines. This is accomplished by considering machine capacity and throughput rates with each lot's remaining processing time and downstream target WIP levels. This dynamic dispatching rule was implemented on the first shift at TSMC in order to compare it with conventional dispatching practices on the second shift. As a result, the first shift's lost machine productive time was 38% lower than the second shift's lost time over three key downstream toolgroups. In addition, total moves increased by over 18% on first shift as a result of the dynamic dispatching rule implementation.

A dispatching approach that was custom-developed for pre-diffusion wet bench operations at a Macronix wafer fab in Hsin-chu, Taiwan is described by Hsieh *et al.* (2002). Wet bench operations at Macronix can accommodate two lots of wafers, while the subsequent diffusion processing step was performed on a batch processing machine that could hold three lots. Senior operators' experience was allowing them to significantly outperform their junior counterparts in terms of maximizing loading efficiency on the diffusion tools. In order to minimize the performance difference between senior and junior operators, Macronix developed a dispatching methodology that evaluates the impact of running a lot on each of the various different wet benches in the fab.

Each wet bench was assigned a priority value for each lot in terms of the amount of processing time that was required for the lot to complete the wet bench operation. In this regard, machines that required the least amount of time to process the lot were given higher priorities. Using this new dispatching approach, junior operators were better equipped to load lots onto more appropriate, faster machines, in order to properly feed the diffusion tools. While junior operators' performance improved 24%, senior operator performance also improved by a total of 4%. This improvement is significant, as diffusion ovens are tools that potentially could have become the Macronix fab's primary bottleneck if they were not loaded properly.

3.2.2 Full wafer fab dispatching rule case studies

This section presents three case studies pertaining to full, wafer fab-wide dispatching studies. They are meant to serve as examples of the two fundamental approaches for manufacturing environment-specific dispatching: 1) create and deploy individual tool-type specific dispatching strategies at each toolgroup which in concert, support the factory management's overarching performance objectives; and 2) identify dispatching rules which support one or more individual factory performance measures, and then blend these rules effectively to create a common, factory-wide dispatching "rule."

SGS-Thomson Fab 8 (Milan, Italy)

The fab management at SGS-Thomson's (now ST Microelectronics) Fab 8 in Milan, Italy's performance objectives include the reduction of mean cycle time and cycle time variability, achieving target fab outs within each planning period, and producing fab outs in a steady, consistent manner across each planning period. Cigolini *et al.* (1999) develop a tool classification-based dispatching strategy for Fab 8 to decrease fab cycle time without adversely affecting tool throughput rates and mean tool utilization levels.

Fab 8's tools are classified according to their wafer processing characteristics and setup time requirements into one of three categories: 1) sequential machines in which lots are loaded sequentially, but subsequent lot processing can only commence once the current lot has completed its processing; 2) batch-processing machines that can process multiple lots simultaneously, provided they share similar processing recipes (requirements); and 3) piggy back machines that operate in a sequential fashion, but can only process a subset of the wafers in a lot at one time.

Prior to the development of the authors' new approach, Fab 8 used a Priority FIFO dispatching scheme for sequential and piggy back machines that do not require significant machine setups and a Priority Setup Minimization approach for tools characterized by large setup times. The authors' new approach extends the fab's tool classification scheme even further by considering each tool's criticality in terms of capacity and flexibility. A tool is deemed capacity critical (non-critical) if its average utilization rate is above (below) some pre-determined threshold utilization level. Similarly, a tool that is (is not) characterized by sequence-dependent setups is termed flexibility critical (non-critical).

Cigolini *et al.* (1999) propose the Maximum Capacity Gain (MCG) dispatching rule for sequential and piggyback tools that are critical both in terms of capacity and flexibility. The MCG rule seeks to minimize tool setup time by sequencing similar lots together in queue. However, individual lot slack calculations ("slack per operation") are also taken into consideration in order to avoid the situation when the only lots selected for processing are of a single product type due to setup avoidance considerations. Tools that are only capacity critical are dispatched according to a shortest processing time (SPT) approach. However, in order to avoid the situation wherein a lot with a long processing time remains in queue for an exceedingly long time, the authors also consider lot slack per operation values, as was the case with the MCG rule in order to force a temporary break ("truncation") in the SPT sequence. Finally, flexibility critical sequential and piggyback tools are dispatched according to a minimum setup rule in concert with lot slack per operation, while totally non-critical tools are dispatching simply according to the minimum slack per operation rule.

Batch-processing tools are also assessed in terms of capacity criticality. However, as setup time is fixed on these tools, the concept of flexibility criticality is no longer of importance. Therefore, the authors present a maximum batch load size-based approach for dispatch rule selection. Whenever a batch-processing tool becomes available to accept another load of wafer lots, some number of lots are in queue of each recipe/process type. Consider the case when B lots of a given recipe type are in queue and available to be batched and processed on a batch-processing tool. The relationship of B to the batch-processing tool's minimum and maximum batch load sizes (B_{min} and B_{max}, respectively) determines whether or not a batch of lots is immediately formed and processed (e.g., when $B \geq B_{max}$), or if the batch-processing tool is held idle for some specified maximum amount of time prior to batching and processing the currently available B lots in the hopes of additional lot arrivals to the tool's queue during the wait time (which thereby would result in a fuller batch and increased batch-processing tool throughput).

Through a large set of simulation experiments, the authors compared the performance of their proposed capacity and flexibility criticality-based dispatching rule ("DEP") to current Fab 8 dispatching practices. Although DEP did increase the number of wafers in WIP by 0.32%, it was able to increase Fab 8's throughput rate by 1.79%. In addition, average lot cycle time was decreased by 0.66% under DEP. However, the authors do point out that DEP does seem to introduce a larger amount of variability in the wafer fab, primarily due to the use of the SPT rule for tools that are only capacity critical.

Motorola MOS 5 (Arizona, United States)

Dabbas and Fowler (2003) present a multi-objective dispatching strategy for front end wafer fabrication at Motorola. The performance measures of interest for the Motorola MOS 5 wafer fab in Mesa, Arizona under study were on-time delivery, the variance of lot lateness, mean lot cycle time, and the variance of lot cycle time. The authors combine three local dispatching policies (critical ratio (CR), throughput (TP), and flow control (FC)) with a fab-wide (global) line balancing (LB) algorithm to create a single, comprehensive dispatching rule. While their approach is applicable for combining any number of dispatching rules, these specific rules were selected due to the respective fab performance objectives that each one addresses individually. For example, LB focuses on minimizing the difference between target WIP levels at each fab operation and actual WIP levels, while FC is concerned with workload balancing. Although being quite effective for maintaining appropriate WIP levels, LB and FC do not promote decreased fab cycle times, increased factory throughput, and on-time product delivery. It is for this reason that the TP and CR dispatching rules are appropriate for pursuing the fab management's performance objectives.

First, each of the four dispatching approaches is individually transformed into a 0-1 scale. Then, the transformed dispatching approaches are aggregated using a linear combination of relative weights. As the efficacy of the comprehensive dispatching rule is directly related to the specification of these relative weights, the authors employ a statistically based desirability function approach in combination with response surface methods to determine appropriate relative weight values (Dabbas *et al.*, 2003). As a result of this aggregation, each lot can be thought of as receiving a single dispatching score that could then be compared with other lot scores to make a dispatching decision. However, the multi-objective dispatching strategy of Dabbas and Fowler (2003) differs from conventional, complex dispatching approaches because the lot with the highest combined dispatching score is not necessarily the lot that is selected for subsequent processing. Instead, this combined dispatching score is considered in concert with a proportional capacity allocation algorithm to establish the final lot priorities and rankings.

The authors allocate machine capacity proportionally to the fab's different products at the different stages of production as a function of their combined dispatching score. Proportional machine capacity allocation is used to balance the production output of a high mix, high volume fab such as Motorola MOS 5 and to counteract the nonlinear, re-entrant product flow present in front end wafer fabrication operations. Once the final lot priorities and dispatching scores have been determined, it is at this point that lots are selected for dispatching in non-increasing final dispatching score order. Ties between lots are broken by CR, although other tie breaking rules could be used.

Once the performance of multi-objective dispatching strategy is verified using a smaller but representative wafer fab model, the authors test their approach on a model full wafer fab model that represents Motorola's MOS 5 facility. Experimental results suggest that the proposed approach improves on-time delivery of customer orders by 22%, mean cycle time by 24%, and the variability of lateness by 53%. Given these promising results, the authors' combined dispatching methodology was implemented in Motorola MOS 5 in October 1998.

As real world wafer fabs frequently experience variable product mix and product starts conditions, key factory parameters must be re-evaluated frequently, such as product WIP targets at various production stages and the weights that are used in the combined dispatching criteria. At Motorola MOS 5, WIP goals are set weekly using simulation model analyses as dictated by the fab's output plan. Other types of fab events, such as adverse changes in fab cycle times and extended down time on key bottleneck tools may also necessitate the need for re-evaluating these factory parameters.

Similar to the results obtained during the experimental studies, Motorola MOS 5 experienced a 20% improvement in on-time delivery performance and a 25% reduction in mean cycle time performance after the combined dispatching rule approach has been implemented. In addition, line WIP balance improved without the need for weekly, manual line balancing decisions by fab supervisory personnel. These results suggest that multi-objective, combined dispatching approaches are capable of supporting multiple fab performance objectives simultaneously in a real world semiconductor manufacturing environment.

Samsung Electronics (Kiheung, South Korea)

With a 20% market share in the areas of static and dynamic random access memory chips, Samsung Electronics is the largest manufacturer of digital integrated circuits in the world in terms of unit volume. Although the company's wafer fabs regularly produced high yielding products via highly productive equipment and labor personnel, Samsung's product cycle times ranked last of 29 different wafer fabs surveyed in the University of California at Berkeley's Competitive Semiconductor Manufacturing study during the early 1990s (Leachman and Hodges, 1996). These excessive cycle times became extremely important to Samsung when the record setting memory chip prices of 1995 encouraged companies like Samsung to invest in additional manufacturing capacity in terms of additional tools and/or wafer fabs. However, the memory chip market downturn in late 1995/early 1996 dramatically reduced the price of memory chips, and in turn, caused the value of memory chip fabs' work in process (WIP, i.e., unfinished chips) to decrease. In an effort to avoid potential lost revenue due to rapidly decreasing sales prices, Samsung Electronics stopped using least slack-based dispatching strategies and implemented a set of methodologies and scheduling applications for managing product cycle time in semiconductor manufacturing known as SLIM (Leachman *et al.*, 2002).

Unlike typical lot-centric dispatching approaches, SLIM focuses on target fab outs for each device type. This "WIP-management paradigm" sets both production targets and device-level priorities at each fab process step. Some of the benefits associated with the "higher level" approach include 1) reduced machine setups/changeovers per shift, as a number of similar device-type lots are processed sequentially prior to a machine changeover; 2) a more consistent fab out schedule due to reducing the number of times an operator mistakenly processes WIP that is ahead of its scheduled due date, rather than a lot that is behind schedule; and 3) effective, dynamic specification of lot priorities and target WIP levels that promote consistent fab outs.

Various components of SLIM exist to perform very specific computations/calculations. For example, SLIM-M sets individual process step-level cycle time targets by allocating all expected buffer/queue time to bottleneck (i.e., constraint) process steps, and then uses Little's Law to establish target WIP levels for these same steps. These two factors are used by SLIM in its determination/setting of short-term production targets and priorities throughout the fab for each device type. Additional SLIM components deal with non-bottleneck equipment (SLIM-L), bottleneck equipment (SLIM-S), batch-processing diffusion ovens (SLIM-D), and lot release (SLIM-I).

By implementing the SLIM methodology, Samsung Electronics reduced cycle times from four days per layer of circuit architecture to a range of 1.3 to 1.6 days per layer. In addition, fab equipment utilization levels were increased and fab WIP levels were redistributed to more appropriate locations. Samsung's president, Mr. Yoon-Woo Lee, stated at the 2001 Franz Edelman Award Competition that as a result of implementing SLIM, the company "increased revenue almost $1 billion through five years without any additional capital investment" (Leachman *et al.*, 2002).

4. SEMICONDUCTOR WAFER FAB SCHEDULING

As mentioned earlier, the authors of this chapter were recently part of a team of researchers that developed a deterministic scheduling approach to scheduling wafer fabrication operations (Fowler *et al.*, 2002). The approach is based on the Shifting Bottleneck (SB) procedure that was developed by Adams *et al.* (1988) as an efficient means to obtain a "good" solution to the $J_m||C_{max}$ problem. The background for the project and the results from it are briefly described below.

4.1 Shifting bottleneck heuristic for classical job shops

While there is a polynomial time algorithm to solve the $J_2||C_{max}$ problem to optimality, almost all other job shop scheduling problems are NP-Hard. Thus, these problems are generally approached with heuristic methods for all but the smallest problem instances. Early documented procedures for the job shop scheduling problem used priority dispatching rules, such as first-in/first-out (FIFO), shortest processing time (SPT), and earliest due date (EDD). Adams *et al.* (1988) developed a procedure to find "better" solutions to the job shop scheduling problems, when compared to available dispatching rules, at the expense of slightly increased computation times. Their Shifting Bottleneck (SB) procedure requires far less computation time for a given problem than completely enumerating all possible solutions in order to find the true optimal set of m machine schedules. The approach basically decomposes the factory into the individual machines, generates candidate schedules for each non-scheduled machine, and selects the one most critical to schedule. This continues until all machines have been scheduled. An overview of the SB procedure is described in Table 9-4.

One of the key aspects of the SB procedure lies in Step Two, wherein the overall job shop scheduling problem is decomposed into separate single machine sub-problems that are solved by Subproblem Solution Procedures (SSP's). In order to fully characterize each unscheduled machine's sub-problem, both the operational ready time and operational due date must be determined for each job. These two quantities are easily determined for node (i, j) by calculating the longest path from node 0 to (i, j) and the longest path from (i, j) to node n, respectively. The determination of the operation ready times and due dates requires $O(n^2)$ steps for a graph with n nodes, although Adams *et al.* (1988) describe a procedure using only "relevant" arcs that requires only $O(n)$ steps. This order of magnitude reduction in complexity greatly reduces the amount of time required to estimate these quantities for large problem instances.

Many different performance criteria have been used as means for determining the "optimal" schedule for a single machine in the SB heuristic's sub-problems. Holtsclaw and Uzsoy (1996) refer to these performance criteria as machine criticality measures (MCM). Adams *et al.*'s MCM is based on Carlier's (1982) algorithm, a branch-and-bound procedure based on the Schrage algorithm, to solve the single machine sub-problems. Carlier's algorithm schedules the operations on a single machine using a pre-emptive, earliest due-date approach and has proven to be both exceptionally fast and efficient at obtaining optimal solutions to single machine scheduling problems. However, Adams *et al.* use a heuristic based on the most work remaining priority dispatching rule to sequence each machine under consideration.

Table 9-4. Shifting Bottleneck Heuristic (Ovacik & Uzsoy, 1997)

Step	Description
1	Let M = the set of all m machines. Initially, the set M_o, the set of machines that have been sequenced or scheduled, is empty.
2	Identify and solve the single sub-problem for each machine $i \in M \backslash M_o$.
3	Identify a critical or bottleneck machine $k \in M \backslash M_o$ from the set of unscheduled machines M_o.
4	Solve the sub-problem associated with machine k and sequence the machine optimally. Set $M_o = M_o \cup \{k\}$.
5	Reoptimize the schedule for each machine $m \in M_o$, considering the newly added disjunctive arcs for machine k. If $M_o = M$, stop. Otherwise, go to Step 2.

Ovacik and Uzsoy (1997) suggest two additional performance criteria for bottleneck machine identification in addition to Carlier's approach. First, the authors consider the $1|r_j|L_{max}$ problem for each machine, identifying the bottleneck as the machine with the largest maximum lateness (L_{max}). This is identical to the approach taken by Balas *et al.* (1995) who use branch-and-bound procedures to solve the single machine sub-problems optimally. Finally, in addition to the previously mentioned maximum lateness and preemptive earliest due-date approaches, Ovacik and Uzsoy also use total workload as a bottleneck identification criterion. In this approach, the machine with the maximum required operation processing time is selected as the bottleneck machine. Pinedo and Singer (1999) developed a shifting bottleneck heuristic for the $J_m \| \sum w_j T_j$ problem. In order to develop an appropriate solution to the this problem, Pinedo and Singer apply the Apparent Tardiness Cost (ATC) index of Vepsalainen and Morton (1987) in the solution procedure of the single-machine subproblems. The ATC index is a composite dispatching rule that blends two heuristics that are effective when used in single-machine scheduling problems that consider job ready times and due dates: weighted shortest processing time and minimum slack.

4.2 Modified shifting bottleneck heuristic for complex job shops

The primary goal of the FORCe scheduling research was to develop solution methodologies for minimizing total weighted tardiness (TWT) in complex job shops. This research built upon the the classical job shop work of Pinedo and Singer (1999) and the research by Mason et *al.* (2002) that developed a modified shifting bottleneck (SB) heuristic for the $FJc \mid r_j, s_{jk}, B, recrc \mid \sum w_j T_j$ problem. The latter paper accounts for batching tools, tool groups (identical machines operating in parallel), sequence-dependent setup times, and recirculating product flow. The heuristic developed in Mason *et al.* (2002) produced lower TWT values than standard first-in/first-out processing for ten different test problems. The authors' heuristic decomposes the factory scheduling problem into smaller, more tractable single tool group problems, which are then iteratively solved using a proposed solution procedure called the Batching Apparent Tardiness Costs with Setup (BATCS) rule. BATCS, which is an extension of the parallel machine version of the Apparent Tardiness Costs with Setups rule (Lee and Pinedo, 1997) blends together different scheduling rules in order to assess the trade-off that may exist between different jobs available for scheduling. These rules include weighted shortest processing time, minimum slack, setup avoidance, and batching efficiency maximization. The BATCS index is calculated as follows:

$$I_{bj}(t,l) = \frac{w_{bj}}{p_{bj}} \exp\left(-\frac{\left(d_{bj} - p_{bj} + \left(r_{bj} - t\right)\right)^{+}}{k_1 \bar{p}} \right) \exp\left(-\frac{s_{bl,bj}}{k_2 \bar{s}} \right) \left(\frac{|bj|}{b_i} \right) \quad (2)$$

where w_{bj} is the average weight of the jobs in batch bj and p_{bj} is the processing time of batch bj, while d_{bj} and r_{bj} denote batch bj's due date and ready time, respectively. Regardless of the type of tool group being scheduled (i.e., single and parallel machines, a subset of which contain sequence-dependent setups, another subset of which are capable of batch processing), the BATCS rule produces a proposed job sequence for each unscheduled tool group.

4.3 Testing the modified shifting bottleneck heuristic

As part of the project, an AutoSched AP-based testing environment to evaluate scheduling approaches in a dynamic, simulation-based environment in order to accommodate real world fab models was developed. Mönch *et al.* (2004) provides additional details. The main purpose of the simulation environment was to emulate the behavior of a real wafer fabrication facility (wafer fab) during scheduler development.

In a real wafer fab, a production planning and control tool obtains its information via the message bus of the Manufacturing Execution System (MES), stores this information in one or several databases, and computes fab control information based on the information found in the database. Then it stores the control information in the database for evaluation purposes, and finally transmits it to the shop floor where the control actions take place at the work centers.

In our design, we mimic the structure found in a real wafer fab (as shown in Figure 9-5). The simulation model generates data, this data is sent to a data model of the fab that stores this information. The message bus is replaced by a set of functions providing a clear separation of data that resides inside the simulation model and data that is going to be transferred to the data model. The scheduler uses only the data stored in the data model to compute the schedule. No internal information from the simulation model is used. The resulting schedule is stored in the data model for testing and to serve as reference data that is required when a decision has to be made upon whether to reschedule or not. The schedule is implemented in the fab by a set of functions.

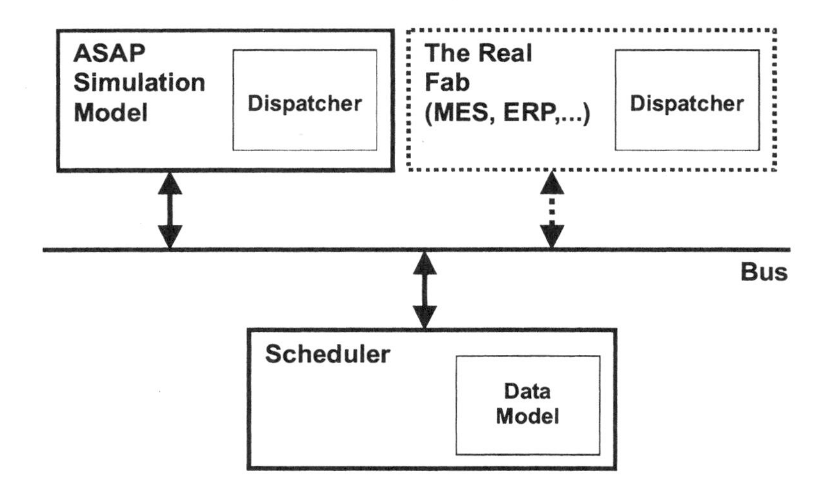

Figure 9-5. Structure of the simulation and testing environment

The center and integrating part of the design is the data model of the fab. It contains both static data, like tool set or product routes, and dynamic data, like tool status information and lot locations. In addition, it is used to store measured and scheduled lot movement instants, like actual and planned ready or completion times. The data model has interfaces to the simulation model, the scheduler, and the rescheduling mechanisms. It is open to future extensions with respect to content and interfaces. We note that the development of the testing environment was not trivial. It took approximately seven person years of effort to develop.

Experimental results demonstrated the efficacy of scheduling wafer fabs to maximize delivery performance of customer orders in an acceptable amount of computation time. For our parameter test studies we considered three wafer fab models:

Fab A (the "MiniFab" model): five tools in three tool groups

Fab B (reduced MIMAC Testbed Dataset 1): 45 tools in 11 tool groups

Fab C (full MIMAC Testbed Dataset 1): 268 tools in 84 tool groups

The goal of the experimentation was to find scheduler parameters that maximize on-time delivery performance of orders from customers of varying importance and priority (i.e., total weighted tardiness or TWT) while running at a bottleneck utilization of 95%. To this end, we developed an experimental design that considered the following.

- Schedule computation frequency—how often should the fab be scheduled?
- Scheduling horizon—how many hours of fab operations should be scheduled?
- Choice of Subproblem Solution Procedure
- Parameters for Subproblem Solution Procedures

In total, we ran more than 1,500 simulation experiments. We compared the TWT results of the scheduler to those obtained using classical dispatching approaches like first in, first out (FIFO); earliest due date (EDD); apparent tardiness cost with setups (ATCS); critical ratio (CR); and operational due date (ODD). In this case, the best scheduler TWT result (MFS) is approximately one seventh (i.e., 13%) of the best dispatching TWT result. For Fab B the TWT produced by the scheduler is 0.0025% of the best dispatching TWT result. Finally, in the case of Fab C (i.e., full MIMAC Testbed Dataset 1), we were able to achieve scheduler TWT values that were 25% of the corresponding best dispatching results. Therefore, we conclude that the scheduler outperforms classical dispatching rules for the three test fab models. The computation time required to schedule the Fab C model took less than 50 seconds on a standard PC. In summary, we have demonstrated that our Shifting Bottleneck-based wafer fab scheduling system has the potential to improve the on time delivery performance of wafer fabs without loss of throughput. See Fowler *et al.* (2005) for additional details on the testing of this approach.

5. SUMMARY AND CONCLUSIONS

Scheduling wafer fabrication facilities is very challenging. These systems are very complex, the equipment highly unreliable, and there are often many jobs to schedule. From the survey described above, it is clear that dispatching is the current state-of-the-practice but that most semiconductor manufacturers would like to employ techniques that consider more than just the current job and machine. The literature discussed in this article suggests that several semiconductor manufacturers have achieved success with sophisticated dispatching systems that have a broader view than traditional dispatching. Finally, this article briefly described an attempt to employ a deterministic scheduling approach for this difficult scheduling problem. The approach shows promise and several key implementation issues were identified. We note that several research teams are currently working on other deterministic scheduling approaches to this challenging problem.

REFERENCES

Adams, J., Balas, E., Zawack, D., 1988, The shifting bottleneck procedure for job shop scheduling, *Management Science* 34, 391-401.

Appleton-Day, K., Shao, L., 1997, Real-time dispatch gets real-time results in AMD's Fab 25, *IEEE/SEMI Advanced Semiconductor Manufacturing Conference and Workshop*, 444-447.

Balas, E., Lenstra, J. K., Vazacopoulos, A., 1995, The one machine problem with delayed precedence constraints and its use in job shop scheduling, *Management Science* 41, 94-109.

Carlier, J. 1982, The one-machine scheduling problem, *European Journal of Operational Research* 11, 42-47.

Cigolini, R., Comi, A., Micheletti, A., Perona, M., Portioli, A., 1999, Implementing new dispatching rules at SGS-Thomson Microelectronics, *Production Planning & Control*, 10 (1), 97-106.

Dabbas, R., Fowler, J., 2003, A new scheduling approach using combined dispatching criteria in wafer fabs, *IEEE Transactions on Semiconductor Manufacturing*, 16 (3), 501-510.

Dabbas, R., Fowler, J., Rollier, D., McCarville, D., 2003, Multiple response optimization using mixture designed experiments and desirability function in semiconductor scheduling, *International Journal of Production Research*, 41 (5), 939–961.

Fowler, J., Brown, S., Carlyle, M., Gel, E., Mason, S., Mönch, L., Rose, O., Runger, G., Sturm, R., 2002, A modified shifting bottleneck heuristic for scheduling wafer fabrication facilities, *12th Annual International Conference on Flexible Automation and Intelligent Manufacturing*, Dresden, Germany, July 15-17, 2002, 1231-1236.

Fowler, J., Gel, E., Mason, S., Mönch, L., Pfund, M., Rose, O., Runger, G., Sturm, R., 2005, Development and testing of a modified shifting bottleneck heuristic for scheduling wafer fabrication facilities, *ASU Working Paper Series*.

Fowler, J., Robinson, J., 1995, Measurement and improvement of manufacturing capacity (MIMAC) project final report, *SEMATECH Technology Transfer #95062861A-TR*.

Holtsclaw, H. H., Uzsoy, R., 1996, Machine criticality measures and sub-problem solution procedures in shifting bottleneck methods: A computational study, *Journal of the Operational Research Society* 47 (5) 666-677.

Hsieh, M.D., Lin, A., Kuo, K., Wang, H.L., 2002, A decision support system of real time dispatching in semiconductor wafer fabrication with shortest process time in wet bench, *2002 Semiconductor Manufacturing Technology Workshop*, 286-288.

ITRS, 2003, International Technology Roadmap for Semiconductors, http://public.itrs.net.

Leachman, R.C., Hodges, D.A., 1996, Benchmarking semiconductor manufacturing, *IEEE Transactions on Semiconductor Manufacturing* 9 (2), 158-169.

Leachman, R.C., Kang, J., Lin, V., 2002, SLIM: short cycle time and low inventory in manufacturing at Samsung Electronics, *Interfaces* 32 (1), 61-77.

Lee, Y. H., M. L. Pinedo. 1997. Scheduling jobs on parallel machines with sequence-dependent setup times. European Journal of Operational Research 100, 464-474.

Mason, S.J., Fowler, J.W., Carlyle, W.M., 2002, A modified shifting bottleneck heuristic for minimizing the total weighted tardiness in a semiconductor wafer fab, *Journal of Scheduling* 5 (3), 247-262.

Mönch, L., Rose, O., Sturm, R., 2004, A simulation framework for the performance assessment of shop-floor control systems, *Simulation: Transactions of the Society for Computer Simulation International*, 79, 163 - 170.

Ovacik, I. M., Uzsoy, R. 1997, *Decomposition methods for complex factory scheduling problems*, Kluwer Academic Publishers, Norwell, MA.

Pickerill, J., 2000, Better cycle time and on-time delivery via real-time dispatching, *Solid State Technology*, June 2000, 151-154.

Pinedo, M. L., M. Singer. 1999. A shifting bottleneck heuristic for minimizing the total weighted tardiness in a job shop. *Naval Research Logistics* 46, 1-17.

Uzsoy, R., C. Y. Lee, L. A. Martin-Vega. 1992, A review of production planning and scheduling models in the semiconductor industry Part I: system characteristics, performance evaluation, and production planning. *IIE Transactions: Scheduling & Logistics* 24, 47-60.

Vepsalainen, A., T. Morton. 1987, Priority rules for job shops with weighted tardiness cost. *Management Science* 33(8), 1035-1047

Yang, T.Y., Huang, Y.F., Chen, W.Y., 1999, Dynamic dispatching model for bottleneck resource allocation, *1999 IEEE International Symposium on Semiconductor Manufacturing Conference Proceedings*, 353-354.

Chapter 10

THE SLAB DESIGN PROBLEM IN THE STEEL INDUSTRY

Milind Dawande[1], Jayant Kalagnanam[2], Ho Soo Lee[2], Chandra Reddy[2], Stuart Siegel[2], Mark Trumbo[2]

[1] *University of Texas at Dallas*
[2] *IBM T. J. Watson Research Center*

Abstract Planners in the steel industry must design a set of steel slabs to satisfy the order book subject to constraints on (1) achieving a total designed weight for each order using multiples of an order-specific production size range, (2) minimum and maximum sizes for each slab, and (3) feasible assignments of multiple orders to the same slab. We developed a heuristic solution based on matchings and bin packing that a large steel plant uses daily in mill operations.

Keywords: steel industry, operations planning, slab design, optimization

Planners in the steel industry design a set of steel slabs (producible units) to satisfy orders in the order book subject to constraints on (1) achieving a total designed weight for each order using multiples of an order-specific range of production sizes, (2) minimum and maximum sizes for each designed slab, and (3) feasible assignments of multiple orders to the same slab. In this paper, we describe a slab design problem for a large steel plant in the Asia-Pacific region. The problem has two optimization objectives: (1) to minimize the number of slabs needed to fill all the orders, and (2) to minimize the total surplus weight. We describe an efficient heuristic for the problem that uses the solution of two classical problems in the operations research literature, namely, matching and bin packing. Our heuristic produced an increase of around 1.3 tons in the average slab weight and a reduction of around three percent of surplus weight. The steel plant estimates the savings from using the heuristic as around $2.5 million per year.

Operations planning in the steel industry typically begins with an order book listing orders to fill. Planners first try to fill the orders using leftover stock from surplus inventory (a problem known as the inventory application problem (IAP)), and second they design slabs (production units) for manufacture for the remaining orders (the slab design problem (SDP)) (Figure 10.1). Usually, they can fill a part of the orders from inventory. The goal of the SDP is to design a minimum number of slabs to fill all the orders and to minimize the surplus weight of the designed slabs.

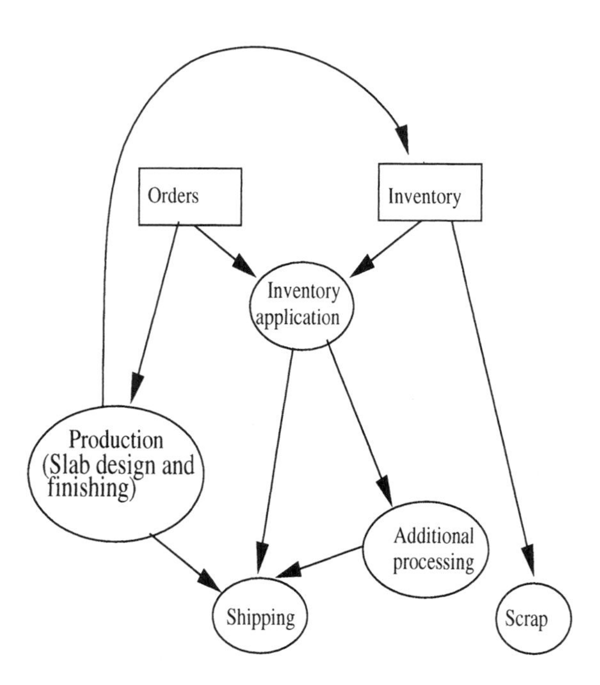

Figure 10.1. Conceptual flow of operations planning in the process industry: a part of the orders are filled from slabs in surplus inventory by solving the inventory application problem. Slabs from surplus inventory may require some additional processing. For the remaining orders, new slabs are manufactured by solving the slab design problem. The complete set of slabs is processed in the steel mill and shipped to customers. Slabs from surplus inventory that cannot be used to fill any order are scrapped.

10.1 The Constraints

The specifications of orders and the rules for packing and processing orders for the construction of slabs in the steel industry have some unique attributes that are essential in understanding the slab design problem.

10.1.1 The Order Book

The order book lists orders from various customers. Each order has a target weight (O_t) to deliver and a minimum weight (O_{min}) and a maximum weight (O_{max}) that the customer will accept at delivery. The size and number of units into which the order can be factored are restricted. Each order specifies a range for the weight of the production units (PUs) delivered, a minimum weight PU_{min} and a maximum PU_{max}. Then, for each order, we must deliver an integral number of production units (PU_{number}) within the weight interval [PU_{min}, PU_{max}] so that the total order weight delivered is in the range [O_{min}, O_{max}]. Equivalently,

$$O_{min} \leq PU_{size} \times PU_{number} \leq O_{max} \quad (10.1)$$

$$PU_{min} \leq PU_{size} \leq PU_{max} \quad (10.2)$$

$$PU_{number} \in \{0, 1, 2, ...\} \quad (10.3)$$

Order #	O_{min}	O_t	O_{max}	PU_{min}	PU_{max}
1	45.3	49.0	54.4	4.1	5.5
2	78.9	87.2	103.0	3.3	5.7
3	20.5	23.8	27.8	2.8	3.6
4	36.0	42.7	52.9	4.5	5.9
5	62.1	70.5	79.0	6.4	8.7

Table 10.1. Sample order characteristics. O_{min}, O_t, and O_{max} denote the minimum weight, target weight, and maximum weight, respectively, for an order. PU_{min} and PU_{max} denote the minimum and maximum production unit size, respectively, for an order.

All the PUs for an order cut from a slab should be of the same size. However, if an order is split between two slabs, the PU sizes from the two slabs can be different. The requirement that all the PUs of an order on a slab should be of the same size is related to control issues: the hot and cold mills transform the slab into a coil or a sheet (Figure 10.2), which is cut into individual PUs. Once we set the cutting machine to handle a certain length of coil, it is preferable to process all the PUs for an order on the slab. If the PUs for an order to be cut from a slab are not of the same

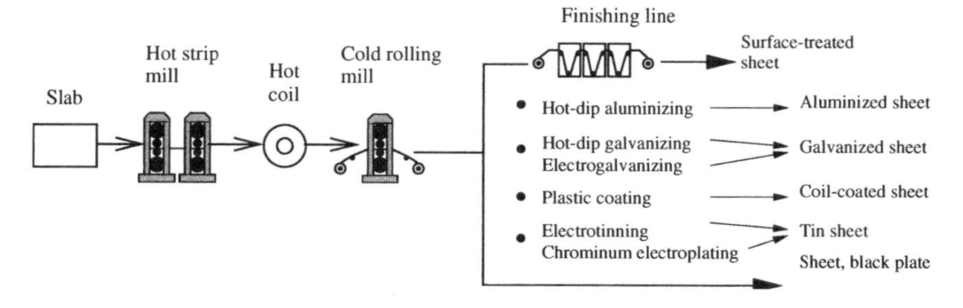

Figure 10.2. A slab goes through the hot strip mill and the cold rolling mill (if required) and subsequently to the finishing line. The hot strip mill or the cold rolling mill or both, transform the slab into a coil or a sheet. The finishing mill provides the coil or sheet with its final surface properties by using such processes as aluminizing, galvanizing, and tinning.

size, we have to reset the cutting machine for another length, which may cause a scheduling bottleneck. Table 10.1 shows the characteristics for a few orders. Typical ranges for the dimensions of a slab are as follows: length between 5.5 meters and 12 meters; width between 1 meter and 2 meters; thickness around 0.2 meters. The weight of a slab typically ranges between 15 tons and 40 tons. The bilinear constraint (10.1), which specifies bounds on the minimum and maximum quantities for an order, is called the order-fulfillment constraint. In general, the PU size for an order can be non-integer because most coil-processing machines are servomotor controlled and can move in a precise and continuous manner.

Order-Fulfillment Rules: It is desirable for a total designed quantity to be the target weight O_t of an order. However, an order is considered filled if the designed quantity, dq, satisfies one of the following conditions (Figure 10.3):

(a) $dq > O_t$ and $(dq - PU_{size}) < O_{min}$ or

(b) $dq < O_{min}$ and $(dq + PU_{size}) > O_{max}$.

Constraint (10.2), which specifies the bounds on the size of a PU, is the production unit size constraint. Constraint (10.3) specifies the integrality of the number of production units.

In addition to the weight requirements, each order has four other classes of attributes: quality requirements, such as grade, surface, and internal properties of the steel; maximum weight of the slab which includes the order; geometric attributes, such as the width and thickness of the product; and the finishing process applied to the production units (for example, car manufacturers often require galvanized steel sheets).

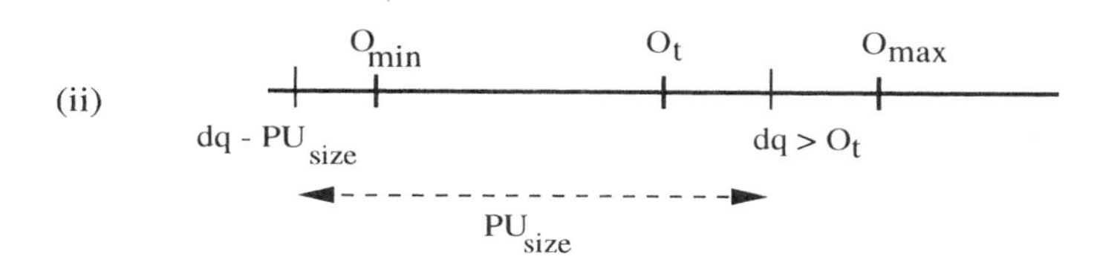

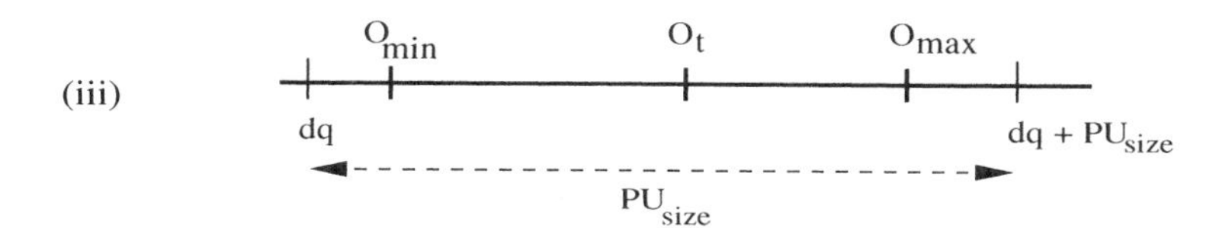

Figure 10.3. It is preferable for the designed quantity, dq, of an order to be its target weight O_t. However, an order is considered filled if dq satisfies either of the conditions in (ii) and (iii).

10.1.2 Packing Constraints

The packing constraints concern packing multiple orders from the same slab (Figure 10.4):

Steel mills produce steel with various chemical compositions (grades). An order specifies a particular range of chemical compositions that map onto a subset of the grades the mill typically produces. Each grade has an associated production cost, so to fill the entire order at the lowest cost, the mill should avoid using a more expensive grade than necessary. Because all orders cut from a slab will have the same grade, planners seek to avoid combining orders for disparate grades. Customers also specify a range of surface quality for orders, and so planners packing orders together try to avoid satisfying an order that requires low surface quality from the same slab used to satisfy an order that requires high surface quality.

Usually, a steel mill can alter the thickness and width of a slab by rolling and trimming it; a corresponding range of constraints identifies limits to such alterations. For example, a slab of width S_w and thickness S_t can be rolled or trimmed or both to create a slab of width in the range $[S_w^{min}, S_w^{max}]$ and thickness in the range $[S_t^{min}, S_t^{max}]$. If planners pack two orders, with widths and thicknesses in this range, on the same slab,

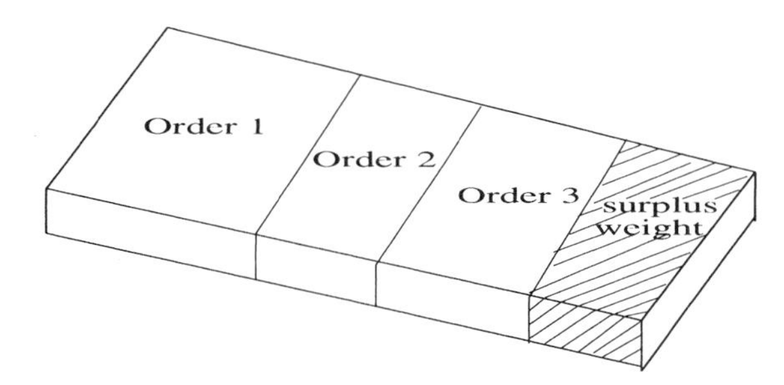

Figure 10.4. A steel slab with three orders packed on it. If the total weight of the orders is less than the minimum weight of the slab, planners add a surplus weight to the slab to increase its weight up to the minimum weight.

the mill uses rolling and trimming downstream to achieve their exact dimensions.

Color Constraints. In passing through a steel mill (Figure 10.2), a slab goes through the hot strip mill and the cold mill (if required) and subsequently to the finishing line. The hot mill or the cold mill, or both, transform the slab into a coil or a sheet. Before sending coils to the finishing line, the mill cuts them to meet different order specifications. It cuts orders with different requirements during finishing from the coil before they go through finishing operations. This makes it possible for planners to assign orders with different finishing requirements to the same slab. However, cutting coils is time consuming and cumbersome. The cutting machine is often the bottleneck in the process flow. Hence the number of allowed cuts per slab is often strongly constrained based on the current state of the cutting machine.

The simplest representation of this constraint is a limit on the number of cuts on a slab or the number of different order types (orders that must be separated before the finishing line). To represent this constraint formally, we assigned a color attribute to each order to indicate the set of finishing operations required. We consider orders that require the same finishing operations to be of the same type (and hence the same color), and they do not need to be separated before the finishing line. Orders that require different finishing operations are of different types (and different colors) and must be separated before the finishing line. By associating a color with each order based on the finishing operations, we can specify a constraint that limits the number of colors allowed on a slab, a *color constraint*. In our application, cuts per slab were limited to one. Thus, the color constraints restrict the number of order types on a single slab to be at most two.

Packing multiple orders on a slab can lead to negative outcomes for a steel producer. To produce orders with different geometries or finishing requirements from a single slab, the mill must add downstream coil-cutting and trimming operations to accommodate these differences. Too many of such added operations can cause scheduling bottlenecks and increase the operator oversight required for the overall steel-making process. Therefore, the packing constraints restrict the orders to be cut from one slab to those with similar geometries and finishing requirements.

10.1.3 Weight Constraints

Each order has a specified maximum allowable slab weight based on its quality requirements. As the quality required increases, the allowable size of the slab to be hot rolled decreases. The stress the slab undergoes in hot rolling affects its internal structure and hence its properties. The orders packed on a slab might have different maximum allowable slab weights; however, the allowable maximum slab weight is determined by the order with the lowest allowable slab weight. This reflects a quality restriction in that if orders of disparate quality are packed together, then the slab must satisfy the requirements of the highest quality order (which corresponds to the smallest allowable slab weight).
A slab cannot exceed 44 tons because the crane that transports slabs to the finishing mill can carry a maximum of 44 tons.

After a slab goes through the hot strip mill and the cold rolling mill,

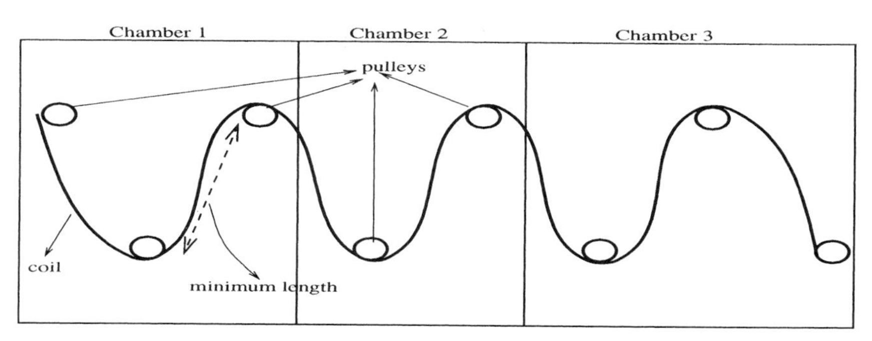

Figure 10.5. In the finishing mill, a coil typically rests on a pulley system. For proper processing, the coil must be of a certain minimum length and, hence, must have a minimum processing weight.

the resulting coil goes to the finishing mill. The finishing mill provides the coil with its final surface properties by using such processes as aluminizing, galvanizing, and tinning. In a typical processing setup in the finishing mill, the coil passes through several chambers, and in each chamber, it rests on a pulley system for a specified amount of time (Figure 10.5). To rest properly on the pulley system, the coil must be a certain minimum length, which means the coil must have a minimum processing weight. This minimum weight requirement on the coil then translates into a minimum weight requirement on the slab.

The constraints on quality and geometry ensure that orders grouped on a single slab are for the same grade and similar thicknesses and widths. The planners adjust the length of the slab to satisfy the weight constraints.

10.1.4 Objectives

The problem has two main optimization objectives: (1) to minimize the number of slabs needed to fill all the orders, and (2) to minimize the total surplus weight.
To demonstrate the trade-off between these objectives, we will use a simple example: consider two orders O^1 and O^2 with target weights of 50 tons and 30 tons respectively. We assume, for simplicity, that $O^1_{min} = O^1_t = O^1_{max} = 50$ tons. Similarly, $O^2_{min} = O^2_t = O^2_{max} = 30$ tons. Let the minimum (resp. maximum) slab weight for both O^1 and O^2 be 15 (resp. 25) tons. Let the PU size of O^1 be fixed at five tons (i.e., $PU^1_{min} = PU^1_{max} = 5$ tons) and the PU size of O^2 be fixed at six tons. Then, one possible design is as follows: slabs S_1 and S_2 each weighing 25 tons (five PUs of five tons) will have order O^1, slab S_3 weighing 24 tons will have order O^2 (four PUs of six tons), and slab S_4 will have six tons of O^2 (one PU). Since the minimum weight of S_4 is 15 tons, it will have 15 - 6 = 9 tons of surplus weight. Thus, this design constructs four slabs with slabs S_1 and S_2 weighing 25 tons, slab S_3 weighing 24 tons and slab S_4 weighing 15 tons. The total surplus weight for this solution is nine tons. An alternative design is to construct five slabs, each having 10 tons of O^1 (two PUs of five tons) and six tons of O^2 (one PU). Here, the number of slabs is more but the surplus weight (zero tons) is less than the first solution.

10.2 Literature Review

Although designing slabs is an essential operation in many steel mills, in a particular instance of the slab-design problem, the complexity of the slab-making operations and the plant's processing restrictions give rise to many constraints. The slab design problems at two different mills are therefore likely to differ. To the best of our knowledge, studies of the slab-design problem we describe have not appeared in the literature. However, Newhart, Stott, and Vasko (1993), Vasko et al. (1996), Vasko and Wolf (1994), Vasko, Wolfe, and McNamara (1989), and Vasko, Wolf, and Stott (1987) have published papers on slab and semi-finished steel size design. Vasko, Newhart and Stott (1999) discussed avoiding combinations of orders with disparate grades in packing. Vasko, Cregger, Newhart, and Stott (1993) discussed handling restrictions on the number of cuts on a slab; Vasko, McNamara, et al. (2000) discussed handling width and length range constraints on orders. Hirayama, Kajihara, and Nakagawa (1996) developed a heuristic based on a genetic algorithm for a much simpler version of the slab-design problem. They did not consider restrictions on PU size, order fulfillment rules, or packing restrictions such as the color constraints. Kalagnanam et al. (2000), Vasko, Cregger, Stott, and Woodyatt (1994), and Vasko, Wolfe, and McNamara (1989) studied the IAP.

The slab-design problem we considered concerns minimizing the number of slabs and minimizing the surplus weight. An equally important issue is minimizing the number of different *slab specifications* used to satisfy the order book (Denton et al. 2003). A slab specification establishes its width, weight, grade, and quality. Given a slab design, the mill uses the *casting* process to produce solid steel slabs meeting the specifications. Large and rapid changes in the specifications of slabs (for example, width changes) reduce the throughput of the caster. This issue is especially significant when caster capacity is a bottleneck. In such cases, production managers may restrict the set of slab specifications to a subset of all potential specifications. Denton et al. (2003) discussed the benefits of made-for-stock slabs (as opposed to make-to-order slabs) and proposed an optimization model for choosing a "good" set of slab specifications. Our client's caster capacity usually exceeds the available orders, so throughput was not a major issue.

Fundamentally, to solve the slab design problem, one designs slabs subject to the various geometrical considerations and assigns orders to those slabs. As such, the problem is related to many well-known and fundamental problems in the literature, such as the bin packing problem (Coffman, Garey, and Johnson (1997), Johnson (1974), Martello and Toth (1990)), the variable sized bin packing problem (Friesen and Langston (1986)), the cutting-stock problem (Gilmore and Gomory (1961), Gimore and Gomory (1963)), and the multiple knapsack problem (Martello and Toth (1981), Martello and Toth (1990)).

10.3 A Heuristic Based on Matching and Bin Packing

We developed a heuristic to produce solutions of good quality within a few minutes. The main building blocks of the heuristic are a nonbipartite matching step and a bin packing step.

10.3.1 Step 1: Nonbipartite Matching

Given the order book, we construct an order-compatibility graph $G = (V, E)$ as follows: For each order O^i, there is a node $v_i \in V$. If two orders, say O^i and O^j, are compatible with each other (satisfy the quality and geometry constraints and can hence be packed on a single slab), we add an edge $e = (v_i, v_j)$. Construction of G requires time $O(|V|^2)$. For an edge $e = (v_i, v_j)$, its weight w_{ij} indicates the maximum assigned weight of a slab that includes orders O^i and O^j. In general, G is nonbipartite. In our application, the graph is quite sparse; usually about 10 percent of the possible number of edges exist. In a graph

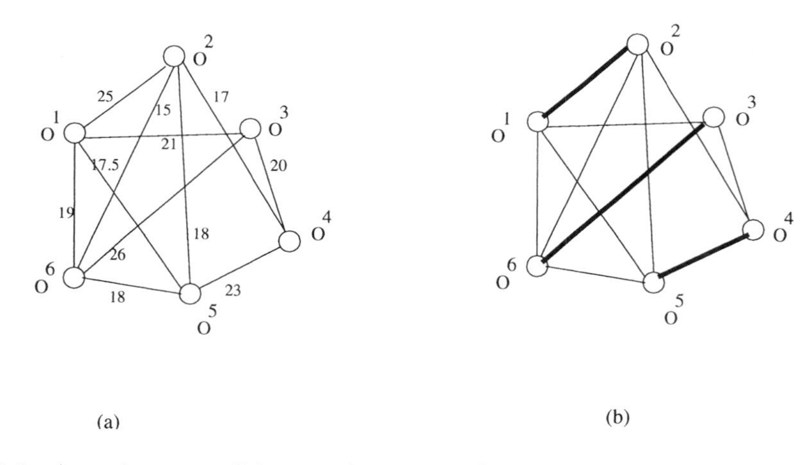

Figure 10.6. An order-compatibility graph corresponding to six orders is shown in (a). For an order pair (O^i, O^j), the weight of the edge joining the corresponding nodes is the maximum weight of the slab that includes O^i and O^j. The production-unit (PU) size ranges of O^1 and O^2 are [8,9] and [6,7], respectively. The maximum slab weights for O^1 and O^2 are 26 tons and 25 tons, respectively. A maximum weight matching is shown in (b).

with six nodes, each corresponding to an order, the edges of the graph indicate the packing feasibility of the corresponding order pairs (Figure 10.6(a)). The maximum weight of a slab that includes orders O^1

and O^2 is Min{26, 25} = 25 tons. We can assign these two orders to the same slab in several ways. For example, we can assign two PUs (each eight tons) of order O^1 and one PU (of seven tons) of order O^2 for a total weight of 23 tons. Alternatively, we can assign one PU (of nine tons) of order O^1 and two PUs (each seven tons) for a total weight of 23 tons. Among all such assignments, we choose the one with the maximum assigned weight as the weight of the edge joining the two orders. We obtain the maximum assigned weight of 25 tons choosing two PUs (each of size nine tons) of order O^1 and one PU (of seven tons) of order O^2. We compute such a weight for each edge in the graph.

Suppose the maximum slab weight associated with order O^i (O^j) is W^i_{max} (W^j_{max}). Then, the maximum weight of a slab that includes both O^i and O^j is $W = \min(W^i_{max}, W^j_{max})$. However, finding the maximum assigned weight is, in general, a nontrivial problem since it requires solving the following problem (P):

$$\begin{array}{lcl}
\max & x_i + x_j & \\
\text{subject to} & & \\
n_i PU^i_{min} & \leq x_i \leq & n_i PU^i_{max} \\
n_j PU^j_{min} & \leq x_j \leq & n_j PU^j_{max} \\
 & x_i + x_j \leq & W \\
 & x_i \leq & O^i_{max} \\
 & x_j \leq & O^j_{max} \\
 & n_i, n_j \in Z^+ &
\end{array}$$

where

$$\begin{array}{rcl}
PU^k_{min} & = & \text{minimum PU size for order } O^k (k = i, j) \\
PU^k_{max} & = & \text{maximum PU size for order } O^k (k = i, j) \\
n_k & = & \text{number of PUs of order } O^k (k = i, j)
\end{array}$$

In this example, we solve $x_1 + x_2 \leq 25$, $6n_1 \leq x_1 \leq 7n_1$, $8n_2 \leq x_2 \leq 9n_2$ with $n_1, n_2 \in Z^+$. (P) is an integer program in two variables. For our application, we easily enumerated the possible solutions and picked the best possible solution: The maximum slab weight, W, is upper bounded by 44 tons (the maximum weight the crane can lift). The minimum PU size is lower bounded by one ton. Typically, the number of PUs on a slab is less than 10.

A maximum matching, M, in G identifies a set of order pairs, S, such that (1) each order pair is compatible (i.e., the two orders in the pair can be packed on the same slab) and (2) the total assigned weight for S is

the maximum among all sets of order pairs with the property that a set contains each order at most once. Ahuja et al. (1993), Lawler (1976), and Papadimitrou and Steiglitz (1982) give an $O(|V|^3)$ algorithm for obtaining a maximum weight matching in a graph with $|V|$ nodes.
The matching M provides us with $|M|$ slabs each with either one or two orders. Each edge (i, j) of the matching corresponds to one slab. This slab contains either one or both the orders O_i and O_j. The order weight assigned to this slab is w_{ij}, the weight of the edge (i, j) in the order-compatibility graph. For example, the maximum weight matching in Figure 6(b) contains three edges $(O^1, O^2), (O^3, O^6)$ and (O^4, O^5). We create three slabs, one for each edge. The slab corresponding to edge (O^1, O^2), for example, will be assigned 25 tons (two PUs of nine tons of O^1 and one PU of seven tons of order O^2). For each slab, we may be able to pack orders in addition to the one or two orders on it. For orders that cannot be packed on any of the $|M|$ slabs, we must construct new slabs.

10.3.2 Step 2: Bin Packing

In this step, we start with the collection of slabs the matching algorithm designed in Step 1. We try to use these slabs better by trying to pack the remaining orders (if any) on them and then, if necessary, construct new slabs. Since we need to check the possibility of assigning a remaining order on the slabs created in Step 1, we visit the slabs in a sequence. From a computational point of view, many strategies can be used to sequence the $|M|$ slabs created in Step 1. For example, we can visit the slabs in a decreasing sequence of their unassigned weight or based on some measure of their likelihood of accommodating further orders. We tried several such strategies but found none to be dominant with respect to our problem's two objectives. In our computations, we tried each remaining order (from a sorted list of remaining orders) for inclusion on all available slabs, in the sequence in which we created the slabs. When we found an order that could be included on a slab, we assigned the maximum possible weight to the slab. If we could not complete the order with applications to existing slabs, we created new slabs to fill the order. We then added these new slabs to the list of available slabs. For example, for an unassigned order O_x, we first tried to assign it to a slab created in the matching step by going through these slabs in the order in which they were created. If we succeeded and assigned all the weight of O_x, we selected another unassigned order and repeated the process. If we could not complete O_x using the available slabs, we created a new slab or slabs for it. We then added these newly created slabs to the set of available slabs and selected another unassigned order. This procedure is similar to the first-fit decreasing heuristic for the bin packing problem, and it has the following steps:

(a) Let LO denote the list of orders and LS, the list of currently available slabs. For order $O^i \in LO$, let RO^i denote the remaining weight of O^i and W^i_{max} denote the maximum allowable weight for a slab on which O^i is applied.

(b) Sort the list LO in ascending order using the function $f(O^i) = \frac{RO^i k_i}{W^i_{max}}$ where k_i is the urgency for O^i. The urgency k_i of an order O^i is the number of orders in LO that can be packed with O^i. We maintain the slabs in LS in the order in which we created them.

(c) For each order $O^i \in LO$ considered in sequence, do the following.

- Consider each slab S_j in LS in sequence. Check whether O^i can be packed on S_j. If O^i can be packed on S_j, apply the maximum amount of RO^i possible on S_j. Update RO^i.
- If O^i is not fulfilled on the slabs in LS, create new slabs and apply O^i on them. In each of these applications, except possibly the last slab, the quantity of O^i applied is equal to W^i_{max}. Update LS by appending the newly created slabs.

This procedure is akin to the first-fit decreasing heuristic for the classical bin packing problem (Coffman, Garey, and Johnson (1997), and Johnson (1974)). In sequencing slabs and orders, we want to maximize the weight of the existing slabs and minimize the number of slabs created. We try already-filled slabs first. If a slab is not filled up to its maximum weight, we can fill it up to its maximum weight. If a slab is filled up to its maximum weight, we can add a new order to extend the maximum allowable weight. By sorting the list of orders according to the function $f(O^i) = \frac{RO^i k_i}{W^i_{max}}$, we first try orders with low urgency, low remaining order quantity, and high maximum allowable weight. By trying the orders with low urgency early, we give the orders with low connectivity a greater chance of being packed with other orders than orders with high connectivity. We give orders with little unassigned weight a better chance of being packed on existing slabs than those orders with a greater unassigned weight. We try to assign orders with high maximum allowable weight early to maximize the size of the existing slabs.

10.3.3 Step 3: Postprocessing to Reduce Surplus

In the matching step and the bin packing step, we try to maximize the slab weight. When maximizing the slab weight, we do not consider the remaining quantity for the orders and whether these quantities are sufficient to create a slab of the required minimum weight. When the order quantities remaining are not sufficient to design a slab of minimum weight, we must add surplus weight to the slab.

Transferring Order Quantities. In this step, we redistribute an order's assigned quantities among the slabs those quantities are assigned to, to reduce surplus weight (Figure 10.7).

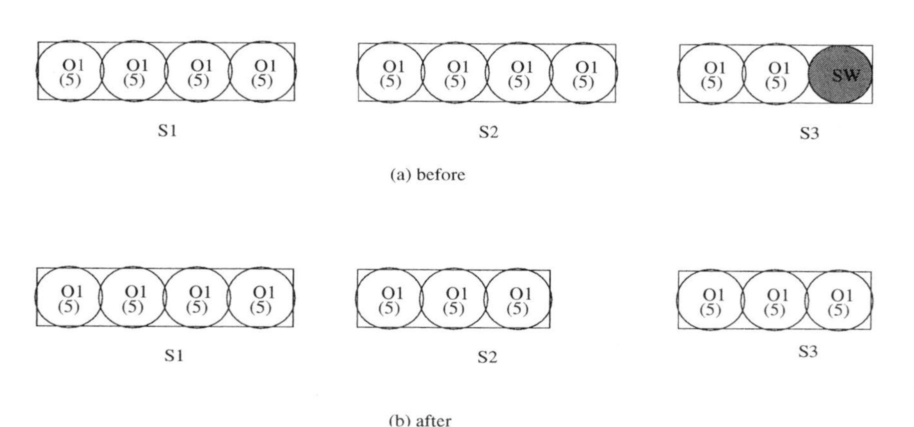

Figure 10.7. Surplus weight can be reduced by redistributing an order's assigned quantity among the slabs those quantities are assigned. In (a), slab $S3$ has a surplus weight of five tons. This surplus weight can be eliminated by transferring five tons from slab $S2$ to slab $S3$.

Suppose that an order, O^1, has a target weight of 50 tons. Let the minimum (W^1_{min}) and maximum (W^1_{max}) allowable slab weights for a slab with O^1 be 15 tons and 20 tons, respectively. Further, suppose we have designed slabs S_1, S_2, and S_3 with a single order, O^1, on them, and they weigh 20, 20, and 10 tons, respectively. Slab S_3 needs five tons more weight to meet its minimum. But S_1 and S_2 each have five tons more order quantity than they need to meet the minimum. We can transfer five tons of order quantity needed for S_3 from S_1 or S_2 or from both (Figure 10.7). The new order quantities on each of the slabs must meet the PU size requirements. That is, the new order quantities must be integral multiples of a valid PU size (constraints (1)-(3)).

This step does not completely remove an order from a slab. Thus, it preserves the packing of orders on a slab. It also preserves the total order quantity applied for each order.

Increase PU Size. In this step, we increase the designed quantity for an order. We exploit the order-fulfillment rule (a) that the designed quantity for an order can exceed its target weight (O_t) up to a maximum weight for the order (O_{max}), provided that removing any PU from the order decreases the designed quantity to fall below the minimum weight for the order (O_{min}). We can exploit this flexibility to reduce surplus weight on a slab.

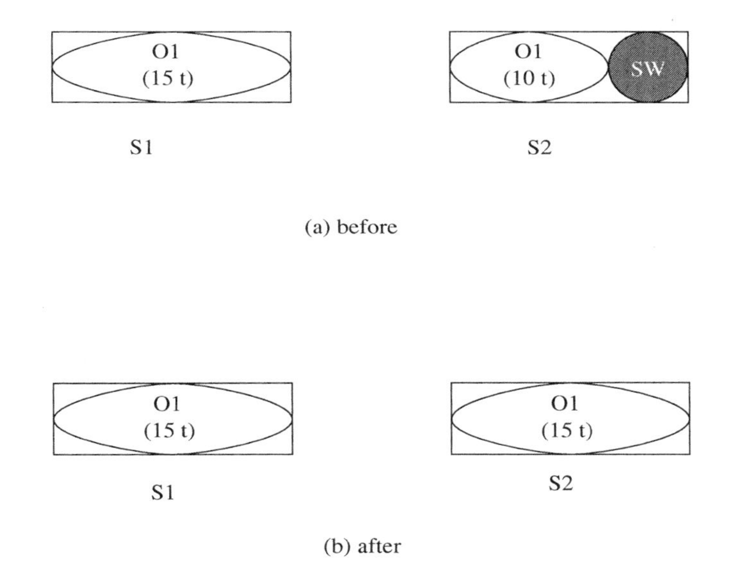

Figure 10.8. Surplus weight can be reduced by increasing the PU size of an order. In (a), slab $S2$ has a surplus weight (SW) of five tons. This surplus weight can be eliminated by increasing the PU size of order $O1$ to 15 tons as shown in (b).

Consider the following example. For an order O^1, suppose that O^1_{min}, O^1_t and O^1_{max} are 20, 25, 30 tons, respectively, and the PU size range is [10, 15]. Let W^1_{min} and W^1_{max} be 15 tons and 20 tons respectively. Let S_1 be a slab with a single PU of O^1 weighing 15 tons, and S_2 be a slab with a single PU of O^1 weighing 10 tons. Slabs S_1 and S_2 together meet the O_t of 25 tons for O^1. If no other order can be packed on S_2, we will incur five tons of surplus weight to meet the minimum allowable slab weight for O^1. We can, however, exploit the order-fulfillment rule (a) and get rid of the surplus weight by increasing the size of the PU on S_2 to 15 tons (Figure 10.8). We thus increase the designed quantity for O^1 from 25 tons to 30 tons. We satisfy rule (a), because we cannot remove any PU without decreasing the designed quantity of O^1 to be below O^1_{min}.

This step, as a side effect of decreasing surplus weight by increasing designed quantity for an order, can also improve average applied slab weight. Figure 10.9 shows the flowchart of the heuristic.

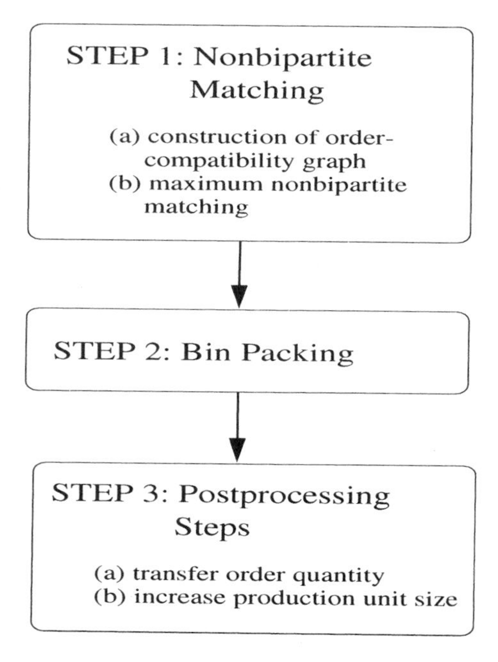

Figure 10.9. The heuristic consists of three steps: nonbipartite matching, bin packing, and postprocessing.

The overall quality of the slab design the heuristic constructs depends largely on Step 1 and Step 2, during which we choose an initial design. In constructing an initial design, we had two main considerations: (1) to obtain an efficient packing of order pairs, and (2) to minimize the number of designed slabs. Packing order pairs via nonbipartite matching is a global optimization step: we consider the entire set of orders and obtain a packing of order pairs with the maximum total assigned weight. In general, this approach is much better than, say, using a greedy algorithm wherein one finds a compatible order pair, packs these orders on a slab, and searches onwards for another compatible order pair. Consider the order-compatibility graph (Figure 10.6(a)) with the weight of the edge joining the order pair (O^1, O^3) changed to 27. A greedy search for the best compatible order pairs without explicit consideration of maximizing the total assigned weight would result in the order pairs (O^1, O^3), (O^4, O^5) and (O^2, O^6), corresponding to three slabs of weight 27 tons, 23 tons, and 15 tons, respectively. The total assigned weight on the three slabs would be 65 tons. The maximum-weight matching (Figure 10.6(b)), however, corresponds to three slabs with a total assigned weight of 74 tons. The second step of our heuristic is based on the first-fit decreasing heuristic for the classical bin packing problem, where the objective is to minimize the number of bins used. The first-fit decreasing heuristic is near-optimal for the bin packing problem with a worst case guarantee of $\frac{11}{9}$ (Coffman, Garey, and Johnson (1997), and Johnson (1974)).

10.4 Computational Experience

A large steel plant has successfully deployed the slab-design system based on our heuristic and uses it daily in the mill operations. The client chose an RS/6000 running AIX as the target platform, but, because the system was implemented completely using standard C++, it is easily portable to other platforms and indeed was tested under Linux with g++ and Windows NT with Visual C++.

10.4.1 Sample Results

We report our experience with solving the slab-design problem on two real world problems, sd_1 and sd_2, based on data from the operations of the steel plant.

First, we decomposed the order-compatibility graph (Figure 10.6(a)) into connected components, where every pair of vertices in a component is connected by a path and no path exists between any two vertices v_i and v_j that belong to different components. We used a simple depth-first search (Horowitz and Sahni (1978)) procedure to find these connected components. Decomposition allowed us to solve independent subproblems separately. Moreover, we needed to decompose the problem because the algorithms we used for bin packing and maximum weight nonbipartite matching would become computationally expensive as the problem got larger.

If a connected component has a balanced and sparse cut, it is possible to further reduce the size of the problems to be considered by branching on the edges of the cut. A cut C of G is a set of edges that, when removed, decomposes the graph into two or more connected components. A sparse cut is a cut with a small number of edges. A balanced sparse cut is a sparse cut that decomposes the graph into two components of roughly the same size. Unfortunately, for our problems, the balanced cuts were of about size 100, which was not small enough to exploit divide-and-conquer strategies.

Problem sd_1 has 1,090 connected components while problems sd_2 has 1,279 connected components. Tables 10.2 and 10.3 show the data characteristics for the 10 largest components for each of these problems. For each component, the number of orders in the component and the number of edges in the order-compatibility graph of the component are indicated in second and third columns respectively. The edge density for a component, calculated as $\frac{|E|}{\binom{|V|}{2}}$ where $|V|$ = number of orders and $|E|$ = number of edges, is provided in column 4. For a component, the minimum (resp. target, maximum) order weight is calculated as the sum of minimum (resp. target, maximum) order weights of the orders in the component. This information is provided in columns 5, 6 and 7. The last row provides the values of these characteristics for the entire problem.

Connected Components	Number of Orders	Number of Edges	Edge Density	Min. Order Weight	Target Order Weight	Max. Order Weight
1	711	21549	8.52%	5585.92	6812.32	8649.09
2	552	18808	12.36%	1998.76	2698.36	3923.65
3	521	10017	7.38%	1630.56	2281.14	3057.19
4	205	321	1.52%	927.78	1277.98	2382.90
5	191	943	5.18%	1015.33	1680.61	3216.04
6	167	1671	12.04%	805.27	1039.32	1334.74
7	153	998	8.58%	651.17	942.18	1230.60
8	137	234	2.50%	4255.15	4892.28	5616.55
9	128	569	7.00%	769.34	907.56	1271.44
10	116	3010	45.12%	1345.65	1603.76	2010.15

Total # of Components	Total # of Orders	Total # of Edges	Edge Density	Total Min. Order Weight	Total Target Order Weight	Total Max. Order Weight
1090	8353	76248	0.21%	144585.39	177409.77	249592.58

Table 10.2. The characteristics for the 10 largest components of problem sd_1. The last row shows the characteristics for the entire problem.

A summary of the computational results for problems sd_1 and sd_2 is shown in Table 10.4. On average, the percentage surplus, calculated as $\frac{\text{Surplus Weight}}{\text{Total Slab Weight}} \times 100$, was 5.7 percent for problem sd_1 and 6.9

Connected Components	Number of Orders	Number of Edges	Edge Density	Min. Order Weight	Target Order Weight	Max. Order Weight
1	205	1165	5.57%	710.82	1012.94	1286.80
2	202	505	2.48%	1640.78	1946.30	2754.48
3	188	1496	8.51%	799.57	1003.69	1193.38
4	184	1144	6.79%	1405.25	1610.80	2362.85
5	156	535	4.42%	795.93	1209.19	2031.62
6	150	4046	36.20%	559.15	727.68	1007.69
7	142	482	4.81%	984.49	1257.00	1476.09
8	137	950	10.19%	415.87	570.87	901.47
9	131	427	5.01%	1114.16	1315.10	1863.53
10	125	1194	15.40%	692.04	823.09	1147.44

Total # of Components	Total # of Orders	Total # of Edges	Edge Density	Total Min. Order Weight	Total Target Order Weight	Total Max. Order Weight
1279	8350	32609	0.09%	131269.73	162887.42	227394.26

Table 10.3. The characteristics for the 10 largest components of problem sd_2. The last row shows the characteristics for the entire problem.

percent for sd_2. The average slab weight was 23.32 for problem sd_1 and 25.05 for problem sd_2. The heuristic required 333 seconds for all the components of problem sd_1 and 162 seconds for all the components of problem sd_2. Across the different data sets we encountered, the time required for constructing and decomposing the order-compatibility graph was about 10 to 15 percent of the total time required by the heuristic.

10.4.2 Savings

By increasing the average slab weight and thereby reducing the number of slabs, we reduce cutting costs and slab-handling costs. Equally important, by reducing surplus weight, we reduce the expected loss resulting from scrapped (never sold) steel. Our client estimated a savings of approximately US$700,000 per year for each ton increase in average slab weight and a savings of approximately US$500,000 per year for each percent decrease in surplus weight. The initial implementation of the system produced an increase of around 1.3 tons in average slab weight and a reduction of around three percent of surplus weight, which together represent a yearly savings of around US$2.5 million. The total cost of development was around US$1.5 million.

Before it adapted our slab-design system, our customer was using a mainframe-based COBOL program. So, executing and maintaining

Problem	Connected Components	Number of Slabs	Applied Order Weight	Total Slab Weight	Surplus Weight	Average Slab Weight	Total Time
sd_1	1	246	6800.32	7251.98	451.66	29.47	82.57
	2	110	2736.75	2942.20	205.45	26.74	68.53
	3	105	2385.44	2644.96	259.52	25.19	47.27
	4	68	1323.34	1521.61	198.27	22.37	3.50
	5	86	1735.06	1976.79	241.73	22.98	7.13
	6	46	1048.83	1105.99	57.16	24.04	6.63
	7	64	958.41	1115.40	156.99	17.42	3.14
	8	207	4957.06	4966.75	9.69	23.99	4.76
	9	40	930.35	1024.78	94.43	25.61	2.72
	10	60	1612.94	1684.89	71.95	28.08	11.18
	Overall	8135	178914.80	189782.42	10867.62	23.32	333.00
sd_2	1	57	1057.64	1236.65	179.01	21.69	5.99
	2	82	1957.69	2104.52	146.83	25.66	4.59
	3	45	1013.14	1129.46	116.32	25.09	6.62
	4	65	1628.06	1758.36	130.30	27.05	5.29
	5	68	1261.23	1520.67	259.44	22.36	4.93
	6	31	719.68	775.17	55.49	25.01	13.95
	7	59	1276.47	1445.61	169.14	24.50	2.95
	8	24	589.01	631.69	42.68	26.32	3.57
	9	54	1324.24	1437.54	113.30	26.62	2.82
	10	29	823.52	856.69	33.17	29.52	4.53
	Overall	7052	164404.39	176721.50	12317.11	25.05	162.00

Table 10.4. Summary of results for problems sd_1 and sd_2. For each component, the number of slabs designed to fill all the orders in the component is shown in column 3. The total weight occupied by the orders (applied order weight), the total slab weight and the total surplus weight (total slab weight - applied order weight) are indicated in columns 4, 5, and 6, respectively. The average slab weight, computed as $\frac{\text{Total Slab Weight}}{\text{Number of Slabs}}$, is shown in column 7. The time required is provided in the last column. For each problem, the last row provides the results for the entire problem.

the slab-design system has turned out to be much less expensive; it runs faster, covers more functionality and hence requires less human intervention, runs on cheaper hardware, and is written in a modern, object-oriented language and so is easier to maintain.

10.5 Conclusions

The steel plant estimates the savings from using the heuristic as around \$2.5 million per year. Although it is unlikely that slab-design operations in other steel mills are exactly the same, we believe that the main ideas of our heuristic can be applied for slab design problems in other similar companies. The problem formulation and the solution method generalize beyond the steel application to other process industries such as paper and metal manufacturing. As such, our approach can also be used for optimizing the design problems in those industries.

References

Ahuja, A.K., T. L. Magnanti, J. B. Orlin. 1993. *Network Flows.* Prentice Hall, Englewood Cliffs, NJ.

Coffman, E.G., M. R. Garey, D. S. Johnson. 1997. Approximation algorithms for bin packing: A survey. D.S. Hochbaum, ed. *Approximation Algorithms for NP-hard Problems.* PWS Publishing Company, Boston, MA.

Denton, B., D. Gupta, K. Jawahir. 2003. Managing increasing product variety at integrated steel mills. *Interfaces* 33(2) 41-53.

Friesen, D.K., M. A. Langston. 1986. Variable sized bin packing, *SIAM Journal of Computing* 15(1) 222-230.

Gilmore, P.C., R. E. Gomory. 1961. A linear programming approach to the cutting stock problem. *Operations Research* 9(6) 849-859.

Gilmore, P.C., R. E. Gomory. 1963. A linear programming approach to the cutting stock problem, Part II. *Operations Research* 11(6) 863-888.

Hirayama, K., H. Kajihara, Y. Nakagawa. 1996. Application of a hybrid genetic algorithm to slab design problem. *Transactions of the Institute of Systems, Control and Information Engineers* 9(1) 395-402.

Horowitz, E., S. Sahni. 1978. *Fundamentals of Data Structures,* Computer Science Press, Inc, Rockville, MD.

Johnson, D. S. 1974. Fast algorithms for bin packing. *Journal of Computer and System Sciences* 8(3) 272-314.

Kalagnanam, J., M. Dawande, M. Trumbo, H. S. Lee. 2000. The inventory matching problem in the process industry. *Operations Research* (Practice Section) 48(4) 505-516.

Lawler, E. L. 1976. *Combinatorial Optimization: Networks and Matroids.* Holt, Rinehart and Winston, New York.

Martello, S., P. Toth. 1990. *Knapsack Problems: Algorithms and Computer Implementations.* John Wiley and Sons, Chichester, U.K.

Martello, S., P. Toth. 1981. A branch and bound algorithm for the zero-one multiple knapsack problem. *Discrete Applied Mathematics* 3(4) 275-288.

Newhart, D. D., K. L. Stott, F. J. Vasko. 1993. Consolidating product sizes to minimize inventory levels for a multi-stage production and distribution system. *Journal of the Operational Research Society,* 44(7) 637-644.

Papadimitriou, C. H., K. Steiglitz. 1982. *Combinatorial Optimization: Algorithms and Complexity.* Prentice Hall, Englewood Cliffs, NJ.

Vasko, F. J., M. L. Cregger, D. D. Newhart, K. L. Stott. 1993. A real-time one-dimensional cutting stock algorithm for balanced cutting patterns. *Operations Research Letters* 14(5) 275-282.

Vasko, F. J., M. L. Cregger, K. L. Stott, L. R. Woodyatt. 1994. Assigning slabs to orders - An example of appropriate model formulation. *Computers and Industrial Engineering.* 26(4) 797-800.

Vasko, F. J., J. A. McNamara, R. N. Parkes, F. E. Wolf, L. R. Woodyatt. 2000. A matching approach for replenishing rectangular stock sizes. *Journal of the Operational Research Society* 51(3) 253-257.

Vasko, F.J., D. D. Newhart, K. L. Stott. 1999. A hierarchical approach for one dimensional cutting stock problems in the steel industry that maximizes yield and minimizes overgrading. *European Journal of Operational Research* 114(1) 72-82.

Vasko, F.J., D. D. Newhart, K. L. Stott, F. E. Wolf. 1996. Using a facility location algorithm to determine optimum cast bloom lengths. *Journal of the Operational Research Society* 47(3) 341-346.

Vasko, F. J., D. D. Newhart, K. L. Stott, F. E. Wolf. forthcoming. A large-scale application of the partial coverage uncapacitated facility location problem. *Journal of the Operational Research Society.*

Vasko, F. J., F. E. Wolf, K. L. Stott. 1987. Optimal selection of ingot sizes via set covering. *Operations Research* 35(3) 346-352.

Vasko, F.J., F. E. Wolf. 1994. A practical approach for determining rectangular stock sizes. *Journal of the Operational Research Society* 45(3) 281-286.

Vasko, F.J., F. E. Wolf, J. A. McNamara. 1989. A multiple criteria approach to dynamic cold ingot substitution. *Journal of the Operational Research Society* 40(4) 361-366.

Chapter 11

A REVIEW OF LONG- AND SHORT-TERM PRODUCTION SCHEDULING AT LKAB'S KIRUNA MINE

Alexandra M. Newman*, Michael Martinez, Mark Kuchta
Colorado School of Mines

Abstract LKAB's Kiruna Mine, located in northern Sweden, produces about 24 million tons of iron ore yearly using an underground mining method known as sublevel caving. To aid in its ore mining and processing system, Kiruna has adopted the use of several types of multi-period production scheduling models that have some distinguishing characteristics, for example: (i) specific rules governing the way in which the ore is extracted from the mine; (ii) lack of an inventory holding policy; and (iii) decisions that are not explicitly cost-based. In this chapter, we review two models in use at Kiruna and three techniques we have employed to expedite solution time, support the efficacy of these techniques with numerical results, and provide a corresponding discussion.

Keywords: Integer programming, production scheduling, underground mining, applications

1. Introduction

Around 1700, two mountains were discovered in northern Sweden above the Arctic Circle which, nearly 200 years later, evolved into LKAB's Kiruna mine. This was when the English metallurgists Thomas and Gilchrist determined how to process high-quality steel from iron ore with a signifi cant phosphorus content. With this innovation, the economic value of the Kiruna site became evident. The mine has been in operation for over 100 years and currently produces approximately 24 million tons of iron ore per year.

The high-grade magnetite deposit at Kiruna supplies several postprocessing mills, which, in turn, send fi nished goods via rail to ports in Narvik, Norway

*Corresponding author

and Luleå, Sweden. Steel mills in Europe, the Middle East and the Far East are recipients of Kiruna's products, which are used to manufacture various items, such as kitchen appliances, automobiles, ships, and buildings.

Once the position and economic value of the orebody are assessed, the appropriate mining method must be determined. The method chosen depends on the depth at which the deposit lies and on its geometry, as well as on the structural properties of the overlying and surrounding earth. Open pit mining is appropriate when deposits lie fairly close to the surface. However, as the pit deepens, open pit mining becomes too costly because gradually sloping pit walls are necessary to prevent waste material from sliding down into the active area of the mine. At this point, the open pit mining operation either ceases or continues underground. There are a variety of underground mining methods, which are categorized as self-supported methods, supported methods, and caving methods. Kiruna currently uses sublevel caving, a technologically-advanced, underground caving method applied to vertically positioned, fairly pure, large, vein-like deposits.

After the mining method has been determined, the mine layout and ore retrieval system must be decided. Miners first drill ore passes that extend vertically from the current mining area down to the bottom of a new mining area where a transportation level is located. They then create horizontal sublevels on which to mine and access routes that run the length of the ore body within a sublevel. Finally, miners drill self-supported horizontal crosscuts through the ore body perpendicular to the access routes. Kiruna places sublevels 28.5 meters apart and crosscuts 25 meters apart. The crosscuts are seven meters wide and five meters high. From the crosscuts, miners drill near-vertical rings of holes in a fan-shaped pattern. Each ring contains around 10,000 tons of ore and waste. The miners place explosives in the holes and blast the rings in sequence, destroying the ceiling on the blasted sublevel, to recover the ore. Miners extract the ore on each sublevel, starting with the overlying sublevels and proceeding downwards. Within each sublevel, miners remove the ore from the hanging wall to the forefront of the mining sublevel, or the footwall. As the miners recover the ore from a sublevel, the hanging wall collapses by design and covers the mining area with broken waste rock.

The Kiruna mine is divided into 10 main production areas which are about 400 to 500 meters long. Each production area has its own group of ore passes, also known as a *shaft group.* Each group of ore passes is located at the center of the production area and extends from the surface to the lowest active level of the mine. One or two 25-ton-capacity electric load-haul-dump units (LHDs), i.e., vehicles that load, haul, and unload the ore, transport the ore on each sublevel within a production area from the crosscuts to the ore passes. (As many as about 20 LHDs can be simultaneously operational in the mine.) Twenty-car trains operating on the lowest active level transport the ore from the ore passes to a crusher, which breaks the ore into pieces small enough to effi ciently hoist to the surface via a series of vertical shafts. Figure 11.1 depicts a sublevel caving operation.

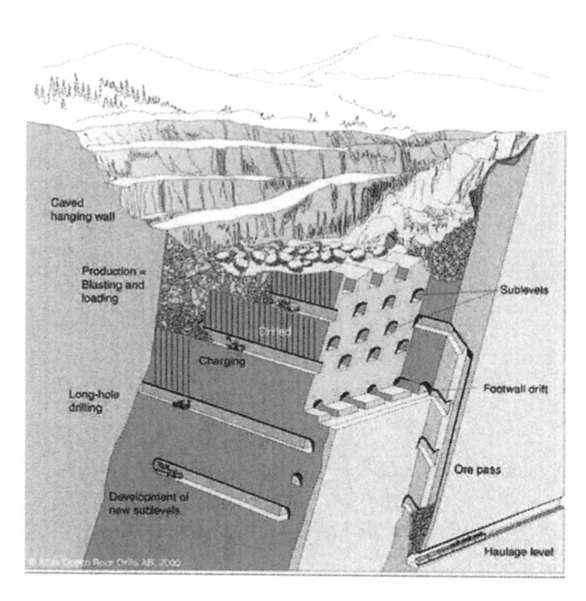

Figure 11.1. This figure depicts a sublevel caving operation in which an orepass extends vertically down to horizontal mining sublevels, access routes (depicted in the figure as the footwall drift) run the length of the ore body within a sublevel, and crosscuts are drilled perpendicular to the access routes. (Source: Atlas Copco 2000)

The site on which each LHD operates is called a *machine placement*, and is equivalent to between 10 and 20 *production blocks*. Machine placements possess the same height as the mining sublevel, are about 200 to 500 meters long, contain from one to three million tons of ore each, and extend from the hanging wall to the footwall. Each machine placement also possesses a series of notional *drawdown lines*, consisting of several production blocks each. The mine is subject to various operational restrictions concerning the relative order in which machine placements, drawdown lines, and the production blocks therein must be mined. We describe, and subsequently mathematically model, these constraints in Section 4. Figure 11.2 depicts the relationship between the machine placements, production blocks and drawdown lines.

After determining the mine layout and ore retrieval system, mine planners must institute an ore mining and processing system. At Kiruna, this system accounts for two main in situ *ore types* that differ in their phosphorus content. About 20% of the ore body contains a very high-phosphorous (P), apatite-rich magnetite known as *D ore*, and the other 80% contains a low-phosphorous,

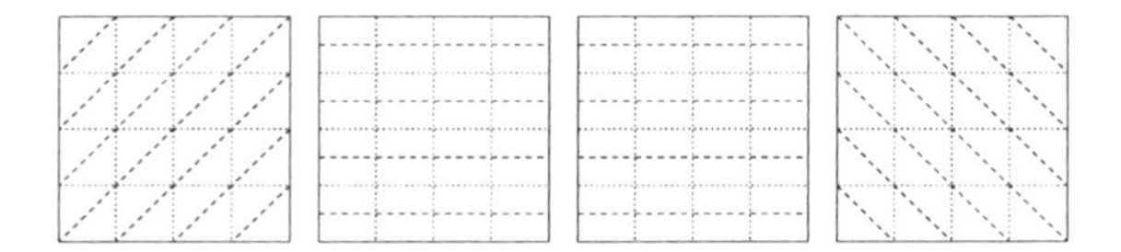

Figure 11.2. Four machine placements, in this example, outlined in solid font, are separated by dotted lines into 16 production blocks each. Four or seven dashed notional drawdown lines lie either horizontally or at 45 degree angles with respect to each machine placement.

high-iron (Fe) content magnetite known as *B ore*. For the entire deposit, the best quality B ore is about 0.025% P and about 68% Fe. The D ore varies considerably and has average grades of about 2% P.

The mine differentiates among the following three raw ore types based on the phosphorus content from its two in situ ore types: (i) B1 ore contains the least phosphorus, (ii) B2 ore contains somewhat more phosphorus, and (iii) D3 ore has the highest phosphorus content. The raw ore is sent to mills that process the B1 ore into high-quality fi nes (of the granularity of fi ne sand) simply by crushing and grinding the ore and removing the contaminants using magnetic separation. The mills process both B2 and D3 into ore pellets approximately spherical in shape by crushing and grinding the ore into a fi ner consistency than the B1 ore, then adding binding agents and other minerals such as olivine and dolomite, and fi nally fi ring the resulting product in large kilns to form hard pellets.

A production schedule is an integral part of the ore mining and processing system. The production schedule requires data regarding production targets, the relative position of each machine placement (and production block and draw-down line), and the iron ore reserves contained in each machine placement. Kiruna calculates iron ore reserves contained in each machine placement in two ways: (i) from an in situ geologic block model developed by drilling and sampling the deposit and then extrapolating estimates from the samples, and (ii) from samples of the production blocks after a production area has fi nished being developed. In the short term, the production schedule consists of determining when each production block should be mined, and, correspondingly, when each drawdown line should fi nish being mined. In the long term, the production schedule dictates the start dates for each machine placement. We seek a set of decisions that minimizes the deviation between extracted and planned production quantities while conforming to the operational restrictions of the mine.

The last stage of the production process consists of restoring the mined site to its original, or, at the very least, an environmentally acceptable state.

Kiruna has adopted our models with success (see, e.g., Kuchta et al., 2004). In this book chapter, we describe several production scheduling models that aid in the fourth stage of mine planning, i.e., the institution of an ore mining and processing system. We explain the signifi cance of our modeling efforts, and mention several techniques we use to expedite solution time. The chapter is organized as follows: In Section 2, we review the relevant literature, focusing primarily on underground mine scheduling models. In Section 3, we describe Kiruna's pre-optimization efforts at scheduling. In Section 4, we contrast our optimization models with those of "classical" production scheduling, and present mathematical formulations. Section 5 outlines several procedures, both exact and heuristic, we use to expedite solution time. Section 6 demonstrates the effi cacy of these procedures, and Section 7 provides a discussion of our solution techniques. Section 8 concludes the chapter.

2. Literature Review

Early optimization work in underground mine scheduling consists of linear programs. Williams, Smith, and Wells, 1972, plan a years' worth of sublevel stoping operations for an underground copper mine using a linear programming model to determine the amount of ore to extract per month from each stope. The objective function minimizes deviations between successive months, seeking to produce a well-balanced schedule. Jawed, 1993, formulates a linear goal program for single-period production planning in an underground room-and-pillar coal operation. The decision variables determine the amount of ore to extract from a given location via a particular method, and the objective function minimizes production deviations from target levels. Tang, Xiong, and Li, 1993, and Winkler, 1998, heuristically integrate linear programming with simulation to address scheduling decisions. The linear program possesses continuous variables, which determine the amount of ore to extract; the simulation model evaluates discrete scheduling decisions. In the latter two examples, the use of simulation to determine optimal values of discrete decisions results in suboptimal schedules.

Trout, 1995, uses integer programming to maximize net present value of multi-time period scheduling decisions in an underground stoping mine for base metals (e.g., copper sulphide). The constraint set incorporates block sequencing, equipment capacity, and backfi ll indicators. However, the algorithm terminates early after 200 hours when the computer reaches memory capacity. Winkler, 1996, models production scheduling in an underground coal mine to minimize fi xed and variable extraction costs, but limits his model to a single time period.

Several years later, researchers formulate more tractable mixed integer programming models. Carlyle and Eaves, 2001, present a model that maximizes revenue from Stillwater's platinum and palladium sublevel stoping mine. The problem focuses on strategic mine expansion planning, so the integer decision variables schedule the timing of development and drilling, and stope preparation. The authors obtain near-optimal solutions after several hours of computing time for a 10-quarter time horizon. Smith et al., 2003, incorporate sequencing relationships, capacities, and minimum production requirements into their lead and zinc underground mine model. However, they significantly reduce the fidelity of the model by aggregating stopes into larger blocks. The resulting model, with time fidelity of one year, maximizes net present value over the life of the mine (here, 13 years). The model generates near-optimal results in less than an hour. The authors note that further research should refine the level of detail to account for ore grade fluctuations and block sequencing considerations. Sarin and West-Hansen, 2005, maximize net present value for an underground coal mine that uses longwall, room-and-pillar, and retreat mining. The primary constraints enforce precedence, smooth production levels between time periods, and capture deviation in coal quality. The authors tailor a Benders decomposition approach to solve some randomly generated problem instances, as well as offer a case study.

Over the last decade, successive efforts at production scheduling for the Kiruna mine sought a schedule of requisite length in a reasonable amount of solution time. Using the machine placement as the basic mining unit, initial attempts significantly shorten the time horizon, sacrificing schedule quality. Almgren, 1994, considers a one-month time frame; hence, in order to generate a five-year schedule, he runs the model 60 times. In a similar vein, Topal, 1998, and Dagdelen et al., 2002, iteratively solve one-year subproblems (with monthly fidelity) in order to achieve production plans for five-year and seven-year time horizons, respectively. These three models provide suboptimal solutions because they disregard part of the planning horizon. Additionally, Kuchta, 2002, develops a computer-assisted manual heuristic scheduling program. However, he admits that it is common to abandon partial schedules and restart the procedure due to the difficulty of satisfying target demands and the inability to assess the future impact of current scheduling decisions.

The literature on open pit (surface) mine production scheduling extends back to Lerchs and Grossmann's, 1965, seminal work, in which the authors develop a graph-theoretic algorithm for optimally calculating the ultimate pit limits, or "break-even" depth of the mine, below which operations should either terminate or switch to underground mining methods. A vein of related literature, e.g., Underwood and Tolwinski, 1998, Hochbaum and Chen, 2000, applies network models to the ultimate pit problem assuming a fixed production rate and cutoff grade (the ore grade that separates the profitable from the unprofitable material). Other work, e.g., Barbaro and Ramani, 1986, Smith, 1998, relaxes these restrictions, which results in integer programming models that are qualitatively similar to underground mining models, but with mathematically different constraint sets. We do not review this body of literature in detail.

3. Previous Manual Scheduler

Before Kiruna adopted a formal optimization model to plan its production, the mine was using a computer-assisted manual heuristic. The program was written in Microsoft Access 97 and included a user interface to allow data entry and program control, and to produce reports. All data are stored in the mine's central relational database, and a schematic overview tracked available machine placements by shaft group and mining sublevel throughout the relevant planning horizon. A scheduler would fi rst establish production targets for the three raw ore types for each month within the planning horizon, and would initialize the schedule by placing all active machine placements into the schedule. For the months after which the ore from the active machine placements could no longer reasonably meet production targets, an available machine placement that would best meet demand while adhering to mine sequencing constraints would be added to the schedule. The scheduling program would then assign start dates to all the production blocks within that machine placement according to various operational restrictions. This process would continue until a schedule of requisite length had been generated. Using this system, the scheduler would require fi ve days, for example, to devise a fi ve-year schedule. Furthermore, these schedules were clearly myopic; that is, they did not incorporate the effects on availability of machine placements even a few time periods into the future. A scheduler could produce schedules that were far from optimal or even infeasible, especially in the "out-years."

4. Optimization Model Formulations

There are distinguishing characteristics between traditional manufacturing settings and Kiruna's underground mining operations, which result in fundamentally different decisions, rules, and goals between the corresponding production schedules. We divide the planning horizon into two parts: the short term and the long term. In the short term, machine placements are currently active, and we are interested in determining how to mine each machine placement. That is, we wish to determine when each production block should be mined. Correspondingly, we wish to determine the time by which each notional drawdown line should fi nish being mined. In the long term, we must determine the time at which various machine placements should start to be mined, and we ignore the additional detail present in the short term concerning production blocks and drawdown lines.

In the long term, there are a variety of operational constraints that dictate the rate at which and the order in which machine placements can be mined. The number of machine placements that can be started in a given time period is restricted due to the availability of the crew that prepares the machine placements for mining. The number of active machine placements, i.e., machine placements currently being mined, is also restricted due to LHD availability. Whether a machine placement can (or must) be mined depends on the relative position of machine placements that have started to be mined. Specifi cally, certain machine placements beneath a given machine placement cannot start to be mined until some portion (typically, 50%) of the given machine placement has been mined, and machine placements to the right and left of a given machine placement must start to be mined after a specifi ed portion (also, typically 50%) of the given machine placement has been mined (to prevent blast damage on adjacent machine placements). These operational constraints are referred to as vertical and horizontal *sequencing constraints*, respectively.

In the short term, within a machine placement, the order in which production blocks must be mined is regulated by a series of notional drawdown lines, each of which spans several blocks either horizontally or at a 45 degree angle from each other within the machine placement. Production blocks in a drawdown line underneath a given drawdown line are precluded from being extracted until all ore in the given drawdown line is extracted. This mining pattern is necessary to correctly execute the sublevel caving method so that the mined out areas do not collapse on top of ore that is yet to be retrieved. Minimum and maximum production levels per month govern the rate at which the blocks within a machine placement are mined. These rates ensure continuous mining of all production blocks within a machine placement until it is mined out, which precludes the need to track partially-mined machine placements and to reblast the rock.

Although there are many costs involved in mining, our production schedule does not consider them explicitly. Analogous in conventional production scheduling to setting up a machine is preparing an area to be mined. The constraints just discussed implicitly consider these setup costs by requiring continuous production of a machine placement once it has started to be mined, and by limiting both the number of active machine placements and the number of machine placements that can start to be mined in each time period.

Kiruna's output can be estimated from the iron ore reserves contained in each machine placement. However, ironically, although the raw materials exist at the manufacturing site *a priori*, these reserves cannot be guaranteed to materialize at the correct time owing to operational constraints that govern the way in which ore is extracted from the mine, and because of uncertainty in ore grade estimation. If the mine falls short, or supplies excess, it has limited ability to extract a single ore type to rectify the loss, not only because of mine sequencing constraints but also because of the heterogeneity of ore types contained in a single machine placement. To exacerbate the problem, company policy prevents the mine from stockpiling ore. In actuality, there is only physical space in which to store about 50 kilotons (ktons) of extracted iron ore. Such a stockpile would be of limited benefi t anyway, because monthly demands are set to regulate the amount of ore processed at the mills, and a shortage in one time period cannot be compensated by a surplus in, say, the following time period.

Because of the three previously-mentioned factors: (i) implicit costs, (ii) the discrete nature of ore retrieval, and (iii) lack of stockpiling, the objective function of our production scheduling model minimizes the difference between the demanded and extracted quantities for each ore type and time period. This objective differs from most usual mine production planning objectives that address precious, rather than base, metals. Precious metals such as gold and silver are traded on, for example, the Commodity Exchange of New York, and mines extracting these metals maximize profi ts by producing as much as is economically viable given current market prices. Generally, these mines also hold stockpiles to synchronize the sale of their products with favorable market conditions. By contrast, markets associated with base metals such as iron ore are regionalized, as transportation costs are high relative to the value of the commodity. Within these markets, steel companies enter into a contract with an iron ore producer, settling on a price commensurate with the chemical and physical characteristics of the iron ore. Large buyers tend to influence contract prices between other buyers and iron ore producers. The negotiated prices generally hold for approximately a year, and iron ore producers are obliged to supply a certain amount of iron ore to each buyer with whom the producers hold a contract. Therefore, iron ore mines such as Kiruna are concerned with meeting contractual demands as closely as possible.

The ability to use an optimization model to generate production schedules for our setting, rather than relying on manually-generated schedules, is particularly important because a schedule cannot be repeated. It is never possible to extract the ore in the same way because a machine placement can never be mined twice, and no two machine placements necessarily look the same or are positioned the same way relative to other machine placements in the mine. Therefore, whereas in some production settings, schedules need only be regenerated when demand or cost data change, in the Kiruna model, schedules are continuously updated.

The temporal fi delity of our production schedules is monthly; this level of resolution corresponds approximately to the amount of time required to mine a production block.

We introduce two formulations: (i) the *long-term* model, which principally determines machine placement start dates, and (ii) the *combined* (short- and long-term) model, which determines not only machine placement start dates, but, for those machine placements already active, the amount mined from each production block in each time period, and fi nish dates for each drawdown line. Both models possess the same temporal fi delity. The difference between the two models is the extra level of detail that determines how the already-active machine placements in the short term should be mined. We introduce all notation fi rst. This is a slightly more detailed restatement of the notation and combined formulation introduced in Newman et al., 2005. A modifi ed version of the long-term formulation appears in Newman and Kuchta, 2005.

SETS:

- K = set of ore types
- V = set of shaft groups
- A = set of machine placements
- A_v = set of machine placements in shaft group v
- IA = set of inactive machine placements, i.e., machine placements that have not started to be mined
- $A_a^{\mathcal{V}}$ = set of machine placements whose start date is restricted vertically by machine placement a
- $A_a^{\mathcal{H}}$ = set of machine placements whose start date is forced by adjacency to machine placement a
- A_t= set of machine placements that can be mined in time period t
- B = set of production blocks
- B_a = set of production blocks in machine placement a
- B_l = set of production blocks in drawdown line l
- B_t= set of production blocks that can be mined in time period t
- L = set of drawdown lines
- L_a = last (i.e., most deeply positioned) drawdown line in machine placement a
- LC = set of drawdown lines constrained by another drawdown line
- L_l = set of drawdown lines that constrain drawdown line l
- $L_a^{\mathcal{V}}$ = drawdown line whose fi nish date vertically restricts starting to mine machine placement a
- $L_a^{\mathcal{H}}$ = drawdown line whose fi nish date forces the start date of machine placement a by adjacency
- L_t = set of drawdown lines that can be mined in time period t

- $\mathcal{T}$ = set of time periods composing the entire time horizon
- T = set of time periods composing the short-term time horizon ($\subset \mathcal{T}$)
- $\mathcal{T}_a$ = set of time periods in which machine placement a can start to be mined (restricted by machine placement location and the start dates of other relevant machine placements)
- $\mathcal{T}_b$ = set of time periods in which production block b can be mined (restricted by production block location and the start dates of other relevant production blocks)
- $\mathcal{T}_l$ = set of time periods in which drawdown line l can fi nish being mined (restricted by drawdown line location and the fi nish dates of other relevant drawdown lines)
- $\hat{\mathcal{T}}_l$ = time period by which all blocks in drawdown line l must fi nish being mined

PARAMETERS:

- p_t = penalty associated with deviations in time period t ($= |\mathcal{T}| + 1 - t$)
- LHD_t = number of machine placements that can start in time period t
- LHD_v = maximum number of active machine placements in shaft group v
- d_{kt} = target demand for ore type k in time period t (ktons)
- $r_{at'tk}$ = reserves of ore type k available at time t in machine placement a given that the machine placement started to be mined at time t' (ktons)
- R_{bk} = reserves of ore type k contained in production block b (ktons)
- $\underline{C}_{at}$ = minimum production rate of machine placement a in time period t (ktons per time period)
- $\bar{C}_{at}$ = maximum production rate of machine placement a in time period t (ktons per time period)
- $\rho_{at't} = \begin{cases} 1 & \text{if machine placement } a \text{ is being mined at time } t \text{ given that} \\ & \text{it started to be mined at time } t' \\ 0 & \text{otherwise} \end{cases}$

DECISION VARIABLES:

- $\bar{z}_{kt}$ = amount mined above the target demand for ore type k in time period t (ktons)

- $\underline{z}_{kt}$ = amount below the target demand for ore type k in time period t (ktons)
- x_{bt} = amount mined from production block b in time period t (ktons)
- $w_{lt} = \begin{cases} 1 & \text{if we finish mining all blocks contained in drawdown line } l \\ & \text{by time period } t \\ 0 & \text{otherwise} \end{cases}$
- $y_{at} = \begin{cases} 1 & \text{if we start mining machine placement } a \text{ at time period } t \\ 0 & \text{otherwise} \end{cases}$

Formulation for the Long-term Model:
(L):

$$\min \sum_{k,t} (p_t)(\underline{z}_{kt} + \bar{z}_{kt})$$

subject to:

$$\sum_{a \in A_t} \sum_{t' \in T_a, \leq t} r_{at'tk} y_{at'} + \underline{z}_{kt} - \bar{z}_{kt} = d_{kt} \quad \forall k \in K, t \in \mathcal{T} \tag{11.1}$$

$$\sum_{a \in A_v \cap A_{t'}} \sum_{t' \in T_a, \leq t} \rho_{at't} y_{at'} \leq LHD_v \quad \forall v \in V, t \in \mathcal{T} \tag{11.2}$$

$$\sum_{a \in IA \cap A_t} y_{at} \leq LHD_t \quad \forall t \in \mathcal{T} \tag{11.3}$$

$$\sum_{t \in T_a, \leq t'} y_{at} \geq y_{a't'} \quad \forall a \in A, a' \in A_a^{\mathcal{V}}, t' \in T_{a'}, a' \neq a \tag{11.4}$$

$$\sum_{t' \in T_{a'}, \leq t} y_{a't'} \geq y_{at} \quad \forall a \in A, a' \in A_a^{\mathcal{H}}, t \in T_a, a' \neq a \tag{11.5}$$

$$\bar{z}_{kt}, \underline{z}_{kt} \geq 0 \quad \forall k, t; \;\; y_{at} \text{ binary } \forall a, t \tag{11.6}$$

We also introduce a model that combines both long- and short-term decisions, (P). This model has the advantage over (L) in that its schedules consider decisions at a finer level of detail than at the machine placement level for the first few time periods; therefore, the quality of solutions it produces, as a function of the deviation from target demands, is better. See Newman et al., 2005, for such a comparison. However, this model has the drawback that it is much larger, thereby limiting the length of the time horizon for which it can be solved. Ideally, (L) is used for strategic planning purposes, whereas (P) is used for operational decision making.

(P) retains the same objective and last four constraints as in (L), modifies the first and second constraints in (L) (now (11.7) and (11.8), respectively), and possesses the following additional constraints, i.e., (11.9) - (11.18):

Formulation for the Combined Model:
(P):

$$\min \sum_{k,t} (p_t)(\underline{z}_{kt} + \bar{z}_{kt})$$

subject to:
(11.3), (11.4), (11.5), (11.6)

$$\sum_{a \in A_t} \sum_{t' \in T_a, \leq t} r_{at'tk} y_{at'} + \sum_{b \in B_t} \frac{R_{bk}}{\sum_{\hat{k} \in K} R_{b\hat{k}}} x_{bt} + \underline{z}_{kt} - \bar{z}_{kt} = d_{kt} \quad \forall k \in K, t \in T \quad (11.7)$$

$$\sum_{a \in A_v \cap A_{t'}} \sum_{t' \in T_a, \leq t} \rho_{at't} y_{at'} + \sum_{a \in A_v} \sum_{l \in L_a \cap L_t} (1 - w_{l\hat{t}}) \leq LHD_v \quad \forall v \in V, t \in T_l, \hat{t} \in \hat{T}_l \quad (11.8)$$

$$\sum_{a \in A_t} \sum_{k \in K} \sum_{t' \in T_a, \leq t} r_{at'tk} y_{at'} + \sum_{b \in B_t} x_{bt} = \sum_{k \in K} d_{kt} \quad \forall t \in \mathsf{T} \quad (11.9)$$

$$\sum_{t \in T_b} x_{bt} \leq \sum_{k \in K} R_{bk} \quad \forall b \in B \quad (11.10)$$

$$w_{lt} \leq w_{l,t+1} \quad \forall l, t \quad (11.11)$$

$$\sum_{b \in B_l} \sum_{u \leq t} x_{bu} \geq \sum_{b \in B_l} \sum_{k \in K} R_{bk} w_{lt} \quad \forall l \in L, t \in T_l \quad (11.12)$$

$$\sum_{u \leq t} x_{bu} \leq \sum_{k \in K} R_{bk} w_{\hat{l}t} \quad \forall l \in LC, b \in B_l, \hat{l} \in L_l, t \in T_{\hat{l}} \quad (11.13)$$

$$\sum_{b \in B_a \cap B_t} x_{bt} \leq \bar{C}_{at} \quad \forall a \in A, l \in L_a, t \in T_l \quad (11.14)$$

$$\sum_{b \in B_a \cap B_t} x_{bt} \geq \underline{C}_{at}(1 - w_{lt}) \quad \forall a \in A, l \in L_a, t \in T_l \quad (11.15)$$

$$w_{lt} \geq y_{\tilde{a}t} \quad \forall a \in A, \tilde{a} \in A_a^{\mathcal{V}}, l \in L_a^{\mathcal{V}}, t \in T_{\tilde{a}} \quad (11.16)$$

$$\sum_{t \in T_{\tilde{a}}, \leq \hat{t}} y_{\tilde{a}t} \geq w_{l\hat{t}} \quad \forall a \in A, \tilde{a} \in A_a^{\mathcal{H}}, l \in L_a^{\mathcal{H}}, \hat{t} \in T_l \quad (11.17)$$

$$x_{bt} \geq 0 \quad \forall b, t; \quad w_{lt} \text{ binary } \forall l, t \quad (11.18)$$

The objective function measures the total weighted tons of deviation, placing more emphasis on meeting demand in the earlier time periods. Not only does the weighting scheme place a greater penalty on deviations from demand known with more accuracy, but it also breaks symmetry which helps to guide the branch-and-bound algorithm.

Constraints (11.1) record for each ore type and time period the deviation between the realized and target demand of ore production for the long-term model. Constraints (11.2) limit the number of active machine placements in each shaft group and time period in the long-term model. Constraints (11.3) limit the number of machine placements that can be started in a time period. Constraints (11.4) and (11.5) enforce vertical and horizontal sequencing, respectively, between machine placements. Constraints (11.6) enforce nonnegativity and integrality, as appropriate.

For the combined short- and long-term model, the left hand side of the deviation tracking constraint, (11.7), is modifi ed to include ore mined in the short term. Similarly, the constraint limiting the number of active machine placements in each shaft group, (11.8), includes those machine placements already being mined, i.e., those machine placements that are considered in the short term. Constraints (11.9) require that, for each time period in the short term planning horizon, the total target amount of ore, regardless of ore type, is met to prevent the postprocessing mills from sitting idle. Constraints (11.10) preclude mining more than the available reserves within a production block. Constraints (11.11) indicate that once a drawdown line has fi nished being mined, it has fi nished for the horizon. Constraints (11.12) establish equivalency between fi nishing to mine a drawdown line and mining all the production blocks within that drawdown line. Constraints (11.13) preclude a production block in a drawdown line from starting to be mined until all blocks in constraining drawdown lines have been mined. Constraints (11.14) and (11.15) enforce monthly maximum and minimum production rates, respectively. Constraints (11.16) and (11.17) enforce vertical and horizontal sequencing, respectively, between drawdown lines modeled in the short term, and machine placements modeled in the long term. Finally, constraints (11.18) enforce nonnegativity and integrality, as appropriate.

Typical model scenarios consist of 65 machine placements, 102 production blocks, 3 ore types, 10 shaft groups and 36 time periods. Long-term model instances contain about 500 binary variables and 1000 constraints while combined model instances contain more than 6000 binary variables and over 13,000 constraints.

5. Solution Methodologies

Because the models are so large, obtaining solutions in a reasonable amount of time, i.e., in a few hours at most, by solving the monoliths, i.e., the full, detailed models, is impossible for more than about 10 time periods. To date, we have developed three different solution techniques, both exact and heuristic, to expedite solution time for either the long-term model, the combined model, or for both models. In this section, we summarize each of the following techniques: (i) variable reduction, (ii) variable aggregation, and (iii) decomposition.

Variable Reduction

We use exact techniques to eliminate all variables that would necessarily assume a value of zero in the optimal solution, and, in fact, in any feasible solution. Specifi cally, for the long-term and combined models, we can determine earliest and latest possible start dates for each machine placement in the scheduling horizon by using the sequencing constraints and the initial condition of the mine. For example, if we know that machine placement a is exactly vertically positioned above machine placement b, then machine placement b cannot start to be mined any earlier than the time at which 50% of machine placement a has been mined. Similarly, if machine placements c' and c'' are directly to the left and right of machine placement a, and machine placement a is active, then we know machine placements c' and c'' must start to be mined once 50% of machine placement a has been mined. After accounting for shaft group constraints, we can use minimum and maximum mining rates to determine these earliest and latest possible start dates, which propagate to machine placements deeper in the mine. Then, rather than defi ning the binary variables y_{at} on the complete set $\{A \times T\}$, for each machine placement a we use the restricted set T_a to denote the eligible time periods in which machine placement a can start to be mined.

For the combined model, we can determine an earliest fi nish date for a drawdown line because each machine placement in the short term is active. Because we know: (i) the maximum and minimum production rates for each machine placement, (ii) the position at which a given drawdown line lies relative to all other drawdown lines whose position could affect accessing the given drawdown line, and (iii) the aggregate tonnage of all blocks comprising each drawdown line, we can compute the earliest date at which a given drawdown line can fi nish being mined; this is the sum of the time at which the fi rst drawdown line in the machine placement fi nishes being mined and the shortest amount of time required for all drawdown lines overlying the given drawdown line to be mined. Similarly, the latest time at which a drawdown line could fi nish being mined is the sum of the time at which the fi rst drawdown line in the machine placement fi nishes being mined and the longest amount of time required for all drawdown lines overlying the given drawdown line to be mined. Then, rather than defi ning the binary variables w_{lt} on the complete set $\{L \times T\}$, for each drawdown line l we use the restricted set T_l to denote the eligible time periods in which drawdown line l can fi nish being mined.

We can use similar principles to establish earliest start and latest fi nish dates for production blocks (in the combined model), thereby using the set T_b to eliminate x_{bt} variables that correspond to mining a production block before its earliest start date or after its latest fi nish date. Although the direct benefi t of eliminating these continuous variables is small, an indirect benefi t of an earliest start date for each production block is its use in establishing an earliest start date for a drawdown line, which is simply the earliest early start date among all blocks in a drawdown line. Earliest start dates for a drawdown line help to eliminate irrelevant terms in constraint (11.8). Details of this procedure, including the early start algorithm, can be found in Martinez et al., 2005.

Variable Aggregation

We use an optimization-based heuristic, which we term the *aggregation procedure*, to eliminate all but a "reasonably good" set of starting times for each machine placement, allowing us to restrict the model to a subset of start date choices beyond the restrictions we determine with the early and late start algorithms. To date, we have found that this procedure is useful only for eliminating the y_{at} variables because the loss of fi delity inherent in the procedure would be unacceptable for short-term decisions.

Our aggregation procedure consists of fi rst "collapsing" the time periods in our long-term production scheduling model to reduce its size. We aggregate demands and the amount of ore in each machine placement that can be mined in a single time period into data corresponding to *phases*, where each phase consists of an equal number of consecutive time periods. We call the model consisting of these aggregated phases the *aggregated model.*

We solve the aggregated model to determine a reasonable interval of start times for each machine placement by noting the phase in which each machine placement a starts to be mined in the optimal solution for the aggregated model (say, τ_a^*). We then limit the start times of machine placement a in the original model to those time periods contained in τ_a^*. For example, consider an aggregated model consisting of two time periods per phase. If, after solving the aggregated model to optimality, the machine placement in the aggregated model starts to be mined in phase 2, we would require that the machine placement either start to be mined in time period 3 or time period 4 (the two time periods contained in that phase). We call this form of the monolith in which we replace T_a by the restricted set of time periods the *restricted problem.*

Because we change the objective function in the aggregated model (from the original model) by eliminating penalties for deviations incurred during consecutive time periods within a phase, we cannot expect to obtain an optimal solution to the original problem by solving the restricted problem. We expect this deviation from optimality to increase with the number of time periods contained in a phase. Therefore, in practice, we actually relax the way in which we use the solution obtained by solving the aggregated model, thereby enhancing the solution quality of the original model. Specifi cally, we allow start times for a machine placement in the restricted model to be: (i) the time periods in the phase in which the machine placement starts to be mined in the aggregated model, and (ii) the time periods contained in the n phases directly before and directly after the phase in which the machine placement starts to be mined in the aggregated model. The larger the value of n, the better the solution, but the longer the solution time for the restricted model. Despite the fact that there is no guarantee of regaining any loss of optimality by adding this flexibility in the restricted model, we have been able to achieve good performance in practice, which we discuss in the next section.

We refer the interested reader to Newman and Kuchta, 2005, for a more detailed description of this procedure, which includes mathematical formulations, as well as method for computing a bound on the worst case performance of the procedure.

Decomposition

Finally, we introduce another heuristic technique, which we apply to the combined model. In this procedure, we decompose the monolithic model into subproblems that represent extreme versions of the original model. Specifically, k subproblems, (P_1), (P_2),... (P_k), consider that the mine consists solely of oretype k. The other two subproblems, (P_o) and (P_u), penalize only overdeviation and only underdeviation, respectively.

Applied to the drawdown line variables (w_{lt}), the heuristic consists of the following steps: (i) solve $k + 2$ subproblems; (ii) from subproblem i solution ($i = 1..k + 2$) and for each drawdown line l, note the time period in which the drawdown line finishes being mined, e_{li}. From this, let a heuristic earliest finish date for drawdown line l considering all $k+2$ subproblems be $\min_i\{e_{li}\}$, and a heuristic latest finish date for drawdown line l considering all $k + 2$ subproblems be $\max_i\{e_{li}\}$; (iii) using the results from (ii), restrict the monolith such that the variables associated with finishing to mine drawdown line l before its earliest heuristic finish date equal zero, and such that the sum of those variables associated with finishing to mine drawdown line l between its earliest heuristic finish date and latest heuristic finish date is greater than or equal to 1. Then, solve this constrained version of the monolith.

Note that the constraint set of each (P_k) is identical to that of (P) with the exception of the first (demand) constraint, which is elasticized. The constraint sets of (P_o) and (P_u) are identical to that of (P). Hence, any solution we obtain for (P_i) is feasible for the original model. Furthermore, (P_i) solves quickly relative to (P) because fewer tradeoffs are necessary when optimizing.

We can also apply this technique to the machine placement variables (y_{at}), establishing heuristic earliest and latest start dates for each machine placement a. Note, however, that whereas all drawdown lines are mined over the course of the model horizon, all machine placements may not be. Therefore, we use the information from (P_i) more loosely, generally indicating whether to start to mine a machine placement over the horizon, rather than when to start mining the machine placement. Martinez et al. (2005) provide details.

6. Numerical Results

We present summary results from implementing the three procedures just described. We conduct numerical experiments on the long-term model with the AMPL programming language (Fourer et al., 2003; and Bell Laboratories, 2001) and the CPLEX solver, Version 7.0 (ILOG Corporation, 2001) using a Sun Ultra 10 machine with 256 MB RAM. We conduct numerical experiments on the combined model with the AMPL programming language and the CPLEX solver, Version 9.0 (ILOG Corporation, 2004) using a Sunblade 1000 Unix workstation with 1 GB RAM. In all cases, we use the CPLEX parameter settings that provide the best performance. All data are taken from LKAB's Kiruna mine.

Our principal data set for the combined model (P) possessing 36 time periods and 3 ore types contains 6084 binary variables, 3888 continuous variables, and 13018 constraints. If we apply the early start, late start, early fi nish and late fi nish algorithms to our model instance simultaneously, the resulting model possesses 1103 binary variables, 687 continuous variables, and 2639 constraints.

We conduct our numerical experiments to test the aggregation procedure using fi ve data sets slightly modifi ed from the one above. Specifi cally, the data sets contain two, rather than three, ore types. The mine is considering recategorizing its ore in this way, and was interested in strategic schedules that reflected this change. In fact, these model instances turned out to be more diffi cult to solve than the three-ore type instances studied in, e.g., Kuchta et al., 2004. A typical monolith contains 500 binary variables and 1000 constraints. We employ two time periods per phase and the best performing relaxation for the way in which we use the solution obtained by solving the aggregated model. That is, eligible machine placement start times in the restricted model are: (i) the two time periods in the phase in which the machine placement starts to be mined in the aggregated model, and (ii) the time periods contained in the two phases adjacent to that in which the machine placement starts to be mined in the aggregated model. A typical aggregated model contains 260 binary variables and 530 constraints; typical restricted model size is 350 binary variables and 740 constraints. Averaging across our fi ve scenarios, we obtain an objective function value about 5% inferior to that provided by the monolith, but in 10% of the time required for the monolith to obtain its optimal solution.

We test the decomposition procedure applied to the combined model, (P), based on the same data set as that to which we apply the early start, late start, early fi nish and late fi nish algorithms, and we specifi cally examine this case after we apply these algorithms. Executing the decomposition procedure on our principal data set, including solving all (P_i) to within at least 1% of optimality, and then solving the resulting, restricted problem (which contains 781 binaries and 2011 constraints) requires 865 seconds of CPU time. Were we to run the monolith for the same amount of time, the objective would be about 3% worse. Were we to run the monolith until the objective matched that obtained by the decomposition procedure, the run time would be more than twice as long.

7. Discussion

In the previous several sections, we have just described, and provided numerical support for, three different methods whereby we expedite solution time of two different models for underground production scheduling, (L) and (P). The variable reduction techniques are exact methods, and should be used to provide a tightly-formulated optimization model in any circumstance. The exact methods we use to determine a restricted set of start and fi nish dates for all entities, i.e., production blocks, machine placements and drawdown lines, can be simply coded and run for a particular data set in a matter of seconds. However, the use of these algorithms alone does not always result in a tractable monolith, particularly if one is interested in production plans that span multiple years.

The aggregation procedure provides a heuristic means through which one can reduce the number of binary, specifi cally, y_{at}, variables virtually as much as necessary. Suppose, for example, that we employ the smallest possible number of time periods contained in each phase and allow no relaxation for the way in which we use the solution obtained by solving the aggregated model. This yields a restricted model with only two time periods in which each machine placement can start to be mined. A solution from such a model is not necessarily of good quality. However, a rough solution of this kind may be adequate for very long-term, e.g., 10-15 year, plans. At least, one can obtain a feasible solution using this method, and the demand and ore reserve data are not known with high accuracy in any case. The aggregation procedure is general enough such that it can be applied in other multi-period production scheduling settings, see, e.g., Martin, 1999, Section 1.3.4, and the references contained therein.

Finally, although the idea of problem decomposition is not new, see, e.g., Newman and Yano, 2000, for a spatial decomposition approach or the early attempts at developing production scheduling models for Kiruna mentioned in the literature review for temporal decomposition, the ability of a decomposition method to produce good-quality solutions is varied. To date, we have found that our procedure works well for determining earliest and latest fi nish dates for drawdown lines, because of the short-term time line on which these decisions are made. The procedure also works reasonably well for establishing heuristic earliest and latest start dates for machine placements. However, solving the decomposed models (in our case, (P_i)) quickly is crucial to the success of any decomposition procedure. We continue to seek ways to solve these subproblems expeditiously.

8. Summary and Future Work

In this chapter, we introduce two multi-period production scheduling models for a large underground mine, and review three techniques we use to reduce solution time for these models. Production scheduling for underground mines is a complex process. Among its important distinguishing characteristics are the existence of specifi c types of sequencing constraints, and, in our case, the lack of inventory held at the mine. Clearly absent from our models are costs, most importantly, setup costs and time-dependent processing costs, whose effects are implicitly considered in the operational constraints and in the contractual agreements which we strive to meet.

Despite the fact that we have developed a variety of techniques to increase tractability, the models remain diffi cult to solve for long planning horizons. Although the models are nonetheless currently useful, future research will involve continuing to develop methodologies to enable us to solve models with an increased planning horizon length. One problem we encounter is a loose lower bound. We hope to be able to develop valid and effective cuts, not only to decrease the solution time of the monolith, but also to enhance our decomposition procedure.

Acknowledgments

The authors would like to thank LKAB for the opportunity to work on this challenging project and for permission to publish these results. The authors would specifi cally like to thank LKAB employees Hans Engberg, Anders Lindholm, and Jan-Olov Nilsson for providing information, assistance, and support. We thank Professors J.K. and F.S. Newman of the University of Illinois at Urbana-Champaign for proofreading this chapter. We also thank the editor of this book, Professor Jeffrey Herrmann, for the invitation to write this chapter.

References

Almgren, T. (1994). *An Approach to Long Range Production and Development Planning with Application to the Kiruna Mine, Sweden*, Luleå University of Technology, Doctoral Thesis number 1994:143D.

AMPL. (2001). Version 10.6.16, Bell Laboratories.

Barbaro, R.W. and R.V. Ramani. (1986). "Generalized Multiperiod MIP Model for Production Scheduling and Processing Facilities Selection and Location," *Mining Engineering* **38**(2): 107-114.

Carlyle, M. and B.C. Eaves. (2001). "Underground Planning at Stillwater Mining Company," *Interfaces* **31**(4): 50-60.

CPLEX. (2001). Version 7.0, ILOG Corporation.

CPLEX. (2004). Version 9.0, ILOG Corporation.

Dagdelen, K., M. Kuchta, and E. Topal. (2002). "Linear Programming Model Applied to Scheduling of Iron Ore Production at the Kiruna Mine, Kiruna, Sweden," *Transactions of the Society for Mining, Metallurgy, and Exploration*, **312**: 194-198.

Fourer, R., D. Gay, and B. W. Kernighan. (2003). AMPL: A Modeling Language for Mathematical Programming, Thompson Learning, Pacifi c Grove, CA.

Hochbaum, D. S. and A. Chen. (2000). "Performance Analysis and Best Implementations of Old and New Algorithms for the Open-Pit Mining Problem," *Operations Research* **48**(6): 894-914.

Jawed, M. (1993). "Optimal Production Planning in Underground Coal Mines through Goal Programming–A Case Study from an Indian Mine," in *Proceedings, 24th International Symposium on the Application of Computers in the Mineral Industry*, Montreal, Quebec, Canada: 43-50.

Kuchta, M. (2002). "A Database Application for Long Term Production Scheduling at LKAB's Kiruna Mine," in *Proceedings, 30th International Symposium on the Application of Computers in the Mineral Industry*, Phoenix, AZ: 797-804.

Kuchta, M., A. Newman, and E. Topal. (2004). "Implementing a Production Schedule at LKAB's Kiruna Mine," *Interfaces* **34**(2): 124-134.

Lerchs, H., I.F. Grossmann. (1965). "Optimum Design of Open Pit Mines," *Transactions, Canadian Mining Institute* **68**: 17-24.

Martin, R. Kipp. (1999). Large Scale Linear and Integer Optimization, Kluwer Academic Publishers, Boston MA.

Martinez, M., A. Newman, and M. Kuchta. (2005) "Using Decomposition to Optimize Long- and Short-term Production Planning at an Underground Mine," Working Paper, Division of Economics and Business and Mining Engineering Department, Colorado School of Mines, Golden, CO, November.

Newman, A., and M. Kuchta. (2005). "Using Aggregation to Optimize Long-term Production Planning at an Underground Mine," *European Journal of Operational Research*, to appear.

Newman, A., M. Kuchta, and M. Martinez. (2005). "Long- and Short-term Production Scheduling at LKAB's Kiruna Mine," *Handbook of Operations Research in Natural Resources*, A. Weintraub and R. Epstein, eds., Springer, to appear.

Newman, A. and C. Yano. (2000). "Scheduling Direct and Indirect Trains and Containers in an Intermodal Setting," *Transportation Science* **43**(3): 256-270.

Sarin, S. and J. West-Hansen. (2005). "The Long-term Mine Production Scheduling Problem," *IIE Transactions*, **37**(2): 109-121.

Smith, M.L. (1998). "Optimizing Short-term Production Schedules in Surface Mining: Integrating Mine Modeling Software with AMPL/CPLEX," *International Journal of Surface Mining* **12**(4): 149-155.

Smith, M.L., Sheppard, I. and G. Karunatillake. (2003). "Using MIP for Strategic Life-of-mine Planning of the Lead/zinc Stream at Mount Isa Mines," in *Proceedings, 31st International Symposium on the Application of Computers in the Mineral Industry*, Capetown, South Africa: 465-474.

Tang, X., G. Xiong, and X. Li. (1993). "An Integrated Approach to Underground Gold Mine Planning and Scheduling Optimization," in *Proceedings, 24th International Symposium on the Application of Computers in the Mineral Industry*, Montreal, Quebec, Canada: 148-154.

Topal, E. (1998). *Long and Short Term Production Scheduling of the Kiruna Iron Ore Mine, Kiruna, Sweden*, Master of Science Thesis, Colorado School of Mines, Golden, CO.

Trout, L.P. (1995). "Underground Mine Production Scheduling Using Mixed Integer Programming," in *Proceedings, 25th International Symposium on the Application of Computers in the Mineral Industry*, Brisbane, Australia: 395-400.

Underwood, R. and B. Tolwinski. (1998). "A Mathematical Programming Viewpoint for Solving the Ultimate Pit Problem," *European Journal of Operational Research* **107**(1): 96-107.

Williams, J., L. Smith, and M. Wells. (1972). "Planning of Underground Copper Mining," in *Proceedings, 10th International Symposium on the Application of Computers in the Mineral Industry*, Johannesburg, South Africa: 251-254.

Winkler, B.M. (1996). "Using MILP to Optimize Period Fix Costs in Complex Mine Sequencing and Scheduling Problems," in *Proceedings, 26th International Symposium on the Application of Computers in the Mineral Industry*, Pennsylvania State University, University Park, PA: 441-446.

Winkler, B.M. (1998). "Mine Production Scheduling Using Linear Programming and Virtual Reality," in *Proceedings, 27th International Symposium on the Application of Computers in the Mineral Industry*, Royal School of Mines, London, United Kingdom: 663-673.

Chapter 12

SCHEDULING MODELS FOR OPTIMIZING HUMAN PERFORMANCE AND WELL-BEING

Emmett J. Lodree, Jr., Bryan A. Norman
Auburn University, University of Pittsburgh

Abstract: Personnel are critical components of many systems. Properly considering human capability and the man-machine interface is essential in order to maximize system effectiveness. The overall performance of a system is often directly related to how system personnel are scheduled. This chapter summarizes research related to scheduling personnel where the objective is to optimize system performance while considering human performance limitations and personnel well-being. Topics such as work rest scheduling, job rotation, cross-training, and task learning and forgetting are considered. For these topics, mathematical models and best practices are described. Additionally, important topics for future research are identified and discussed.

Key words: Scheduling theory, human performance, human factors, ergonomics

1. INTRODUCTION

Over the years, many manufacturing and service organizations have reaped several benefits such as shop floor efficiency, reduced costs, and increased responsiveness as a result of effective planning and scheduling. To this end, the role of theoretical and applied scheduling research has been paramount in the advancement of the steadily evolving manufacturing enterprise. For example, scheduling research methods contribute mathematical models and algorithmic solution procedures that empower decision-aiding mechanisms that often underlie Material Requirements Planning (MRP), Enterprise Resource Planning (ERP), and other scheduling oriented software packages. Although motivated primarily by situations encountered on the shop floor in manufacturing systems, the theory of scheduling has emerged as a viable entity in the research literature and is well equipped for modeling and solving scheduling problems that arise in other environments, most notably in computer science. The purpose of this chapter is to demonstrate the potential benefits that can be realized by applying the principles of sequencing and scheduling theory in the realm of human performance.

Human performance, safety, and well-being have historically been the subject matter of human factors engineering, industrial psychology, exercise and work physiology, and medicine, among others. These disciplines have well justified the benefits associated with establishing mechanisms to help the human at work. Such benefits include reducing the number of employees suffering from work-related injuries, reducing or eliminating the magnitude of injury suffered by a given employee, minimizing job turnover rates, and increasing the productivity for an employee or group of employees. The economic impact of these benefits is also significant. For instance, as the number and magnitude of employee injuries increase, the firm's costs associated with insurance premiums and lawsuits also increase. Additionally, injuries lead to either an understaffed workforce if injured workers are not temporarily replaced, or an underperforming workforce because of untrained or unqualified temporary replacements. In either case, the firm's level of productivity is impaired. Also, poorly designed jobs result in a high probability of injury and lead to high turnover, which leads to substantial costs associated with maintaining a properly trained workforce.

The abovementioned disciplines are actively involved in mitigating potentially adverse effects associated with human labor-intensive environments and exploring ways to improve human productivity. On the other hand, the conventional impetus of production scheduling models and methods has been shop floor efficiency, but not specifically human efficiency with explicit considerations of human characteristics. This chapter is concerned with (i) reviewing research in which remedies have some form of scheduling implication, (ii) identifying scheduling oriented guidelines for improving human performance, and (iii) describing promising new areas of multi-disciplinary research. Section 2 surveys existing approaches to human performance improvement that involve scheduling decisions, and summarizes best practices. It turns out that none of the approaches discussed in Section 2 are based on the fundamental scheduling models that are often applied to manufacturing systems and computer processing environments. Therefore, Section 3 proposes a framework that integrates scheduling theory, human factors engineering, operations research, and potentially other disciplines. Section 3 also describes research opportunities based on the proposed framework. Section 4 provides a summary and closing remarks.

2. GENERAL SCHEDULING MODELS

Human factors engineering, industrial psychology, and other disciplines prescribe a host of antidotes to ensure the health, safety, and productivity of humans in the workplace. This section discusses such approaches that have a significant scheduling component. For a comprehensive presentation of other approaches (those that do not necessarily involve scheduling decisions), refer to Chengalur et al. (2004). The approaches that will be discussed in this section include *work-rest scheduling*, *job rotation scheduling*, and *personnel scheduling*. Other issues inherent in human task sequencing and scheduling such as group/team work and cross-training are also discussed

2.1 Work-rest scheduling

Perhaps the most common engineering approach to addressing human performance is work-rest scheduling. A work-rest schedule specifies time intervals during a work shift in which workers are to engage in work activities, and time intervals in which workers should engage in rest breaks to facilitate recovery. An example schedule over a four-hour shift is 45-15, in which an employee works during the first 45 minutes of each hour and spends the last 15 minutes of each hour resting. The motivation behind work-rest scheduling is that the effects of fatigue, stress, motivation and other human characteristics facilitate the need for humans to engage in recovery activities periodically during a work shift.

2.1.1 Literature review

The majority of the work-rest scheduling literature involves empirical studies in which human subjects are exposed to at least two different work-rest cycles and the most preferred schedule based on the human subjects' responses is determined. For example, Van Dieen and Vrielink (1996a, 1996b) determined that a 60-15 work-rest cycle is least preferred for poultry inspectors when compared to 45-15, 30-15, and 30-30. Respondents also indicated no significant difference between the latter three schedules. The reader is referred to Konz (1998) for a comprehensive survey of an empirical approach to determining work-rest schedules for various occupations and tasks.

One limitation of the above-mentioned approach to specifying work-rest schedules is that the different schedules tested in empirical experiments are usually based exclusively upon subjective criteria and lack a mathematically justifiable framework. This motivated a few researchers to investigate objective measures based on operations research approaches to generate optimal work-rest schedules. Eilon (1964) introduced a mathematical framework of work-rest cycles in which linear functions were used to represent both human performance depreciation due to fatigue during work periods, and recovery from fatigue during rest periods. The optimal start time and duration of exactly one rest period were determined in closed form such that total productivity during the work shift was maximized. Later, Gentzler et al. (1977) considered the start times of several resting periods during a work shift, but the duration of each break was fixed and ensured full recovery. Optimal start times that maximized productivity were determined based on linear and exponential decrements in work rate. Bechtold et al. (1984) generalized the modeling frameworks of Eilon (1964) and Gentzler et al. (1977) by developing a mathematical programming model and sophisticated solution procedures that generate the number, duration, and start times of rest breaks based on linear decay and recovery such that the worker's productivity during the shift is maximized. Their solution procedure was used to generate schedules for employees of a major international airline resulting in a 37.3% increase in work output when compared to the airline's previous scheduling procedures. Similarly, Bechtold and Sumners (1988) developed a mathematical programming model and solution procedures that yield work-rest schedules based on exponential decay in work rate and linear recovery. Bechtold (1991) introduced a quadratic programming formulation for determining work-rest schedules where some of the rest periods are predetermined and fixed due to union contracts and management policies. All of these models are concerned with work-rest schedules associated with an individual worker and do not consider the implications of group work. Thus, Bechtold and Thompson (1993) developed a mixed-binary cubic programming formulation that specifies work-rest cycles for a group of workers that share a common work-rest schedule. The primary difference is that their model incorporates varying work and recovery rates (both linear) for employees in a work group resulting in work-rest schedules that have different effects on employees in the work group. The objective is to maximize the work output of the group.

2.1.2 Guidelines

From the above survey of the literature, two mainstream approaches to generating effective work-rest schedules are identified: (i) empirical analysis and (ii) operations research oriented approaches. We recommend a hybrid approach similar to Bechtold et al. (1984) that incorporates both components. But first, the two existing approaches are summarized:

<u>Empirical approach</u>

Step 1. Determine the performance indicators that will be measured (e.g., total productivity during the test period and heart rate).
Step 2. Identify two or three work-rest patterns to be tested.
Step 3. For each of the work-rest patterns identified in step 1, assemble a group of human subjects that will implement the corresponding work-rest schedule.
Step 4. Appropriately measure the performance indicators identified in Step 1.
Step 5. Recommend the most effective schedule based on Step 4.

Step 2 is subjective and can be accomplished by interviewing workers and getting input based on their experience. Also in Step 4, measurements can be taken at the end of the testing period, but it may also be necessary to take measurements continuously or periodically throughout the testing period depending upon the performance criteria decided upon in Step 1. For instance, total productivity is measured at the end of the test period. However if the objective is to minimize the maximum heart rate, then measurements must be taken during the testing period.

Now the analytical approach is described.

<u>Analytical approach</u>

Step 1. Determine the performance indicators that will be measured.
Step 2. Select the form of the performance decay function (for example, linear or exponential).
Step 3. Select the form of the recovery function (again for example, linear or exponential).
Step 4. Determine the appropriate parameters for the type of function selected in Step 2 and Step 3.
Step 5. Formulate an appropriate mathematical model.
Step 6. Solve the optimization problem formulated in Step 5 to determine the optimal schedule.

The proposed hybrid approach for identifying the best work-rest scheduling essentially integrates the empirical and analytical approaches just described.

<u>Proposed Hybrid Approach</u>

Step 1. Determine the performance indicators that will be measured.
Step 2. Determine the parameters of the performance decay function as follows.

a. Assemble a diverse group of human subjects and allow them to work the maximum possible time without taking a break, W_{max}.
b. Measure the appropriate performance indicators identified in Step 1 periodically during the time interval [0, W_{max}].
c. Use an appropriate statistical regression to determine the parameters.

Step 3. Determine the parameters of the recovery function as follows.
a. Assemble a diverse group of human subjects.
b. From Step 2, identify the points in time (call them critical points, C_t) where there are noticeable differences in performance.
c. For each of the break points from Step 3b, allow workers to rest the maximum allowable time, B_{max}.
d. Measure the appropriate performance indicators identified in Step 1 periodically during each time interval [C_t, $C_t + B_{max}$], for $t = 1, \ldots, n$, where n is the number of critical points.
e. Use an appropriate statistical regression to determine the parameters.

Step 4. Formulate an appropriate mathematical model.

Step 5. Solve the optimization problem represented in Step 4.

Step 6. Evaluate the performance of the sequence generated by Step 5 with respect to the measures specified in Step 1 by testing the solution on human subjects.

The reader should be advised that the above hybrid approach is a recommendation based on integrating existing analytical and empirical methods, but the hybrid approach itself has not been validated. Also, the reader should not expect an effective "one-size-fits-all" guideline that is appropriate for all occupations, tasks, and demographics. For examples of specific guidelines (primarily based on empirical approaches), refer to Konz (1998) and Chengalur et al. (2004).

2.2 Personnel scheduling

Personnel scheduling involves assigning employees to work shifts such that minimum staff level requirements are satisfied, labor costs are minimized, and labor laws and agreements are upheld. More generally, personnel scheduling (also referred to in the literature as rostering, workforce scheduling, employee scheduling, crew scheduling, labor scheduling, and tour scheduling among others) entails demand modeling, days off scheduling, shift scheduling, line of work construction, task assignment, and staff assignment (Ernst et al., 2004a). *Demand modeling* is the process of translating demand information into staffing requirements. *Days-off scheduling* involves allocating off days among a workforce, which is common in industries that implement flexible and part-time work schedules. *Shift scheduling* is particularly applicable to firms that operate on a 24-hour basis, which has traditionally been associated with service industries such as police departments and hospitals, but is now also common in the manufacturing sector. Shift schedules assign the workers of a workforce to the different work shifts such that demand during each shift is satisfied. *Line of work construction* specifies work schedules (i.e., sequences of duties or tasks) to be performed by individual employees, while *task assignment* ensures that the necessary skill levels are available to carry out the tasks corresponding to lines of work. Finally, *staff assignment* involves assigning individual staff members to lines of work. Personnel scheduling has been a very active area of research since the 1950s. Ernst et al. (2004a) and Ernst et al. (2004b) have prepared recent comprehensive literature surveys in this area, and also cite more than ten earlier surveys. However, these reviews do not discuss personnel scheduling research that explicitly addresses human performance, health, and characteristics, which is the focus of our discussion.

2.2.1 Literature review

In the personnel scheduling literature, human characteristics are emphasized primarily through the psychological, sociological, and ergonomic effects associated with *shift work*. More specifically, researchers over the years have verified that shift work can lead to impaired productivity, coronary heart disease, psychosocial disorders (depression), and sleep deprivation (Chengalur et al. 2004). Additional problems include gastrointestinal disorders, cardiovascular disorders, and fatigue (Kostreva et al., 2002). These adverse effects of shift work are often attributed to circadian rhythm disruption. *Circadian rhythms* are the natural fluctuation of physical and mental conditions that are governed by the Earth's day-night cycle (Wickens et al., 2004). Therefore from the perspective of human performance and well-being, the objective of shift schedule design is to minimize the degree of circadian rhythm disruption. The decision parameters involved in constructing shift rotation schedules include speed of rotation, direction of rotation, shift duration, start time of morning shift, and distribution of days off (Czeisler 1982, Knauth 1993, Monk 2000). Another decision parameter associated with constructing shift schedules is shift duration (e.g., Smith et al., 1998), which usually involves comparisons between 8-hour and 12-hour work shifts. *Speed of rotation* refers to how often an employee rotates from one work shift (for example, morning) to another work shift (for example, evening). *Direction of rotation* is either forward or backward. If an employee rotates from the morning shift to the evening shift (or from the evening shift to the night shift), this is considered *forward rotation*. If the rotation is from the evening shift to the morning shift (or the night shift to the evening shift), this is considered *backward rotation*. Similar to the work-rest scheduling literature, the methods used to identify the most effective shift schedules based on human performance metrics and the abovementioned decision parameters are dominated by empirical analysis, and rarely utilize optimization methods (refer to Chengalur et al. 2004 and Konz 1998 for more detailed discussion and a literature review of these mainstream methodologies). The exception is Kostreva et al. (2002), who construct a computer simulation of circadian rhythms based on sinusoidal equations and evaluate the effectiveness of shift schedules in the simulated environment.

2.2.2 Guidelines

Based on the results of several studies, a number of guidelines have been reported with respect to constructing shift schedules that optimize human performance and health. Examples are presented below. Items 1 to 4 are general guidelines while the remainder of the list (items 5 to 8) applies to 8-hour shifts for control rooms. For more comprehensive presentations of shift design guidelines including applications to a variety of other occupational settings, refer to Konz (1998) and Chengalur et al. (2004).

1. Use forward rotating schedules (i.e., avoid backward rotating schedules).
2. Consider short rotation cycles as opposed to longer rotation cycles (e.g., rotation every month is preferable to rotation every two months).
3. Shifts should not begin too early in the morning.
4. Avoid assigning employees permanently to the nightshift.
5. The number of consecutive workdays should not exceed seven.
6. The number total number of workdays should not exceed 21 during any 4-week period.
7. Two consecutive full days off should be assigned after working 9 consecutive days.
8. A series of night shifts should be followed by a minimum of two full days off.

Again, the above recommendations are not necessarily optimal for all industries, occupations, or individual workers. Employers should conduct appropriate studies to determine shift design guidelines that are specific to the industry, occupation, and task type.

2.3 Job rotation scheduling

Job rotation is another important facet of personnel scheduling. Whereas work-rest scheduling often assumes that a worker is performing the same task throughout a workday, job rotation examines having workers do different tasks throughout the day. This method can be particularly effective for reducing the potential for worker injury when different tasks have very different severities with regard to their potential for workplace injury.

In addition to reducing the potential for worker injuries, job rotation can also help increase productivity by reducing monotony associated with performing the same task throughout the entire workday. It can also reduce errors in inspection and other vigilance or demanding tasks.

2.3.1 Literature review

One of the principal reasons that job rotation is employed is to reduce worker injuries in repetitive task environments. As Hagberg et al. (1995) indicated, performing the same task for every hour of every workday can lead to a significant potential for occupational illness and injury, in particular, cumulative trauma disorders. One example of a repetitive task is lifting resulting from manual material handling. Manual material handling is one of the primary sources of back injuries and back injuries are a leading cause of lost days in the workplace as noted by Liles and Deivanayagam (1984). Moreover, there are many settings where the physical demands of different tasks (lifting or other) vary significantly. For example, in manual material handling environments some work areas may require moving an average of two 50-pound loads every minute while other work areas require only moving two 10-pound loads every minute. Similarly, in a sawmill setting some workstations have significant exposure to noise while others have much less noise exposure.

Generally, the likelihood of occupational injury is reduced by using one of three methods: (1) create an engineering solution, such as redesigning the job or using automation or mechanization; (2) using an administrative control such as work-rest scheduling, job rotation, or worker screening; and (3) utilizing personal protective equipment (Tayyari and Smith, 1997). Ideally, the first method is used because this often eliminates the source of injury potential. However, this is often not practical either due to constraints imposed by the inherent nature of the task or due to cost considerations. The third method is applicable to some tasks but for many tasks personal protective equipment can reduce but not eliminate the potential for injury. Job rotation is often an effective administrative control that can be used in conjunction with the other two methods to provide maximum reduction in worker injury potential.

The effectiveness of job rotation has been shown in many settings. A few examples of where it has been applied include Kuijer et al.'s (1999) study of refuse collecting, Hinnen et al.'s (1992) study of cashiering, and Henderson's (1992) study of poultry processing. All of these studies indicate that job rotation reduces the likelihood of worker injuries and also reduces task errors and leads to improved employee satisfaction.

It is important to note that much of the research related to job rotation is found in the human factors literature where the focus is generally on comparing two or more operational policies to determine their effect on worker performance (similar to the empirical approach shown in Section 2.1.2). Recently, more research has been conducted to apply mathematical models to the problem of job rotation to try and find "optimal" job rotation schedules. Examples of this work include Carnahan et al. (2000) who investigate a manual lifting task and utilize a genetic algorithm to find a family of near optimal solutions and then create job rotation rules or guidelines based on patterns found within this family of solutions. Tharmmaphornphilas and Norman (2004b) investigate a similar problem and utilize heuristic methods to find robust solutions for the case where the task demands are stochastic. There has also been related work in the realm of cellular manufacturing to determine the appropriate level of cross-training and then considering the ensuing worker to task assignment problem. Molleman and Slomp (1999) and Slomp and Molleman (2000) consider the problem of assigning workers to tasks considering skill requirements and how to reduce the workload of the bottleneck worker. Campbell and Diaby (2002) develop an assignment heuristic for assigning workers to tasks at the start of a shift.

There are several aspects of job rotation that make it a challenging problem and these need to be considered in future research. First, in many settings the task demands vary over time rather than being static. While this has been considered in some of the current research (one example being Tharmmaphornphilas and Norman, 2004b) this topic requires further analysis. Second, it is important to consider workforce heterogeneity. That is, different workers have different capabilities and thus, cannot be treated as being completely interchangeable with regard to task assignment. This has important implications for worker training which will be discussed in more detail in Section 2.4 in the discussion of worker cross-training. The third aspect is to consider how frequently to rotate workers. In many studies, the rotation interval is simply assumed. However, it is better to treat this as a decision variable rather than as a constraint. Initial work by Tharmmaphornphilas and Norman (2004a) has indicated that in many settings a rotation interval of two hours is short enough to obtain most of the benefits of job rotation with regard to reducing injury potential. However, further study would be beneficial in order to investigate more settings and provide more comprehensive guidelines.

2.3.2 Guidelines

Job rotation can be a useful tool for both reducing worker injury potential and increasing productivity. In particular, job rotation can be effective when there is significant variation in task demands or significant variation in worker capabilities. Job rotation can be applied in virtually any setting and should be considered as one method for improving the performance of a work area. Some guidelines for applying job rotation are summarized below: (Note that these have been written assuming the focus is to reduce injury potential but the same guidelines can be modified to consider rotating workers for productivity reasons too.)

1. Assess the injury potential of the work area and determine the injury criteria.
2. Determine relationships for how the work content of the various tasks relate to the injury criteria. For example, use the Job Severity Index (Liles and Deivanayagam, 1984) for a lifting environment.
3. Consider the application of engineering controls and the use of personal protective equipment (if applicable.)
4. Determine if there is significant variation in the task demands that relate to the injury criteria within the work area. If so, then specify the amount of variation as precisely as possible. If not, then job rotation will likely not provide significant benefit with regard to injury potential or productivity. However, it may still provide benefits with regard to reducing monotony.
5. Examine the heterogeneity of the workforce. Depending on the severity of the task requirements consider using a screening procedure to eliminate workers that would be at too high a level of risk if assigned to tasks in the work area.
6. Develop a mathematical model for the resulting job rotation model.
7. Solve the model and implement the recommended schedules.
8. Evaluate the performance of the rotation sequence generated by Step 5 with respect to the measures specified in Step 1 by testing the solution on human subjects.

2.4 Cross-training

Cross-training is an important facet of workforce scheduling and is closely related to job rotation since workers can be rotated only to jobs that they are capable of performing. In recent years a number of papers have been written concerning different facets of cross-training. In this section we will consider why cross-training is often used in practice, how it relates to promoting employee well-being in the workplace, and how to determine appropriate levels of cross-training.

2.4.1 Literature review

There are several reasons that cross-training is employed in practice. The primary reasons center on flexibility. If workers can perform several different tasks then it is easier to create feasible worker assignments that will satisfy the work requirements for a given production planning period. In particular, in recent years companies have worked to reduce inventories and still have rapid response times. In order to do this it is necessary to produce parts, products or services to satisfy customer demands rapidly and this can only be done if the workforce has the flexibility to perform different tasks in order to respond to changing weekly, daily, or hourly demand. Having a flexible workforce helps provide an organization with a capacity buffer in order to respond better to variable demands (Hopp et al., 2004).

Similar to the production flexibility benefits of having a cross-trained workforce there are significant benefits with regard to worker well-being. As noted in Section 2.3, job rotation can reduce the potential for worker injuries in physically demanding work environments and can also create job enlargement in all settings. However, job rotation is practical only if workers have been sufficiently cross-trained that they can perform several different tasks without sacrificing productivity.

A key question is how much or what level of cross-training is necessary? This issue has been addressed by several authors in the literature although they are primarily concerned with system productivity and determining cross-training levels that will meet certain system throughput requirements. We now discuss a few of these papers. Molleman and Slomp (1999) developed a linear goal programming model that considered worker skill requirements. In their extended study (2000), they formulated linear programming models and presented a hierarchical procedure for worker cross-training in order to reduce the workload of the bottleneck worker. Norman et al. (2002) developed a mixed integer programming model to assign workers to tasks in manufacturing cells. The model considers both technical and human skills with the objective to maximize an organization's effectiveness. Slomp et al. (2005) studied the need for cross-training workers in a cellular manufacturing environment. They developed an integer programming model that can be used to select workers to be cross-trained for particular machines. Their study has shown that cross-training decisions in a cellular manufacturing environment should support the forming of effective "chains" between workers and tasks in order to shift loads from a loaded worker to a less loaded worker. This model is helpful when management is making decisions regarding the trade-off between training costs and workload balance among workers. Further discussion of chaining can be found in Jordan and Graves (1995) and Jordan et al. (2004). Campbell and Diaby (2002) developed an assignment heuristic for allocating cross-trained workers to multiple departments at the beginning of a shift.

Quantitative studies in the area of workgroup selection in cellular manufacturing have also been found in the literature including Askin and Huang (1997), Askin and Huang (2001), and Bhaskar and Srinivasan (1997). Askin and Huang (1997) compared two integer programming models for assigning workers to cells and evaluated the training program for each worker. In their extended study (Askin and Huang, 2001), they developed a multi-objective model to create work teams for cellular manufacturing systems. Dynamic task assignment in the traditional serial line model with partially cross-trained workers is addressed by Askin and Chen (2004) and the objective was to maximize throughput. Hopp and Van Oyen (2004) outline approaches for accessing and classifying manufacturing and service operations in terms of their suitability for use of cross-trained workers. They define production agility as the ability to achieve heightened levels of efficiency and flexibility while meeting objectives for quality and customer service.

A common theme seen in these and other papers in the literature is that it is not necessary to fully cross-train the workforce in order to gain many of the benefits of having a flexible workforce. In fact, Slomp and Molleman (2002) show that there are diminishing returns after workers are cross-trained on a few tasks. However, it is important to consider what the impact of these various jobs is on the well-being of the worker and to make sure that those considerations are included in these models. For example, chaining workers across two or three jobs may be sufficient in terms of providing robustness from a productivity perspective but may not be sufficient to insure that workers do not suffer from overexertion or cumulative trauma disorders. In particular, it is important to consider the physical demands imposed by the jobs that comprise the chain.

2.4.2 Guidelines

Cross-training is an important part of workforce scheduling and can permit greater flexibility in worker assignment. This flexibility can foster both greater productivity and greater worker satisfaction and safety. The proper implementation of cross-training is related to the use of both job rotation and work teams since both affect how workers will move between different tasks. Some guidelines concerning the use of cross-training are summarized below.

1. Determine how much workforce assignment flexibility is necessary considering both productivity and worker well-being objectives. This will be a basis for deciding what degree of cross-training is necessary. Additionally, one can then investigate if the setting requires full cross-training, partial cross-training or if chaining should be used. Recall that there is generally a decreasing marginal benefit for cross-training workers for more than a few tasks.
2. Recognize if there are any impediments to utilizing cross-trained workers, such as union rules or pay grade differences in jobs.
3. Investigate how expensive is it to cross-train workers. It is important to consider the training costs, time lost due to training, and productivity that will be lost as workers climb their respective learning curves (there is more discussion of learning in Section 3).

2.5 Group and team work

Group and team work have become increasingly important in the realm of personnel scheduling in recent years. More companies are moving towards using cellular manufacturing and other production strategies where workers are divided into teams that operate within certain cells or production areas within a facility (Gordon, 1992). The concept of using teams is often combined with cross-training because companies find it provides more flexibility with regard to production capacity and also lessens the effects of worker absenteeism and turnover.

2.5.1 Literature review

When implementing a group or team concept in the workplace there are many factors to consider. Bidanda et al. (2005) note the following important factors that affect group or team effectiveness: training, communication, autonomy, conflict management, and teamwork. Training is important because the operators typically must be able to perform multiple tasks in order to gain the flexibility benefits from using a team concept. Communication is essential because the members of the team must share information related to production schedules and quantities, quality, and other topics. Self-directed teams have an even greater need for good communication to insure that they are able to operate effectively (Forza and Salvador, 2001). While workers in almost any position require some degree of communication skills, the skill requirements for highly interactive work teams are significant and therefore workers generally benefit from receiving specialized training related to communications skills. In addition, successfully implementing any type of change in a manufacturing system requires effective communications (Axley, 2000). The degree of autonomy of the workers is also an important factor. Providing workers with more self-control, accountability, and ownership are ways to increase their autonomy. Conflict management relates to the team's ability to resolve differences that arise within the team itself or between the team and other teams or management within the facility. It is important that the team members be able to handle conflict effectively and look for "win-win" solutions rather than "win-lose" or "lose-lose" solutions. Effective conflict management is often dependent on having good communication skills.

There are also many facets of teamwork in general to consider. It is important to note that many entire books have been written on teamwork so only a few ideas are touched on here. First, teams need to have clear, unambiguous goals to be effective (McComb et al., 1999 and Sweeney and Lee, 1999). Second, the organizational culture must be unified from top to bottom in the support of the team concept and must create an atmosphere of trust (Sweeney and Lee, 1999 and Groesbeck and Van Aken, 2001). Third, effective human resource management including the use of *team* performance incentives is important (Hellinghausen and Myers, 1998). Fourth, team members must be carefully selected. There are many different tools and methods to use for doing this. One method suggested by Hut and Molleman (1998) considers the following four principles: (1) the principle of requisite variety: the team as a whole must contain all of the skills needed to complete its tasks, (2) the principle of redundancy: this insures that more than one team member can perform each required task to that the team can function even if one member is absent, (3) the principle of minimal critical specification: the team should be given as much flexibility as possible to determine how to respond to its task demand requirements, and (4) the principle of double loop learning: both the input and output processes relating to the team can be changed over time in response to the team's experience and learning.

Teams can work together in several different ways. One way is for team members to rotate among the different tasks assigned to the team. This type of rotation results in systems that are similar to those described under the job rotation and cross-training sections. Another way that teams can share work that has grown in interest over the last ten years is through dynamic work balancing or bucket brigades. Bucket brigade systems are now described in more detail.

A significant problem in managing serial flowlines or other team based production processes is balancing the workload among the workers on the line so that the line is maximally productive. This can be particularly challenging in settings where the workload is changing over time. To deal with these types of settings, a variation of the classic serial flowline has been introduced which is called a bucket brigade. The basic idea behind a bucket brigade is that workers are ordered along a serial line and each worker carries a product towards completion. When the last worker finishes his product, he sends it off and then walks back to take over the work of his predecessor, who is now released and walks back to his predecessor. This process repeats until the first worker walks back and starts a new product (Bartholdi and Eisenstein, 1996). Note that the ordering of the workers is preserved. In the classic bucket brigade when a worker comes to a busy station he waits because no passing is allowed.

Bartholdi and Einstein (1996) analyzed systems using the bucket brigade policy under the assumption of deterministic processing times and non-identical workers, each with a processing rate that depends on the particular task performed. When workers are assigned on a production line from slowest to fastest, and move according to bucket brigade rules, the production rate will converge to a value that is maximum possible among all ways of organizing workers and stations (Bartholdi and Eisenstein, 1996). The weakness of the model is that it assumes constant worker velocities over time. Bartholdi et al. (1999) suggest that it is better for management to sequence workers from slowest to fastest and include very different workers (fast and slow) on the same team in order to achieve the maximum production rate. They also suggest that the greater the range in velocities on a team, the greater the rate of convergence. For a line with three workers, it is sufficient for the last worker to be the fastest in order for there to be convergence to a fixed point. However, they conclude that this is not true for lines with more than three workers.

One drawback of the bucket brigade approach as generally modeled in the literature is the fact that all the workers must be trained for all tasks. An important question is how much can the system deviate from this assumption and still have the bucket brigade remain effective. Bucket brigades are most successful in applications where the skills required to perform the operations on the line are very similar, such as warehouse picking, fast food preparation, and textile sewing operations (Hopp and Van Oyen, 2004). A second common assumption is that faster workers dominate slower workers across all stations. However, in practice, this may not be the case. A two worker bucket brigade is studied by Gel and Armbruster (2004) where one worker is faster than the other over some part of the production line and slower over another part of the line. Two environments were analyzed with passing and blocking. Their conclusion is that it may not always balance itself on a fixed point but rather to two stable positions where workers pass jobs. Workers would hand over jobs at exactly two fixed locations that they visit periodically.

There are many opportunities for investigating other aspects of dynamic work-sharing or bucket brigade systems. For example, one could consider letting workers' productivity rates change over time and consider tasks that have different complexity levels. It is also important to analyze systems with more than two or three workers to see how the bucket brigade results generalize to larger systems.

2.5.2 Guidelines

When considering the use of work teams there are many important factors to consider. There is no universally correct answer for each item considered below but rather each setting or context must be examined individually.

1. Determine the degree of autonomy and responsibility the team will have. For example, will the team itself determine which workers do which tasks and how to sequence work that is assigned for the shift?
2. Select team members that fit with the required autonomy and responsibility that were established in Step 1.
3. Develop team based incentives that will reinforce the team concept rather than focusing on only individual performance.
4. Properly train the team members. This includes training not only on the technical skills that relate to performing the various tasks that need to be done but also on the human or soft skills such as communication, conflict management and overall teamwork.
5. Determine how work will be shared among the team members. Will job rotation be employed? Will dynamic work sharing methods such as bucket brigades be used?

3. RESEARCH OPPORTUNITIES: SEQUENCING HUMAN TASKS

The classic sequencing/scheduling problem involves a set of jobs and a set of resources, where resources perform operations and each job requires one or more operations for successful completion. Job sequences for each resource (or equivalently, resource routes for each job) must be determined such that some combination of objectives is optimized and relevant constraints are satisfied. Common objectives include (i) makespan minimization, (ii) flow time minimization, and (iii) minimization of the number of tardy jobs. Example constraints include (i) job preemption, (ii) precedence relationships, and (iii) each resource can process at most one job at a time. The most common applications of the above scheduling framework are manufacturing systems in which resources are machines and computer systems in which the resource is the computer's Central Processing Unit (CPU). In theory however, the framework is relevant to a variety of real-world scenarios such as gate assignment at an airport (Pinedo, 2002) and sequencing human tasks (e.g., Dessouky et al., 1995). This section describes the implications of applying the classic scheduling framework to sequence tasks performed by humans. More specifically, this section addresses the *human task-sequencing problem.*

The Human Task Sequencing Problem: *A group of human workers is confronted with a set of tasks. These tasks are available simultaneously or can arrive over a specific planning horizon. Task sequences for each worker in the group are to be determined such that some performance criterion is optimized.*

Because manufacturing and computer applications involving inanimate resources (machines and computer CPUs) inspire the vast majority of scheduling theory and practice, human characteristics such as fatigue, motivation, and stress are often overlooked as resource attributes in traditional scheduling models. However, in order to use scheduling models effectively as a decision aid for scheduling decisions with respect to tasks carried out by human resources, it is imperative that ergonomic factors be accounted for. Therefore we now survey areas of the scheduling literature that are appropriate for the abovementioned human task-sequencing problem (note that Section 2 does not specifically address task sequencing). More specifically, we briefly review scheduling models with (i) sequence dependent processing times, (ii) learning (i.e., the human learning curve) and (iii) rate-modifying activities. This section also outlines a framework for integrating human characteristics into scheduling theory and describes research opportunities based on the framework.

3.1 Sequence-dependent processing times

In deterministic models of classical scheduling theory, job processing times p_j are fixed parameters. Now consider job processing times of the form (1), where s_j is the start time of job j in some job sequence and $f_j : \Re \rightarrow \Re$ is called a *processing time function*.

$$p_j = f_j(s_j) \tag{1}$$

Scheduling models with job processing times that satisfy (1) are said to have *sequence-dependent (or variable) processing times*. If the function f_j is increasing, then job j is said to be a *deteriorating job*. This framework is useful for modeling the effects of performance variations due to the effects of fatigue, stress, motivation and other factors. For example, it is logical that if a processor accumulates fatigue over time, then job processing times will take longer if scheduled later in a sequence than if scheduled earlier. Conversely, if learning occurs or motivation increases with time, then job processing times are likely to improve (be reduced) if scheduled later in a sequence than if scheduled earlier.

Just as the empirical experiment conducted by Moray et al. (1991) demonstrates that excluding human behavior from scheduling algorithms will lead to suboptimal or infeasible results when applied to sequencing human tasks, analytical models also demonstrate that the properties of classical scheduling problems do not necessarily generalize to scheduling problems with variable processing times. Consequently, solution algorithms that optimize classical scheduling problems will potentially lead to suboptimal solutions to the corresponding problem with sequence dependent processing times. For example, it is well known that the single machine makespan problem is sequence independent. However, the single machine makespan problem with linearly deteriorating jobs (see Equation (2) below) is not sequence independent and is actually optimized by sequencing jobs in non-decreasing order of their original processing times (e.g., Browne and Yechiali, 1990).

The research literature addresses various forms of processing time function (1) including linear deterioration, piecewise linear deterioration, and nonlinear deterioration (Alidaee and Womer, 1999). For example, the most general form of linear deterioration is given by (2) where $\alpha_j \geq 1$ is the *growth rate* of job j and X_j is the "original" processing time of job j (the processing time if job j is scheduled first).

$$p_j = X_j + \alpha_j s_j \tag{2}$$

For comprehensive surveys of scheduling models with sequence-dependent processing times, the reader is referred to Alidaee and Womer (1999) and Cheng et al. (2004).

3.2 Human learning curve

A special case of sequence-dependent processing times is when job processing times are of the form (3), where p_{ir} is the processing time of job i if it processed in sequence position r, p_i is the original processing time of job i, $a = \log_2 s \leq 0$ is the constant *learning index*, and s is the standard *learning rate*.

$$p_{ir} = p_i r^a \tag{3}$$

This framework represents a unique generalization of classical scheduling theory because it is motivated purely by applications to human performance. Biskup (1999) introduced the processing time function (3) within the single machine context, which lead to a series of related papers (Mosheiov 2001, Lee and Wu 2003, Biskup and Simons 2004). However, scheduling with learning effects (not in the form of Equation (3)) was first addressed by Meilijson (1984).

3.3 Rate-modifying activities

Lee and Leon (2001) introduced the concept of rate-modifying activity to the scheduling literature. *A rate modifying activity (RMA) is any activity that alters the speed in which a resource executes tasks*. More specifically, the framework of RMA scheduling (on a single machine) is characterized by a set of jobs J where each job $j \in J$ has processing time p_j if scheduled before the RMA and processing time $\alpha_j\, p_j$ ($\alpha_j \in \Re \backslash \{0\}$) if scheduled after the RMA. Example RMAs include maintenance activities and rest (or break) periods. This generalizes models that integrate job and maintenance scheduling in that the latter assumes job processing times remain unaltered both before and after maintenance takes place. The RMA concept is also related to work-rest scheduling in the human factors literature. Work-rest scheduling involves determining the number, placement, and duration of rest breaks such that human productivity, comfort, and safety are optimized.

The RMA scheduling problem involves determining job sequence as well as the sequence position that one RMA of fixed length should assume. Lee and Leon (2001) determined the optimal policy for the single machine makespan problem, proved complexity results for total completion time and maximum lateness objectives, and constructed dynamic programming based algorithms that optimally solve the latter problems. Lee and Lin (2001) derived optimal policies for scheduling fixed length RMAs and task sequencing in an environment characterized by random machine breakdowns. Finally, Mosheiov and Sidney (2003) derived complexity results and algorithms for minimizing the number of tardy jobs on a single machine with an RMA that only reduces processing times (that is, $\alpha_j \leq 1$ in equation 2), and also developed a $O(n^4)$ algorithm that minimizes makespan where job processing times are affected by both learning and an RMA. The latter problem and solution is appropriate for scheduling human tasks and a break when the learning curve is considered.

3.4 A framework for sequencing human tasks

Lodree et al. (2005) introduce a framework for incorporating human characteristics into single machine scheduling models, which is briefly reviewed here. The framework addresses the single resource version of the human task-sequencing problem defined earlier in this section. One component of the framework includes scheduling models with sequence-dependent processing times (Sections 3.1 and 3.2), in which processing time functions such as Equations (1), (2), and (3) are used to characterize the effects of human learning, fatigue, stress, and other human relevant characteristics. Realistically, there are attributes other than time that impact performance related human characteristics such as the number of times a particular task type has been performed by time t, the amount of time since a particular type of task has been performed, and the amount of energy exerted by time t. Therefore we generalize the processing time function to one with multi-variable input.

$$p_j = f_j(\mathbf{v}_j) \tag{4}$$

In (4), $\mathbf{v}_j \in \mathbf{R}^\alpha$ is a vector and α is the number of attributes that affect performance (i.e., processing time).

Another component of the framework includes rate-modifying activities (RMAs, Section 3.3). As discussed in Section 2.1, the effects of human fatigue necessitate recovery during and between work shifts. Furthermore recovery oriented activities impact human performance just as RMAs affect machine speeds. Given that existing RMA scheduling research involves job sequencing and determining the sequence position of an RMA of fixed length, the following generalization based on the work-rest scheduling literature (Section 2.1) is suggested: in addition to specifying job sequences, determine the number, duration, and placement of RMAs (breaks).

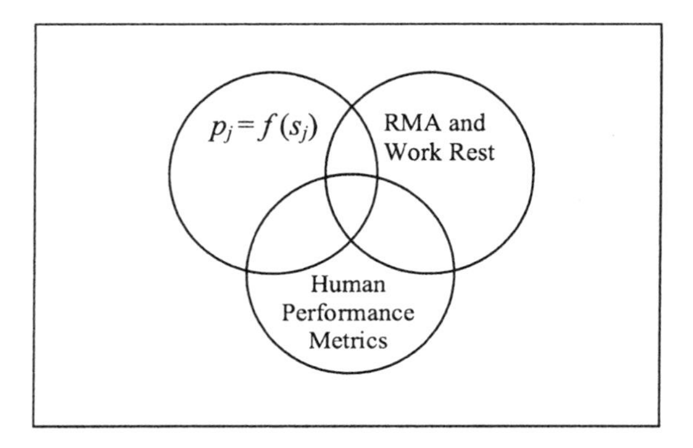

Figure 12-1. Proposed scheduling theoretic framework for sequencing human tasks.

Lastly, appropriate performance metrics should be incorporated. In manufacturing systems and computer processing, metrics such as minimizing makespan, flow time, and number of tardy jobs all relate to operational efficiency. However human factors engineering concerns reducing error, improving productivity, enhancing safety, and enhancing comfort (Wickens et al., 2004). Therefore performance metrics such as minimizing cumulative physiological burden, maximum heart rate, and maximum blood pressure are also relevant. The framework is summarized by Figure 12-1.

3.5 Research needs

In order to effectively apply the framework of scheduling theory to make human task sequencing decisions, human characteristics such as fatigue and motivation should be explicitly represented in scheduling models. The two primary implications of incorporating human characteristics into scheduling models are (i) variation of human performance over time and (ii) the need to schedule breaks. Representative scheduling models should then include both of these characteristics as described in the framework proposed in Section 3.4. To date, it seems that only Mosheiov and Sidney (2003) have done so by modeling task processing times that are affected by both sequence dependent processing times (learning curve) and a rate-modifying activity. Therefore from a modeling perspective, there are many opportunities to develop mathematical formulations, exact and heuristic solutions approaches, and computational complexity results for α / RMA, $p_j = f_j(t)$ / γ type scheduling problems, where α / β / γ is the Graham et al. (1979) scheduling notation. There are also many opportunities to examine the effects that human performance metrics have on sequencing decisions. For example, which task sequence minimizes the number of errors during a work shift? Which task sequence minimizes the total physiological burden acquired while carrying out a job? There are also several multi-disciplinary opportunities that engage production engineers, operations research analysts, exercise and work physiologists, human factors engineers, industrial psychologists, and others that arise from the fact that there is no "one size fits all" solution. In particular, the nature of processing time and recovery functions associated with various occupations, tasks, and demographics should be determined to generate effective scheduling models. Finally the proposed framework (Section 3.4) should be generalized to accommodate the many issues associated with groups and teams of human workers. Given the complex interactions among humans, we expect that the generalization will be more complex than generalizations from single machine to multiple machine environments.

4. CONCLUSION

Scheduling research has made a significant impact in production systems. This chapter has described how the production scheduling concepts and models that have brought so much success to the shop floor can be applied to task sequencing issues that impact human performance. Existing approaches to human performance improvement including work-rest scheduling, job rotation scheduling, cross-training, teamwork and shift-work scheduling have been surveyed along with practical implementation guidelines. The literature survey revealed that classical scheduling research has done little to specifically address human task sequencing. Consequently, a framework for characterizing the human element in scheduling models has been described as well as promising new multi-disciplinary research opportunities. Consideration of human characteristics as resource attributes in scheduling models will potentially inspire a new generation of scheduling research and practice.

REFERENCES

Alidaee, B. and Womer, N.K. (1999). Scheduling with time dependent processing times: review and extensions. *Journal of the Operational Research Society*, Vol. 50, No. 7, pp. 711-721.

Armbruster, D., Gel, E.S., and Murakami J. Bucket brigades with worker learning. European Journal of Operational Research, to appear.

Askin, R. G. and Huang, Y. (1997). Employee Training and Assignment for Facility Reconfiguration. *Institute of Industrial Engineers 6th Industrial Engineering Research Conference Proceedings*, 426 – 431.

Askin, R. G. and Chen, J. (2006). Throughput maximization in serial production lines with worksharing. *Int. J. Production Economics,* **99**, 88-101.

Askin, R. G. and Huang, Y. (2001). Forming Effective Worker Teams for Cellular Manufacturing. *International Journal of Production Research*, 39(11), 2431 – 2451.

Axley, S. R. (2000). Communicating change: Questions to consider. Industrial Management, 18–22.

Bartholdi III, J. J. and Eisenstein D. (1996). A production line that balances itself. *Operations Research,* **44** (1), 21-34.

Bartholdi J., Bunimovich L. A. and Eisenstein D. (1999). Dynamics of two- and three-worker "bucket brigade" production lines. *Operations Research* Vol. 47, No.3, 488-491.

Bhaskar, K. and Srinivasan, G. (1997). Static and Dynamic Operator Allocation Problems in Cellular Manufacturing Systems. *International Journal of Production Research*, 35(12), 3467 – 3418.

Bechtold, S.E. (1991). Optimal work-rest schedules with a set of fixed-duration rest periods. *Decision Sciences*, Vol. 22, pp. 157-170.

Bechtold, S.E., Janaro, R.E., and Sumners, D.L. (1984). Maximization of labor productivity through multi-rest break scheduling. *Management Science*, Vol. 30, pp. 1442-1458.

Bechtold, S.E. and Sumners, D.L. (1988). Optimal work-rest scheduling with exponential work-rate decay. *Management Science*, Vol. 34, No.4, pp. 547-552.

Bechtold, S.E. and Thompson, G.M. (1993). Optimal scheduling of a flexible-duration rest period for a work group. *Operations Research*, Vol. 41, No. 6., pp. 1046-1054.

Bidanda, B., Ariyawongrat, P., Needy, K.L., Norman, B.A. and Tharmmaphornphilas, W. (2005). Human Related Issues In Manufacturing Cell Design, Implementation, And Operation: A Review & Survey. *Computers and Industrial Engineering*, Vol. 48, 507–523.

Biskup, D. (1999). Single-machine scheduling with learning considerations, *European Journal of Operational Research*, Vol. 115, Vo. 1, pp. 173–178.

Biskup, D. and Simons, D. (2004). Common due date scheduling with autonomous and induced learning. *European Journal of Operational Research*, Vol. 159, pp. 606-616.

Browne, S. and Yechiali, U. (1990). Scheduling deteriorating jobs on a single processor, *Operations Research*, Vol. 38, No. 3, pp. 495-498.

Campbell G. M. and Diaby M. (2002). Development and Evaluation of an assignment heuristic for allocating cross-trained workers. *European Journal of Operational Research* Vol. 138, 9-20.

Carnahan, B. Norman, B.A. and Redfern, M.S. (2000). Designing safe job rotation schedules using optimization and heuristic search. *Ergonomics*, 43(4):543–560.

Cheng, T.C.E., Ding, Q., and Lin, B.M.T. (2004). A concise survey of scheduling with time-dependent processing times. *European Journal of Operational Research*, Vol. 152, pp. 1-13.

Chengalur, S.N., Rodgers, S.H., and Bernard, T.E. (2004). *Kodak's Ergonomic Design for People at Work.* 2nd Edition, John Wiley & Sons, Inc., Hoboken, NJ.

Czeisler, C.A., Moore-Ede, M.C. and Coleman R.C. (1982). Rotating shift work schedules that disrupt sleep are improved by applying circadian principles. *Science*, Vol. 217, pp. 460-463.

Dessouky, M.I., Moray, N., and Kijowski, B.A. (1995). Taxonomy of scheduling systems as a basis for the study of strategic behavior. *Human Factors*, Vol. 37, No. 3, pp. 443-472.

Eilon, S. (1964). On a mechanistic approach to fatigue and rest periods. *International Journal of Production Research*, Vol. 3, pp. 327-332.

Ernst, A.T., Jiang, H., Krishnamoorthy, M., and Sier, D. (2004a). Staff scheduling and rostering: A review of applications, methods and models. *European Journal of Operational Research*, Vol. 153, pp. 3-27.

Ernst, A.T., Jiang, H., Krishnamoorthy, M., Owens, B. and Sier, D. (2004b). An annotated bibliography of personnel scheduling. *Annals of Operations Research*, Vol. 27, pp. 21-144.

Forza, C., & Salvador, F. (2001). Information flows for high-performance manufacturing. *International Journal of Production Economics,* 70, 21–36.

Gentzler, G.L., Khalil, T.M., and Sivazlian, B.B. (1977). Quantitative models for optimal rest period scheduling. *Omega*, Vol. 5, pp. 215-220.

Gordon, J. (1992). Work teams: how far have they come? Training , 59–65.

Graham, R.L., Lawler, E.L., Lenstra, J.K. and Rinnooy Kan, A.H.G. (1979). Optimization and approximation in deterministic sequencing and scheduling theory: a survey. *Annals of Discrete Mathematics*, Vol. 5, pp. 287-326.

Groesbeck, R., and Van Aken, E. M. (2001). Enabling team wellness: Monitoring and maintaining teams after start-up. *Team Performance Management: An International Journal*, 7(1/2), 11–20.

Hagberg, M., Silverstein, B., Wells, R., Smith, M.J., Hendrick, H.W., Carayon, P. and P′erusse, M. (1995). *Work Related Musculoskeletal Disorders (WMSDs): A Reference Book for Prevention.* Taylor&Francis, Great Britain, 31

Hellinghausen, M. A., & Myers, J. (1998). Empowered employees: A new team concept. Industrial Management, 40(5), 21–23.

Henderson, C.J. (1992). Ergonomic job rotation in poultry processing. *Advances in Industrial Ergonomics and Safety*, 4:443–450.

Hinnen, U., Laubli, T., Guggenbuhl, U. and Krueger, H. (1992). Design of check-out systems including laser scanners for sitting work posture. *Scandinavian Journal of Work, Environment and Health*, 18:186–194.

Hopp, W. J. and Van Oyen, M. P. (2004). Agile Workforce Evaluation: A Framework for Cross-training and Coordination. *IIE Transactions,* Vol. 36, 919-940

Hopp, W.J., Tekin, E., and Van Oyen, M.P. (2004). Benefits of skill chaining in production lines with cross-trained workers. *Management Science,* Vol. **50:1**, 83-98.

Hut, J., & Molleman, E. (1998). Empowerment and team development. Team Performance Management Journal, 4, 53–66.

Jordan, W.C., Inman, R.R. and Blumenfeld, D.E. (2004). Chained cross-training of workers for robust performance. *IIE Transactions,* Vol. 36, 953-967.

Jordan, W.C., and Graves, S.C. (1995). Principles on the benefits of manufacturing process flexibility. *Management Science,* Vol. **41:4**, 577-594.

Knauth, P. (1993). The design of shift systems. *Ergonomics*, Vol. 36, pp. 15-28.

Konz, S. (1998). Work/rest: Part II – The scientific basis (knowledge base) for the guide. *International Journal of Industrial Ergonomics*, Vol. 22, pp. 73-99.

Kostreva. M., Mcnelis, E., and Clemens, E. (2002). Using a circadian rhythms model to evaluate shift schedules, *Ergonomics*, Vol. 45, No. 11, pp. 739 -763.

Kuijer, P. P. F.M., Visser, B. and Kemper, H.C.G. (1999). Job rotation as a factor in reducing physical workload at a refuse collecting department. *Ergonomics*, 42(9):1167–1178.

Lee, C.-Y. and Leon, V.J. (2001). Machine scheduling with rate-modifying activity, *European Journal of Operational Research*, Vol. 128, pp. 119-128.

Lee, C.-Y. and Lin, C.-S. (2001). Single machine scheduling with maintenance and repair rate-modifying activities. *European Journal of Operational Research*, Vol. 135, pp. 493-513.

Lee, W.-C. and Wu, C.-C. (2004). Minimizing total completion time in a two-machine flowshop with a learning effect. *International Journal of Production Economics*, Vol. 88, pp. 85-93.

Liles, D. and Deivanayagam, S. (1984). A job severity index for the evaluation and control of lifting injury. *Human Factors*, 26(6):683–693.

Lodree, E.J., Geiger, C.D., and Jiang, X. (2005). Taxonomy for integrating scheduling theory and human factors: Review and research opportunities. *Journal of Scheduling,* Under Review.

McComb, S. A., Green, S. G., and Compton, W. D. (1999). Project goals, team performance, and shared understanding. *Engineering Management Journal*, 11(3), 7–12.

Meilijson, I. and Tamir, A. (1984). Minimizing flow time on parallel identical processors with variable unit processing time. *Operations Research*, Vol. 32, No. 2, 440-448.

Molleman E. and Slomp J. (1999). Functional Flexibility and Team Performance. *International Journal of Production Research*, 37(8), 1837 – 1858.

Monk, T.H. (2000). What can the chronobiologist do to help the shift worker? *Journal of Biological Rhythms*, Vol. 15, pp. 86-94.

Moray, N., Dessouky, M.I., Kijowski, B.A., and Adapathya, R. (1991). Strategic behavior, workload, and performance in task scheduling. *Human Factors*, Vol. 33, No. 5, pp. 607-629.

Mosheiov, G. (2001). Scheduling problems with a learning effect. *European Journal of Operational Research*, Vol. 132, pp. 687-693.

Mosheiov, G. and Sidney, J.B. (2003). New results on sequencing with rate modification, *INFOR*, Vol. 41, No. 2, pp. 155-163.

Norman, B. A., Tharmmaphornphilas, W., Needy, K. L., Bidanda, B., and Warner, R. C. (2002). Worker Assignment in Cellular Manufacturing Considering Technical and Human Skills. *International Journal of Production Research*, 40(6), 1479 – 1492.

Pinedo, M. (2002). *Scheduling theory, algorithms, and systems*. 2nd Edition, Prentice Hall, Upper Saddle River, NJ.

Slomp J. and Molleman, E. (2000). Cross-Training Policies and Performance of Teams. *Group Technology/Cellular Manufacturing World Symposium*, San Juan, Puerto Rico, 107-112.

Slomp J. and Molleman, E. (2002). Cross-training policies and team performance. *International Journal of Production Research*, 40(5), 1193-1219.

Slomp J., Bokhorst, J.A. and Molleman, E. (2005). Cross-training in a cellular manufacturing environment. *Computers & Industrial Engineering,* Vol. 48, 609-624.

Smith, L., Hammond, T., Macdonald, I., and Folkard, S. (1998). 12-h shifts are popular but are they a solution? *International Journal of Industrial Ergonomics*, Vol. 21, pp. 323-331.

Sweeney, P. J., & Lee, D. R. (1999). Support and commitment factors of project teams. *Engineering Management Journal*, 11(3), 13–17.

Tayyari, F. and Smith, J.L. (1997). *Occupational Ergonomics Principles and Applications.* Chapman and Hall, London, UK.

Tharmmaphornphilas, W. and Norman, B. A. (2004a). A Quantitative Method for Determining Proper Job Rotation Intervals. *Annals of Operations Research* special issue on *Staff Scheduling and Rostering: Theory and Applications*, Vol. 128, 251-266.

Tharmmaphornphilas, W. and Norman, B. A. (2004b). Robust Job Rotation Methodologies to Reduce Worker Injuries. Technical Report 04-03, Department of Industrial Engineering, University of Pittsburgh, 15261.

Van Dieen, J. and Vrielink, H. (1996a). Toward an optimal sampling strategy of EMG and EMG spectral parameters, when using test contractions to monitor muscle fatigue, In: Mital et al. (Ed.), *Advances in Occupational Ergonomics and Safety, International Society for Occupational Ergonomics and Safety*, Cincinnati, OH, 534–539.

Van Dieen, J. and Verielink, H., 1996b, Evaluation of workrest schedules with respect to postural workload in standing work, In: Mital et al. (Ed.), *Advances in Occupational Ergonomics and Safety, International Society for Occupational Ergonomics and Safety*, Cincinnati, OH, 394–399.

Wickens, C.D., Lee, J.D., Liu, Y., and Gordon-Becker, S.E. (2004). *An Introduction to Human Factors Engineering*. 2nd Edition, Prentice Hall, Upper Saddle River, NJ.

Index

Early Titles in the
INTERNATIONAL SERIES IN
OPERATIONS RESEARCH & MANAGEMENT SCIENCE
Frederick S. Hillier, Series Editor, *Stanford University*

Saigal/ *A MODERN APPROACH TO LINEAR PROGRAMMING*
Nagurney/ *PROJECTED DYNAMICAL SYSTEMS & VARIATIONAL INEQUALITIES WITH APPLICATIONS*
Padberg & Rijal/ *LOCATION, SCHEDULING, DESIGN AND INTEGER PROGRAMMING*
Vanderbei/ *LINEAR PROGRAMMING*
Jaiswal/ *MILITARY OPERATIONS RESEARCH*
Gal & Greenberg/ *ADVANCES IN SENSITIVITY ANALYSIS & PARAMETRIC PROGRAMMING*
Prabhu/ *FOUNDATIONS OF QUEUEING THEORY*
Fang, Rajasekera & Tsao/ *ENTROPY OPTIMIZATION & MATHEMATICAL PROGRAMMING*
Yu/ *OR IN THE AIRLINE INDUSTRY*
Ho & Tang/ *PRODUCT VARIETY MANAGEMENT*
El-Taha & Stidham/ *SAMPLE-PATH ANALYSIS OF QUEUEING SYSTEMS*
Miettinen/ *NONLINEAR MULTIOBJECTIVE OPTIMIZATION*
Chao & Huntington/ *DESIGNING COMPETITIVE ELECTRICITY MARKETS*
Weglarz/ *PROJECT SCHEDULING: RECENT TRENDS & RESULTS*
Sahin & Polatoglu/ *QUALITY, WARRANTY AND PREVENTIVE MAINTENANCE*
Tavares/ *ADVANCES MODELS FOR PROJECT MANAGEMENT*
Tayur, Ganeshan & Magazine/ *QUANTITATIVE MODELS FOR SUPPLY CHAIN MANAGEMENT*
Weyant, J./ *ENERGY AND ENVIRONMENTAL POLICY MODELING*
Shanthikumar, J.G. & Sumita, U./ *APPLIED PROBABILITY AND STOCHASTIC PROCESSES*
Liu, B. & Esogbue, A.O./ *DECISION CRITERIA AND OPTIMAL INVENTORY PROCESSES*
Gal, T., Stewart, T.J., Hanne, T. / *MULTICRITERIA DECISION MAKING: Advances in MCDM Models, Algorithms, Theory, and Applications*
Fox, B.L. / *STRATEGIES FOR QUASI-MONTE CARLO*
Hall, R.W. / *HANDBOOK OF TRANSPORTATION SCIENCE*
Grassman, W.K. / *COMPUTATIONAL PROBABILITY*
Pomerol, J-C. & Barba-Romero, S. / *MULTICRITERION DECISION IN MANAGEMENT*
Axsäter, S. / *INVENTORY CONTROL*
Wolkowicz, H., Saigal, R., & Vandenberghe, L. / *HANDBOOK OF SEMI-DEFINITE PROGRAMMING: Theory, Algorithms, and Applications*
Hobbs, B.F. & Meier, P. / *ENERGY DECISIONS AND THE ENVIRONMENT: A Guide to the Use of Multicriteria Methods*
Dar-El, E. / *HUMAN LEARNING: From Learning Curves to Learning Organizations*
Armstrong, J.S. / *PRINCIPLES OF FORECASTING: A Handbook for Researchers and Practitioners*
Balsamo, S., Personé, V., & Onvural, R./ *ANALYSIS OF QUEUEING NETWORKS WITH BLOCKING*
Bouyssou, D. et al. / *EVALUATION AND DECISION MODELS: A Critical Perspective*
Hanne, T. / *INTELLIGENT STRATEGIES FOR META MULTIPLE CRITERIA DECISION MAKING*
Saaty, T. & Vargas, L. / *MODELS, METHODS, CONCEPTS and APPLICATIONS OF THE ANALYTIC HIERARCHY PROCESS*
Chatterjee, K. & Samuelson, W. / *GAME THEORY AND BUSINESS APPLICATIONS*
Hobbs, B. et al. / *THE NEXT GENERATION OF ELECTRIC POWER UNIT COMMITMENT MODELS*
Vanderbei, R.J. / *LINEAR PROGRAMMING: Foundations and Extensions, 2nd Ed.*
Kimms, A. / *MATHEMATICAL PROGRAMMING AND FINANCIAL OBJECTIVES FOR SCHEDULING PROJECTS*
Baptiste, P., Le Pape, C. & Nuijten, W. / *CONSTRAINT-BASED SCHEDULING*
Feinberg, E. & Shwartz, A. / *HANDBOOK OF MARKOV DECISION PROCESSES: Methods and Applications*
Ramík, J. & Vlach, M. / *GENERALIZED CONCAVITY IN FUZZY OPTIMIZATION AND DECISION ANALYSIS*
Song, J. & Yao, D. / *SUPPLY CHAIN STRUCTURES: Coordination, Information and Optimization*
Kozan, E. & Ohuchi, A. / *OPERATIONS RESEARCH/ MANAGEMENT SCIENCE AT WORK*
Bouyssou et al. / *AIDING DECISIONS WITH MULTIPLE CRITERIA: Essays in Honor of Bernard Roy*

Early Titles in the
INTERNATIONAL SERIES IN
OPERATIONS RESEARCH & MANAGEMENT SCIENCE
(Continued)

Cox, Louis Anthony, Jr. / *RISK ANALYSIS: Foundations, Models and Methods*
Dror, M., L'Ecuyer, P. & Szidarovszky, F. / *MODELING UNCERTAINTY: An Examination of Stochastic Theory, Methods, and Applications*
Dokuchaev, N. / *DYNAMIC PORTFOLIO STRATEGIES: Quantitative Methods and Empirical Rules for Incomplete Information*
Sarker, R., Mohammadian, M. & Yao, X. / *EVOLUTIONARY OPTIMIZATION*
Demeulemeester, R. & Herroelen, W. / *PROJECT SCHEDULING: A Research Handbook*
Gazis, D.C. / *TRAFFIC THEORY*
Zhu/ *QUANTITATIVE MODELS FOR PERFORMANCE EVALUATION AND BENCHMARKING*
Ehrgott & Gandibleux/ *MULTIPLE CRITERIA OPTIMIZATION: State of the Art Annotated Bibliographical Surveys*
Bienstock/ *Potential Function Methods for Approx. Solving Linear Programming Problems*
Matsatsinis & Siskos/ *INTELLIGENT SUPPORT SYSTEMS FOR MARKETING DECISIONS*
Alpern & Gal/ *THE THEORY OF SEARCH GAMES AND RENDEZVOUS*
Hall/*HANDBOOK OF TRANSPORTATION SCIENCE - 2nd Ed.*
Glover & Kochenberger/ *HANDBOOK OF METAHEURISTICS*
Graves & Ringuest/ *MODELS AND METHODS FOR PROJECT SELECTION: Concepts from Management Science, Finance and Information Technology*
Hassin & Haviv/ *TO QUEUE OR NOT TO QUEUE: Equilibrium Behavior in Queueing Systems*
Gershwin et al/ *ANALYSIS & MODELING OF MANUFACTURING SYSTEMS*

**** A list of the more recent publications in the series is at the front of the book ****